PHYSICS OF ATOMS AND MOLECULES

Recent volumes in this series:

COINCIDENCE STUDIES OF ELECTRON AND PHOTON IMPACT IONIZATION
Edited by Colm T. Whelan and H. R. J. Walters

DENSITY MATRIX THEORY AND APPLICATIONS, Second Edition
Karl Blum

IMPACT SPECTROPOLARIMETRIC SENSING
S. A. Kazantsev, A. G. Petrashen, and N. M. Firstova

INTRODUCTION TO THE THEORY OF X-RAY AND ELECTRONIC SPECTRA OF FREE ATOMS
Romas Karazija

PHOTON AND ELECTRON COLLISIONS WITH ATOMS AND MOLECULES
Edited by Philip G. Burke and Charles J. Joachain

POLARIZED ELECTRON/POLARIZED PHOTON PHYSICS
Edited by H. Kleinpoppen and W. R. Newell

PRACTICAL SPECTROSCOPY OF HIGH-FREQUENCY DISCHARGES
Sergei A. Kazantsev, Vyacheslav I. Khutorshchikov, Günter H. Guthöhrlein, and Laurentius Windholz

SELECTED TOPICS ON ELECTRON PHYSICS
Edited by D. Murray Campbell and Hans Kleinpoppen

THEORY OF ELECTRON–ATOM COLLISIONS, Part 1: Potential Scattering
Philip G. Burke and Charles J. Joachain

VUV AND SOFT X-RAY PHOTOIONIZATION
Edited by Uwe Becker and David A. Shirley

A Chronological Listing of Volumes in this series appears at the back of this volume.

A Continuation Order Plan is available for this series. A continuation order will bring delivery of each new volume immediately upon publication. Volumes are billed only upon actual shipment. For further information please contact the publisher.

Impact Spectropolarimetric Sensing

S. A. Kazantsev

St. Petersburg State University
St. Petersburg, Russia, and
Paris Observatory
Meudon, France

A. G. Petrashen

St. Petersburg Institute of Fine Mechanics and Optics
St. Petersburg, Russia

and

N. M. Firstova

Irkutsk Institute of Solar and Terrestrial Physics
Irkutsk, Russia

Kluwer Academic / Plenum Publishers
New York, Boston, Dordrecht, London, Moscow

Library of Congress Cataloging-in-Publication Data

Kazantsev, S. A.
Impact spectropolarimetric sensing / S.A. Kazantsev, A.G. Petrashen, and N.M. Firstova.
p. cm. -- (Physics of atoms and molecules)
Includes bibliographical references and index.
ISBN 0-306-45892-6
1. Solar flares--Measurement. 2. Remote sensing. 3. Solar radiation--Polarization--Measurement. 4. Astronomical spectroscopy. I. Petrashen, A. G. II. Firstova, N. M. III. Title. IV. Series.
QB526.F6K39 1999
523.7'5'028--dc21 98-48668
CIP

ISBN 0-306-45892-6

233 Spring Street, New York, N.Y. 10013

10 9 8 7 6 5 4 3 2 1

A C.I.P. record for this book is available from the Library of Congress.

Printed in the United States of America

Preface

This book arises from the significant quantity of original results in the field of impact spectropolarimetry, mainly for solar flare studies, and the evident potential applications of this technique to environmental physics, astrophysics, and hot plasma. This is a practically oriented book, describing theoretical fundamentals and implementation of this new interdisciplinary remote sensing technique.

A basic phenomenon for impact spectropolarimetric sensing is the polarization of the electron shells of an ensemble of free atomic particles (atoms, ions, molecules) due to collisional interaction. Slow collisions in the presence of internal anisotropies in the relative velocity space inside any ionized medium, as well as bombardment of a diluted gas by fast precipitating light projectiles, are analyzed in detail as principal impact polarization mechanisms. Impact spectropolarimetric sensing incorporates state-of-the-art theoretical methods of collisional physics in combination with reliable polarimetric measurements. This technique is illustrated by the new quantitative sensing of energy transport to the upper region of the chromosphere during solar flares, making use of ground-based solar spectropolarimetric observations. Apart from general astrophysical and solar-terrestrial significance, a solar flare, the brightest nonthermal phenomenon, is a good candidate for the demonstration of new opportunities of impact spectropolarimetry. New astrophysical achievements of quantitative spectropolarimetry are particularly considered in this book.

The theoretical part presents a quantum mechanical description of impact polarization and the spectropolarimetric manifestations for different energies of the colliding partners. Special attention is drawn to the new polarization effects induced by anisotropic relaxation within an array of closely spaced quantum atomic states and by the charge exchange process. Different techniques for observing the polarization of solar optical radiation are described, representing results from the most eminent research groups. The ground-based observations of optical polarization of solar flare emissions, as well as of other nonequilibrium entities in the solar atmosphere, are reviewed, and spectropolarimetric diagnostics for energy transport, using data for hydrogen lines, are described in detail.

This book presents an extensive description of theoretical and observational spectropolarimetric methods and will be useful not only to astrophysicists, but to all researchers, PhD students, and optical engineers utilizing spectroscopic methods of sensing for a number of problems of terrestrial, laboratory, and technical physics of ionized gases.

The authors take special pleasure in extending their gratitude to Prof. Dr. H. Kleinpoppen for a discussion of the general structures of the book and for scientific advice and support.

S. Kazantsev

1

Physical Introduction: Concepts, Phenomenology, Retrospectives, Applications

1.1. Polarization of Atomic Ensembles in Ionized Media

Spectropolarimetry offers new means for sensing ionized media. Polarimetric remote sensing is based on the polarization spectroscopy of ionized gases that was actively developing during the last two decades in connection with different problems of atomic and collisional physics. The general physical principles governing the polarization spectroscopy of ionized gases, the quantitative spectropolarimetric sensing of various gas discharges, and the first steps toward exploiting this method in diagnostic problems of the solar atmosphere have been presented in the book *Polarization Spectroscopy of Ionized Gases* by S. A. Kazantsev and J.-C. Hénoux.[1]

Rising interest in applications of spectropolarimetry to the study of solar flares, one of the most intriguing nonthermal phenomena in the solar atmosphere, and other recent results in solar physics justify the need for a comprehensive profound description of the theoretical fundamentals of impact spectropolarimetry. Presentation of quantum mechanical methods of the theory of atomic collisions, review of observations on linear polarization in solar flare emissions, and interpretation of polarization effects in terms of energy transport by precipitating fast particles should be very important for many problems of remote sensing. The development of quantitative spectropolarimetric techniques for the upper atmosphere of the Earth (for which there exist a number of reliable qualitative spectropolarimetric observations) for sensing hot plasmas and many other ionized media with a well-expressed structure can be envisaged in the near future.

A quantum state of an excited atomic particle moving in an ionized gas is determined by its total energy and angular momentum. Properties of the linear spectrum acquired as a result of spontaneous transitions to lower levels reflect the characteristics of an ensemble of radiating particles and internal interaction processes. Studies of the optical emission of a plasma and features of the frequency distribution of a spectral line intensity create a basis for well-developed and widely used remote spectroscopic diagnostic techniques. Information on the mechanisms governing the excitation and ionization of particles is inferred by means of measurement and theoretical analysis of the spectral distribution of the emission or absorption intensity.

The polarization of a spectral line is effectively a basic spectroscopic characteristic which, within the simplest approach, is connected to the nonequilibrium distribution of populations of magnetic substates of excited atomic particles (atoms, ions, molecules) and reflects an ordering of the total angular momenta of electron shells in an ensemble, or, in other words, the polarization of an ensemble.(2) In many cases the polarization of an emission spectrum is directly related to the stucture of an ionized medium, which is reflected in anisotropic properties of an ensemble of the excited atomic particles. Structural features of an ionized medium, such as the anisotropy of the resonance optical field, distribution of internal fields within a medium, energy transport by precipitating particles or by heat conduction, gradients of parameters, and boundary effects, give rise to anisotropy in the velocity space of free atomic particles and electrons. In this way, collision-induced polarization of the emission spectrum stems from anisotropies in the space of relative velocities of atoms, ions, and light charged projectiles, such as electrons or protons, responsible for excitation and optical emission.

Until recently, the impact polarization of a spectral line as a source of information on the physical properties of ionized media of various physical natures was not widely exploited, but usually polarization spectroscopic effects have been safely detected in many ionized natural and artificial objects. The reasons were the following: the complexity of the spectral line polarization creation, even for a simple gas discharge, difficulties of experimental observation and detection, lack of basic theoretical relations between the polarization of radiation and the kinetic characteristics of the plasma, as well as indispensable cross sections and atomic parameters. These difficulties have been overcome during last two decades in the course of developing quantum collision theory, with extensive analysis of polarization effects under different conditions. As a result, principles of polarization spectroscopy as a diagnostic tool and as means for remote sensing have been proposed.

1.2. Polarization of Solar Flare Emission

A special place in the history of spectropolarimetric sensing is occupied by ongoing astrophysical studies of solar flares. Therefore, polarization effects in an ionized medium under natural conditions are illustrated, and implementation of impact spectropolarimetric sensing is presented based on solar flare studies. According to present-day concepts, a solar flare is a consequence of the liberation of magnetic free energy in the outer coronal region of the solar atmosphere as a result of the reconnection of the magnetic field lines. Usually, secondary processes due to energy transformation and transport from the primary source of the flare are observed. One of these processes is the formation of the ionized medium of the upper layer of the solar chromosphere. The energy source for the creation of the nonthermal chromospheric region of a solar flare, where optical emission lines are excited, was studied for a long time because the chromospheric effects are secondary with respect to the flaring corona, where the principle energy release takes place.

A number of energy transport modes from the coronal flare location have been treated, principally heat conduction, X-ray irradiation, and accelerated particles, propagating within the magnetic loop of a solar flare. In most cases the energy is transported from the corona to the lower chromosphere, either by means of heat transfer from the high temperature plasma heated in the coronal region to a temperature of about 10^8 K, or by beams of charged particles that propagate along the coronal loop and bombard the chromospheric base of the loop. Of all known diagnostic methods, only optical observations and, more specifically, spectropolarimetric analysis of the chromospheric emissions could shed light on the relative contribution of each energy transport mode. As the solar flare is a rather complicated object with a highly expressed structure and involving many physical processes, detailed analysis of the observational spectropolarimetric results and application of the quantitative methods of impact spectropolarimetric sensing are extremely useful for the general presentation of this new technique.

Systematic observations started at the beginning of the 1980s and revealed the linear polarization of different chromospheric emission lines (hydrogen Balmer α,[3] SI, 1437 Å[4]). These first observations and new fundamental possibilities provided by linear polarization led to some idea of the quantitative polarimetric sensing of the conductive flux at the top of the flaring chromospheric layers and, later on, yielded some quantitative estimates for this flux.[4] Successive analysis of the spectropolarimetric information enabled the second energy transport mode associated with accelerated particles to be revealed. The orientation of the polarization direction was the important parameter which could distinguish the energy transport modes.[5] Later, spectropolarimetric studies of solar flares were related to quantitative sensing of corpuscular modes of energy transport. Special analysis of impulse relaxation for different particles as candidates for linear polarization of chromospheric emissions showed that only protons could create impact polarization at the chromospheric level, because they could conserve the directivity of their motion at larger distances. Hence quantitative spectropolarimetry offered an excellent opportunity for solving the principle problem of the detection and sensing of accelerated low energy protons during solar flares.[150] A straightforward approach to the optical detection and sensing of low energy protons was provided only by the impact polarization of the chromospheric lines.

1.3. Impact Polarization

Generally, linear polarization subject to collisional excitation by a beam of charged particles provides information on the nature of the incident particles, their number flux, and direction of propagation. For a monoenergetic beam of particles of energy E, the polarization of emission of an atomic ensemble undergoing collisional excitation is defined by[6]

$$P(\beta, E) = \frac{I_{\parallel} - I_{\perp}}{I_{\parallel} + I_{\perp}}$$

where β is the angle between the line of sight and the beam propagation direction, while $I_{\parallel}$ and $I_{\perp}$ are the intensities with linear polarizations, parallel and perpendicular to the plane, defined by the beam and the line of sight.

The degree of polarization depends on the particle energy and on the angle of observation. The maximum polarization is observed at an angle of 90° to the beam direction. The angular dependency of the degree of linear polarization $P(\beta, E)$ for the dipole radiation is given by[6]

$$P(\beta, E) = P_{90}(E)\frac{\sin^2\beta}{1 - P_{90}(E)\cos^2\beta}$$

With this basic approach, the angular and energy dependencies of the observed polarization degree depend on the atomic characteristics of the optical transition under study and were exploited later in problems dealing with the sensing of protons during solar flares on the basis of hydrogen emission lines. These works have been reviewed elsewhere.[7]

In spite of impressive solar physics applications, polarization of excited atomic ensembles and spectropolarimetric effects in low temperature plasma emission have a long history of laboratory studies. The first qualitative studies in simple laboratory gas discharges, performed in the mid-1970s[2] indicated that the spectropolarimetric effects created by direct electron impact excitation are very informative and reflect directly the structural features of a plasma entity. Making use of these fundamentally new possibilities, spectropolarimetric studies allowed one to detect the very fine kinetic effect of the formation of loss cones in the velocity distribution of fast electrons, resulting from interaction of the low temperature diluted plasma with the wall of a discharge tube. Moreover, it was possible to sense some orientation of these loss cones and the radial profile of the internal plasma electric field.[8] Analysis of spectropolarimetric effects in the simplest case of low temperature gas discharge plasma, in the absence of external electromagnetic fields, beams of precipitating particles, and gradients of environmental parameters showed that the observed polarization is determined by the self-alignment of particles or the quadrupole orientation of momenta, induced by internal excitation and relaxation processes in an ionized gas. Self-alignment of atomic ensembles reflects one of the fundamental properties of a real ionized gas entity, such as deviations from thermal equilibrium and spatial confinement, and thereby it could be regarded as a general feature of ionized gas entities.[9]

Most promising in respect of diagnostics is the self-alignment caused by impact processes in a plasma, due to anisotropies in the relative velocity space of the colliding particles. Historically, the most evident way to understand and demonstrate practical potentialities of impact self-alignment was to analyze this effect under the direct excitation of atomic particles by fast electrons in a simple laboratory gas discharge. Electron–atom collisions represent a typical mechanism of excitation and ionization of atoms in low temperature diluted plasmas, formed by means of electric discharges in gases. Understanding the basic properties of impact self-alignment of atomic ensembles, and formulating the basic integral equations connecting the observable spectropolarimetric characteristics with the kinetic parameters of fast exciting electrons, allowed polarization spectroscopy to be regarded as a principally new method of quantitative remote sensing.[10] The most important result in view of a new contribution of spectropolarimetry to remote sensing of ionized gases is that the Stokes parameters of an emission spectral line by a plasma are expressed in terms of the quadrupole moment of the velocity distribution function of fast electrons, or $f^{(2)}(v, \mathbf{r})$. The quadrupole moment is the third coefficient of the multipole representation of the velocity distribution function in velocity space:

$$f(\boldsymbol{v}, \mathbf{r}) = \sum_{k=0}^{\infty} \sum_{q=-k}^{k} Y_q^{(k)}(\mathbf{n}) f_q^{(k)}(v, \mathbf{r})$$

where $Y_q^{(k)}(\mathbf{n})(\mathbf{n} = \boldsymbol{v}/v,\ v = |\vec{\mathbf{v}}|)$ are spherical harmonics determined in the velocity space of electrons.

By its nature, the quadrupole moment of the velocity distribution function is the anisotropic pressure tensor of the electron component and characterizes the local value of the momentum flow density of electrons, and hence the quantitative characteristics of the energy exchange. Indeed, the principle processes indispensable for understanding the nature of an ionized gas entity and related to energy exchange with the environment are directly reflected in the kinetics of the most mobile electron component. The polarization of an ensemble of atomic particles in a plasma produced by the electron impact excitation, $\rho_q^{(2)}(J, \mathbf{r})$, is directly related to the quadrupole moment of the velocity distribution function via the basic relation

$$\rho_q^{(2)}(J, \mathbf{r}) \sim \int_0^\infty dv v^3 \sigma^{(2)}(J, v) f_q^{(2)}(v, \mathbf{r})$$

where $\sigma^{(2)}(J, v)$ is the cross section for electron impact excitation.

1.4. Gas Discharge Studies

The energy aspect of the maintenance of an ionized gas entity was studied. It was concluded that energy transport into an ionized medium is implemented, as a rule, in a nonthermal form and accompanied by gradients of parameters and the spatially located anisotropy of the electron motion, especially in the fast part of the velocity distribution. Interaction of fast electrons with other plasma components transforms the energy associated with the ordered motion into the energy of the chaotic thermal motion. Energy dissipation at the boundaries of an ionized medium is also reflected by the anisotropy of the electron motion and provides an increase in the local value of the momentum flow tensor.[11] All the anisotropies in the velocity space of electrons directly connected to the structural features of an ionized entity are thereby reflected in the polarization of emission (degree and polarization plane orientation). In the course of spectropolarimetric studies of laboratory gas discharges, it was possible to obtain sufficient observational data to reach these general conceptual considerations.

Retrospectively, the principle results for understanding the role of impact spectropolarimetric effects as a diagnostic tool were connected to laboratory experiments on high-frequency discharge. These were also very illustrative in regard to the universal nature of polarization effects. High-frequency discharge in different forms is a well-known laboratory source of low temperature plasma with well-expressed structural features. The capacitive high-frequency discharge, which is formed inside the capacitor of the oscillating circuit of a high-frequency generator due to the potential high-frequency alternating electric field, is very important, because it could enable us to understand and even simulate processes of energy deposition in many plasmas under different conditions.

Boundary electrode processes in capacitive discharge are very important because, under certain conditions, the stochastic Fermi process of heating electrons, of fundamental physical significance, takes place. If the frequency of the electric field lies between the ion and electron plasma frequencies, the potential barrier of the boundary of the electrode sheath is modulated with a high amplitude at the frequency of the external field. For conditions of low temperature diluted plasma, when the velocity of this boundary is comparable to the mean electron velocity, most of the bulk plasma electrons will be reflected from this boundary with enriched energy, due to Fermi acceleration. These concepts have been confirmed by the anomalous polarization effect detected at the boundary of the electrode sheath for low pressure discharges in noble gases.[12,143] In the course of spectropolarimetric studies of high-frequency discharge, polarization spectroscopy enabled us to undertake for the first time remote sensing of the most important structural parameter: the energy deposition in the plasma through the electrode sheath and the alternating electric field strength in the middle of the discharge, while maintaining the low temperature plasma.

The principal possibilities of impact spectropolarimetry, apart from other techniques of polarization spectroscopy, have been illustrated by some experimental investigations and remote sensing of low temperature laboratory plasmas.[13] Hence there is an evident need for a more substantial presentation of the theoretical fundamentals of this universal technique with a view to many prospective applications to laboratory and technical plasmas.

The theoretical basics of impact polarization spectroscopy are very closely related to the quantum mechanical theory of atoms and to collision theory, employing the general formalism of the quantum theory of angular momentum and of group theory. The physical principles of basic quantum mechanical theory of the collisional ordering of atomic states and of its spectroscopic manifestations appeal to the fact that, as a result of nonelastic collisions, an ensemble of excited atomic particles, acquires a memory of the velocity distribution anisotropies of the exciting particles. In other words, the angular momentum vectors of the electron shells of excited particles obtain a particular ordering that corresponds to the kinetic aspects of the collisional excitation. On the mathematical side, this ordering is characterized by spherical tensors of different ranks or by polarization moments of the atomic density matrix. Polarization moments are the coefficients of expansion of the density matrix of an excited atomic ensemble in terms of the irreducible representation of the three-dimensional rotation group. The orderings of the excited atomic ensembles reflected by the polarization moments are expressed directly by the spectropolarimetric characteristics of the emitted light.[1]

1.5. Collisional Kinetics of Polarization

Collisions between free atomic particles in ionized media play an important role in the kinetics of polarization. Collisions between isotropically moving particles play a purely destructive role by increasing chaos in an ensemble, which is related to the thermal properties of an ionized entity, and destroying orderings of angular momenta. In this case isotropic relaxation of atomic polarization moments takes place. This relaxation is represented by an exponential decay of the polarization moments with different ranks or different relaxation rates. The corresponding rate constants characterize the process of collisional depolarization of linearly and circularly polarized light emitted by an ensemble. The quantitative theory of the collisional relaxation of polarization moments is based on the quantum mechanical description of an elementary collision. This problem is treated traditionally on the basis of the system of equations governing the impact parameter method, with a detailed account of the splitting of degenerate atomic states during a collision and collisionally induced transitions between these degenerate states.

For an anisotropic velocity distribution of atomic particles with a preferential direction of collisions, the collisional relaxation of the atomic polarization moments undergoes crucial changes. In this case inelastic collisions play a constructive role by creating an ordering of the angular momenta. Such a principle feature stems from a change in the type of interaction symmetry from spherical to axial. As a result, multiexponential temporal decay and mutual transformation of polarization moments may take place. Anisotropic collisional relaxation leads to the collisional creation of the quadrupole orientation of momenta from the nonequilibrium population of atomic and ionic states.(14) This effect has a very high diagnostic significance. Polarization created by anisotropies of slow collisions of atomic particles reflects internal anisotropies of the studied object, providing a tool for remote sensing of many plasma parameters.(15,16)

When observing this effect under laboratory conditions, it should be noted that, among gas discharges, the discharge in a hollow cathode is special in that practically all the dc electric field applied to it is located within a very narrow sheath near the inner surface of the metallic cathode, known as the dark cathode space. The electric field strength inside the dark cathode space is rather high, so the ion drift motion in the vicinity of the boundary of this cathode sheath at low pressures is very pronounced. Due to the high anisotropy of motion of ions relative to the isotropic environment of neutral atoms, this new collisional polarization effect resulting from anisotropic collisional relaxation was observed.(15)

Different probabilities of collision along the radial ion drift velocity and in the perpendicular direction provide a preferential direction of collisional interaction within a plasma. Such partial anisotropy of collisional interaction of ions with an isotropic atomic environment is one reason for the anisotropic collisional relaxation within the system of the closely spaced fine structure levels of an energy structure of most atomic ions. The polarization moments of an ensemble of ions undergoing anisotropic collisional relaxation are described by the kinetic equation

$$\frac{d}{dt}\rho_0^k(J,\mathbf{r}) = \sum_{k_1}\frac{4a^{3/2}}{\sqrt{\pi}}e^{-av^2}\sum_{q,l}(-1)^q\begin{bmatrix} k & k_1 & l \\ 0 & 0 & 0\end{bmatrix}\begin{bmatrix} k & k_1 & l \\ q & -q & 0\end{bmatrix}$$

$$\cdot\int dv\,\tilde{R}_q^{kk_1}(JJ_1v,\mathbf{r})v^2e^{-av^2}\left(\frac{v}{2avv_0}\right)^l\left(\frac{1}{v}\frac{d}{dv}\right)^l\frac{sh\,2avv_0}{2avv_0}\rho_0^{k_1}(J_1,\mathbf{r})$$

where

$$a = \frac{m_i m_a}{2k(m_i T_a + m_a T_i)}$$

Here m_a, m_i, T_a, and T_i are the masses and temperatures of atoms and ions, while $\tilde{R}_Q^{kk_1}(JJ_1 v)$ is the matrix of the total anisotropic collisional relaxation.

Due to a linkage of momenta with different ranks under anisotropic collisional relaxation, the drift of ions is accompanied by the creation of longitudinal alignment from the populations of the fine structure levels with the axis parallel to the drift velocity vector, known as drift self-alignment. The process of drift self-alignment is specific in that the projection of the momentum on the drift axis is conserved, but not the rank of the tensor, k. The study of drift self-alignment of ions, first undertaken on the system of close multiplets of noble gas ions excited in a hollow cathode discharge, revealed a number of specific features of this phenomenon. The efficiency of drift self-alignment depends on the degree of anisotropy of motion and on the energy of the ion. The general requirement for drift self-alignment is a deviation in the fine level populations from their equilibrium values. Naturally, this process is only possible if the kinetic energy exceeds the multiplet splitting:

$$m_1 v_1^2/2 > \Delta E$$

This condition may be fulfilled for a number of excited states of different ions under ordinary low temperature plasma conditions.[(16)]

The transverse profiles of the degree of polarization for ion emission lines from a cylindrical hollow cathode discharge have been measured. Also, the drift self-alignment of ions with allowance for the interaction energy of colliding particles was computed, incorporating the polarization interaction of the ion with the neutral atom, dispersive interaction, and combined polarization interaction of the neutral atom with the charge and quadrupole moment of the ion. These results enabled the radial profile of the drift velocity of argon ions to be determined.[(17)] These drift velocities were obtained for two different doublets of the argon ion in the boundary region of the negative glow and the mobility of ions in the gas, and they enabled the internal electric field profile in the plasma to be estimated. Reliability of these data was confirmed by comparing the spectropolarimetric results with independent electric measurements of the potential drop on the hollow cathode. The main advantage of the spectropolarimetric technique based on drift self-alignment is that it can be employed to measure low drift velocities, not being limited by doppler broadening.

1.6. Spectropolarimetric Sensing

The theory of impact polarization for an ensemble of atomic particles under collisional interaction with light projectiles bombarding this ensemble yields an important result, namely, general integral relations connecting the polarization parameters of an emission spectrum with the kinetic parameters of the projectiles. These general theoretical relations provide the means for quantitative diagnostics of the structural and nonthermal features of a medium based on the spatial distribution of the polarization of its emission spectrum. In order to develop and put in perspective quantitative methods of impact spectropolarimetric sensing, theoretical chapters of the book comprise a detailed description of the quantum mechanical formalism of the theory of angular momenta ordering of an ensemble of atomic particles. The formalism of the density matrix in the representation of polarization moments enables direct relations to be established between the internal anisotropic and nonthermal features of an ionized gas medium and the spectropolarimetric characteristics of its optical emission.

Special attention is paid to the alignment creation parameters for the processes which govern the polarization of an ensemble and to all the other characteristic parameters of atomic physics that are indispensable for the solution of inverse problems of impact spectropolarimetric diagnostics. These characteristics are analyzed in more detail for the hydrogen states, which have prior astrophysical and ionospheric significance. The theoretical description of the elementary impact process, using different approximations and mathematical methods, has received much attention. Original theoretical results presented in the book refer to anisotropic collisional relaxation of angular momenta ordering caused by slow collisions of atomic particles, which provide the means for polarimetric diagnostics. An important part of the theoretical description of impact polarimetric effects is related to charge exchange collisional effects, which could be regarded as one of the most promising approaches for diagnostic tasks. The charge exchange process seems to be extremely important for various astrophysical objects in which jets and beams of charged particles play an important role, determining their structural nature and acting as dominant energy transport modes.

The diagnostic possibilities of impact spectropolarimetry have been demonstrated more effectively in the chapters of the book devoted to spectropolarimetric techniques used in different observatories and the results of spectropolarimetric studies of solar flares, by playing a principal role in the development of impact spectropolarimetric sensing. The optical emission of a solar flare, observed mainly in the H_α hydrogen line, characterize the low temperatrure chromospheric part of the flare. Special spectroscopic studies of solar flares indicated a presence of a linear polarization of hydrogen emission lines from the upper chromosphere of several percent with the orientation of the polarization toward the center of the disk.[5] These qualitative studies of the polarized emission of the upper chromosphere and special considerations on the conservation of the directivity of motion of different charged particles inside a flare loop have enabled us to infer that the radiation of the chromosphere during a flare is caused by suprathermal protons.

A significant part of the book is devoted to the application of spectropolarimetric sensing to obtain quantitative estimates of energy transport by charged projectiles. This work is based on observations of the hydrogen H_α emission line ($n = 3$) and other emisson lines, and has been performed in leading astrophysical observatories. Much of the material is the practical realization of the impact spectropolarimetric sensing of solar flares. Special attention is paid to the analysis of observational instrumental methods and techniques of spectroscopic observation of the linear polarization of solar flare line emission. Different instruments and polarimetric facilities are described, showing the present state of this technique in different research groups. Observations of the spectropolarimetric effects of solar flares, undertaken in different leading observatories, are analyzed in detail. The last chapters of the book deal with practical implementation of impact spectropolarimetric sensing to obtain precise conditions of energy transport to the chromospheric location of the flare.

1.7. Polarization Effects in the Earth Atmosphere

Turning to possible new prospective applications of polarization spectroscopy, we should mention the ionized diluted medium of the upper terrestrial atmosphere, which interacts with precipitating charged particles from outer space. There are many interesting phenomena in the outer ionosphere of the Earth, where impact spectropolarimetric sensing can contribute to an understanding of environmental effects at a fundamental level.[18] No quantitative spectropolarimetric studies have so far been made in the atmosphere of the Earth. Therefore, we shall review here known observational data on the polarization of emission from different ionized entities in the outer shells of the Earth atmosphere, showing the range of prospective applications of impact spectropolarimetric sensing to geophysical and environmental investigations.

In the course of studies into optical emissions from the upper atmosphere of the Earth, there were many attempts to observe and analyze the polarization. These observations were extremely complicated owing to the specific nature of optical effects and observational difficulties. In most observations of the polar auroras, for example, the polarization of emission was no higher than several percent or even not undetected at all.[19] However, some observations showed significant degrees of polarization, attaining 30–60%.[20] Theoretical estimates[21] confirmed that for some events subject to electron impact excitation, the polarization of the red oxygen line could be about 14% while the polarization of the hydrogen lines excited during proton polar auroraes under collisional interaction of a beam of fast protons with the atoms of the upper atmosphere could reach 20%.[(22)]

Qualitative studies showed that spectropolarimetric effects in the outer atmosphere are very different and depend on the nature of the precipitating particles which create the polarization of ensembles. The polarization of the red oxygen 630.0 nm line was observed[(20)] during an extensive series of zenith spectropolarimetric observations of polar auroras in Australia. The degree of polarization was detected to be 30% and the magnetic vector was oriented in the north–south direction. At the same time, spectropolarimetric studies of the 557.7 and 630.0 nm lines of oxygen emitted by the polar aurora showed a degree of polarization of about 1% or less.[(23)] A low degree of polarization of emission of auroras was connected to the fact that the emission of the polar auroras in this case is determined by slow electrons, possessing a rather isotropic velocity distribution function in the outer atmosphere, while when auroras were excited by precipitating proton beams the degree of polarization is usually high. A middle-latitude auroras is quite a rare geophysical phenomenon, which happens only during solar activity periods.[(24)] It is a dynamic and short-term phenomenon, reflecting different phases of a geomagnetic storm and the degree of excitation anisotropy of the studied levels; in addition, the observed polarization should be time-dependent. Theoretical computations and predictions show that the 630.0 nm oxygen line could be polarized by collisions with slow electrons, and due to different charge exchange reactions as leading mechanisms of upper level excitation.

There exist other geophysical phenomena such as optical flares of the night sky, which were discovered in the auroral zones, and later in subauroral zones and at middle latitutes. Detected polarization degrees[(25)] of about 30% have been observed for 17 optical flares with the aid of a photometer-polarizer during 341 hours of observation. The degree of polarization of optical flares was subsequently confirmed to be in the range of 18–56%, while the polarization plane was inclined at an angle not exceeding 40° with respect to the meridian. Even night sky emissions showed linear polarizations of 2–4% for some events. Observations[(26)] showed the degree of polarization to be 14%. Later works reported a degree of polarization equal to no more than 1–2%.

Polarization effects could be anticipated in the outer atmosphere of the Earth in the course of different experiments on the artificial modification of ionospheric and magnetospheric media. This could take the form of the injection of electron or ion beams, formation of a local plasma medium and plasma jets, irradiation by ultraviolet radiation, radioheating of the ionosphere, and other techniques. Special artificial effects in the outer ionosphere, such as launches and flights of satellites and spacecrafts, are accompanied by optical emissions in which spectropolarimetric effects could be anticipated and used. All the impact spectropolarimetric methods developed in this book could be employed directly for prospective applications in studies of environmental phenomena because, methodologically, the problem of spectropolarimetric sensing of the outer atmosphere of the Earth is close to the same problem in the solar atmosphere.

Contrary to classical optical intensity methods, the spectropolarimetric technique based on the impact polarization of atomic ensembles clearly demonstrates its basic ability for quantitatively sensing many astrophysical, extraterrestrial, and laboratory ionized media with highly expressed structural features, and is suitable for studies of nonthermal effects. Apart from the quantitative spectropolarimetric sensing of solar flares, which is presented in the book as a basic illustrative phenomenon, many problems of gas discharge and technical physics related to a diagnostic of plasma boundaries, the transport of high current charged particle beams in a gas medium, energy deposit and transformation in ionized gases, as well as a number of fundamental questions of environmental physics related to corpuscular interactions within the upper atmosphere of the Earth and to solar–terrestrial communications, could be effectively resolved with the help of quantitative impact spectropolarimetry. This book presents theoretical and experimental methods of impact spectropolarimetric sensing and provides a unique possibility for extensive interdisciplinary applications of this technique to remote sensing of ionized media of various physical natures.

The contributions of the coauthors are as follows: The Physical Introduction, and Chapters 2–6 and 12 were written by Drs. S. A. Kazantsev and A. G. Petrashen, Chapters 7–11 were written in Russian by Dr. N. M. Firstova, while Chapter 13 was written by all three coauthors. The translation, scientific editing, and preparation of the manuscript as a final computor file were carried out by Dr. S. A. Kazantsev.

This book was written at the DASOP - L.P.S.H. Department of the Meudon Section of the Paris Observatory under the general scientific supervision of its Director, Dr. J.-C. Hénoux. The authors sincerely acknowledge the scientific cooperation, valuable scientific research contributions, the many remarks that improved the presentation of the material, and the help proferred in the preparation of the manuscript by Dr. J.-C. Hénoux. The authors are also grateful to Mme. G. Gelugne for assistance in the preparation of the manuscript. Support from the Paris Observatory, and funds from International Joint Projects of INTAS, EC, and NATO are highly appreciated.

2

Polarization of a Photon Beam

2.1. Introduction

The aim of this chapter is to demonstrate a relation between the physically clear concept of electromagnetic wave polarization and the photon characteristics. We shall analyze various aspects of the phenomenon of light polarization but not follow a complete and mathematically rigorous description. Rather, we try to focus on problems of special importance in this book, related to remote sensing of ionized media. This methodological approach is used in order to provide physical clearness, simplicity of presentation, and connect directly to the classical models with which to present and study new opportunities offered by spectropolarimetry as an effective remote sensing technique for investigating ionized gases. Most of the material in this chapter is of a general nature and can be found in well-known textbooks on quantum mechanics and quantum electrodynamics, such as by Berestetsky *et al.*[33] The material in Section 2.7 is based, to a certain degree, on the results of original work.[137,138]

2.2. Wave Function of a Photon

Consider some volume V, the electromagnetic field at every point of which is characterized by the electric field $\mathbf{E}(\mathbf{r}, t)$, magnetic field $\mathbf{H}(\mathbf{r}, t)$, and vector potential $\mathbf{A}(\mathbf{r}, t)$. The vector potential can be expanded in terms of a Fourier series as follows:

$$\mathbf{A}(\mathbf{r}, t) = \frac{c}{\sqrt{V}} \sum_k \mathbf{e} q_k(t) e^{i(\mathbf{k},\mathbf{r})} \tag{2.2.1}$$

where $\mathbf{e}$ is a unit vector specifying the direction of vector $\mathbf{A}$ and $q_k(t)$ are the coefficients of the expansion.

Due to the transversality of the electromagnetic field we write:

$$\operatorname{div}\mathbf{A}(\mathbf{r}, t) = i\frac{c}{\sqrt{V}}\sum_k (\mathbf{e}, \mathbf{k})\exp[i(\mathbf{k}, \mathbf{r})] = 0 \tag{2.2.2}$$

Hence, the vector $\mathbf{e}$ is perpendicular to the direction of propagation, which is defined by vector $\mathbf{k}$ and therefore this vector depends on $\mathbf{k}$.

Introducing two perpendicular orths $\mathbf{e}_\lambda$, $(\lambda = 1, 2)$, as shown in Figure 1, such as $(\mathbf{e}_\lambda, \mathbf{k}) = 0$, the vector $\mathbf{e} = \mathbf{e}(\mathbf{k})$ may be expanded in terms of these orths and, as a result, the vector potential assumes the form:

$$\mathbf{A}(\mathbf{r}, t) = \frac{c}{\sqrt{V}}\sum_{k,\lambda} \mathbf{e}_\lambda(\mathbf{k})q_{k\lambda}(t)e^{i(\mathbf{k},\mathbf{r})} \tag{2.2.3}$$

The vector potential is a real vector function, so it is possible to write

$$\mathbf{A}(\mathbf{r}, t) = \frac{c}{\sqrt{V}}\sum_{k,\lambda} \mathbf{e}_\lambda(-\mathbf{k})q^*_{-k\lambda}(t)e^{i(\mathbf{k},\mathbf{r})} \tag{2.2.4}$$

Special requirements for the q values will have the form

$$q^*_{-k1} = q_{k2} \qquad \text{and} \qquad q^*_{-k2} = q_{k1} \tag{2.2.5}$$

From the principles of electrodynamics it follows that, knowing the vector potential, it is possible to calculate the electric and magnetic field strengths by the following formulae:

$$\boldsymbol{E} = -\frac{1}{c}\frac{\partial A}{\partial t} = \frac{1}{\sqrt{V}}\sum_{k,\lambda} \mathbf{e}_\lambda(\mathbf{k})\dot{q}_{k\lambda}(t)e^{i(\mathbf{k},\mathbf{r})}$$

$$\boldsymbol{H} = \operatorname{curl}\mathbf{A} = i\frac{c}{\sqrt{V}}\sum_{k,\lambda} [\mathbf{e}_\lambda(\mathbf{k}) \times \mathbf{k}]q_{k\lambda}(t)e^{i(\mathbf{k},\mathbf{r})} \tag{2.2.6}$$

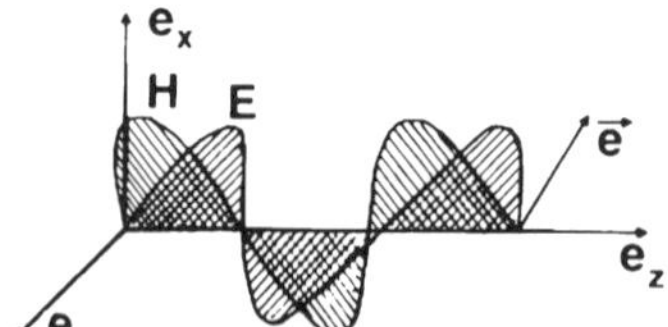

Figure 1. Orientation of the electric vector of an electromagnetic wave.

The energy of the electromagnetic field in volume V, according to basic relations involving the electric and magnetic field strengths, with the help of Eq. (2.2.5), takes the form

$$W = \frac{1}{8\pi}\int_V (H^2 + E^2) d\mathbf{r} = \frac{1}{8\pi}\sum_{k,\lambda}\left[\omega_k^2 q_{k\lambda} q_{k\lambda}^* + \dot{q}_{k\lambda}\dot{q}_{k\lambda}^*\right] \tag{2.2.7}$$

where $\omega_k = c|\mathbf{k}|$.

Formally, the electrodynamic relation (2.2.7) is analogous to the expression for the energy of an ensemble of harmonic oscillators of frequency ω_k.[36] It provides for the possibility that an electromagnetic field within some volume V can be considered classically as an ensemble of such harmonic oscillators. The transition to a quantum electrodynamic description of an electromagnetic field can be made by using this harmonic oscillator analogy.

By introducing the lowering operator $\hat{a}_{k\lambda}$ and its adjoint, the raising operator $\hat{a}_{k\lambda}^\dagger$, with the help of relations

$$\hat{q}_{k1} = \frac{\hat{a}_{k1} + \hat{a}_{-k2}^\dagger}{\sqrt{2\omega_k/\hbar}} \quad \text{and} \quad \hat{q}_{k2} = \frac{\hat{a}_{k2} + \hat{a}_{-k1}^\dagger}{\sqrt{2\omega_k/\hbar}} \tag{2.2.8}$$

we have

$$\mathbf{A}(\mathbf{r}, t) = \sqrt{\hbar}\frac{c}{\sqrt{V}}\sum_{k,\lambda}\frac{1}{\sqrt{2\omega_k}}\left[\mathbf{e}_\lambda(\mathbf{k})a_{k\lambda}(t)e^{i(\mathbf{k},\mathbf{r})} + \mathbf{e}_\lambda(-\mathbf{k})a_{k\lambda}^\dagger(t)e^{-i(\mathbf{k},\mathbf{r})}\right] \tag{2.2.9}$$

where the operators $\hat{a}_{k\lambda}$ and $\hat{a}_{k\lambda}^\dagger$ can be interpreted as operators responsible for creation and destruction of an oscillator, characterized by wave vector $\mathbf{k}$ and polarization vector $\mathbf{e}_\lambda$.

This means that, if $|N_{k\lambda}\rangle$ is the state vector of an ensemble of $N_{k\lambda}$ particles, we can formally write:

$$\hat{a}_{k\lambda}^\dagger|N_{k\lambda}\rangle = \sqrt{N_{k\lambda}}|N_{k\lambda} - 1\rangle \quad \text{and} \quad \hat{a}_{k\lambda}^\dagger|N_{k\lambda}\rangle = \sqrt{N_{k\lambda+1}}|N_{k\lambda} + 1\rangle \tag{2.2.10}$$

and hence

$$\hat{a}_{k\lambda}^\dagger\hat{a}_{k\lambda}|N_{k\lambda}\rangle = N_{k\lambda}|N_{k\lambda}\rangle \tag{2.2.11}$$

Therefore the operator $\hat{n}_{k\lambda} = \hat{a}^{\dagger}_{k\lambda}\hat{a}_{k\lambda}$ can be regarded as an operator acting on a number of particles of an ensemble with given wave vector and polarization. The relation

$$\hat{a}_{k\lambda}\hat{a}^{\dagger}_{k\lambda}|N_{k\lambda}\rangle = (N_{k\lambda} + 1)|N_{k\lambda}\rangle \tag{2.2.12}$$

is also valid, so the operators $\hat{a}^{\dagger}_{k\lambda}$ and $\hat{a}_{k\lambda}$ should obey the commutation relations:

$$\hat{a}_{k\lambda}\hat{a}^{\dagger}_{k'\lambda} - \hat{a}^{\dagger}_{k'\lambda}a_{k\lambda} = \delta_{kk'} \tag{2.2.13}$$

If Eq. (2.2.13) is substituted into (2.2.7) with allowance for (2.2.8), we have

$$W = \sum_{k,\lambda} \hbar\, \omega_k\, (\hat{a}^{\dagger}_{k\lambda}\hat{a}_{k\lambda} + \tfrac{1}{2}) \tag{2.2.14}$$

It follows from the latter relation that every oscillator possesses a constant energy quantum $h\omega_k/2$. A divergence problem arises while summing over an infinite number of oscillators, because expression (2.2.14) tends to infinity. Nevertheless, as we are interested only in thc cigcnvalues of the energy operator of the field, this difficulty can be overcome by just removing the term $\frac{1}{2}$ in expression (2.2.14).

By employing quantum electrodynamic language, Eq. (2.2.14) enables one to represent the electromagnatic field as an ensemble of particles (photons) with energy $h\omega_k$. Therefore, by rewriting relation (2.2.9) in the form

$$\hat{A} = \frac{c\sqrt{\hbar}}{\sqrt{V}} \sum_{k\lambda} (\hat{a}_{k\lambda}\mathbf{A}_{k\lambda} + \hat{a}^{\dagger}_{k\lambda}\mathbf{A}^{\dagger}_{k\lambda}) \tag{2.2.15}$$

the vector function

$$\mathbf{A}_{k\lambda} = \sqrt{\frac{2\pi}{\omega_k}}\, e^{i(\mathbf{k},\mathbf{r})}\mathbf{e}_{\lambda}(\mathbf{k}) \tag{2.2.16}$$

can be considered as the wave function of a photon with impulse $\hbar\mathbf{k}$ and direction of polarization given by the vector $\mathbf{e}_{\lambda}(k)$.

2.3. Momentun, Spin, and Parity of a Photon

The state of every particle in terms of a quantum mechanical description can be characterized by the total angular moment, which is the sum of the orbital l and spin s moments. The orbital moment reflects the coordinate properties of the wave function of a particle. The eignfunctions of a quantum state with given orbital moment l are a set of spherical functions of order l, which will be introduced in the next section. The orbital moment disappears within a reference frame in which the particle remains at rest.

The spin moment characterizes the internal symmetry of a system. The value of the spin moment does not depend on the choice of reference frame. It is connected to the fact that the set of $2s+1$ eignfunctions of the spin moment are transformed under rotations of the reference frame. If there exist some relations, connecting the spin variables and the spatial variables, the state of a particle can be described in this case not in terms of the separate orbital and spin moments, but only in terms of the total momentum.

In order to analyze the possibility of describing a photon state with the aid of its orbital and spin moments, we consider the expression for the photon wave function within the coordinate representation obtained at the end of the previous section. Transformation to the impulse representation yields the following expression for the wave function:

$$\mathbf{A}_{k\lambda} = \int d\mathbf{r}\, \mathbf{A}_{k,\lambda}(\mathbf{r}) e^{-i(\mathbf{k},\mathbf{r})} \tag{2.3.1}$$

The condition of transversality imposed on the vector components will take the form

$$(\mathbf{k}, \mathbf{A}_{k,\lambda}) = 0 \tag{2.3.2}$$

The wave function of a photon is a vector transformed under the reference frame rotation obeying the three-dimensional rotation group $D^{(1)}$, so the photon spin should be prescribed as the unit value. The transversality condition (2.3.2) imposes limitations on the allowed values of the impulse, as the impulse dependence corresponds to the coordinate dependence of the wave function within the impulse representation. These limitations, as well as the fact that there exist no system of coordinates in which the photon is at rest, lead to the conclusion that the state of the photon should be described only in terms of the total momentum j and, rigorously speaking, a separate description in terms of orbital or spin moments is impossible. Nevertheless, we shall formally retain the terms involving the spin and orbital moments of a photon. As the wave function of a photon is transformed as a vector under space rotations, we can allow the photon spin to equal unity. The orbital moment will then be determined by the order of the wave functions, representing every component.

An important photon characteristic is its parity, or the quantity reflecting the behavior of the wave function under inversion of the frame of reference. As the photon wave function is a vector, the inversion transformation not only reverses all the signs of the impulse components, but also reverses the signs of the coordinate axes. Therefore, if l is the orbital moment of a photon, or the quantity characterizing the order of the spherical functions of the wave function of a photon, the latter, due to inversion ($\theta \to \pi - \theta$; $\varphi \to \varphi + \pi$), will possess a factor $(-1)^{l+1}$. For a fixed total angular moment j and spin momentum $s = 1$, the orbital moment is strictly determined as $l = j; j \pm 1$. Three states with different parity are only possible in this case: one with parity $(-1)^{j+1}$ and two with parity $(-1)^{j}$. For $j = 0$, only one value of the orbital moent, $l = 1$, is possible and consequently only one even state can exist.[33]

All the above is valid for a vector function of type (2.3.1) with independent components. For a photon these components are not independent. They are connected by the transversality condition, which prohibits states with longitudinal orientation of the vector **k**. Therefore, of the total number of formal solutions, the states corresponding to a longitudinal orientation should be excluded. The general structure of the wave function corresponding to a longitudinal orientation can be represented as $\mathbf{k}f(\mathbf{k})$. The transformational features of a vector are linked to a certain coordinate or some value of **k** in the impulse representation, therefore the quantity $\mathbf{k}f(\mathbf{k})$ is transformed under rotations as a scalar. Hence the "longitudinal" states of a particle should be described by a zero value of the spin moment. This means that the value of the total moment of the wave function of the longitudinal state should be the same as the order of the spherical functions.

Transformation of the "transversal" wave function under a reference frame inversion can be written in the form

$$\hat{I}\mathbf{k}f(\mathbf{k}) = -(-\mathbf{k})f(-\mathbf{k}) = (-1)^{j}\mathbf{k}f(\mathbf{k}) \tag{2.3.3}$$

This means that the longitudinal state is of parity $(-1)^{j}$, so of all the states obtained above with parity $(-1)^{j}$ (two for $j \neq 0$ and one for $j = 0$), one should be excluded. Therefore, for $j \neq 0$ there exist two states with opposite parity. No one real state corresponds to a value $j = 0$, because this state is described by a spherically symmetric function and this contradicts the transversality condition. A photon with momentum j and parity (-1^{j}) is called an electric photon, and with parity $(-1)^{j+1}$ a magnetic photon.

2.4. Rotation Matrix

The spin of a particle characterizes the internal symmetry of a particle, describing the transformation of the wave function within the frame of reference in which this particle remains at rest. Let us introduce an infinitely small rotation of angle φ around the axis OZ. The coordinate transformation $x' =$

$x + \varphi y$; $y' = -\varphi x + y$ corresponds to such a rotation. For the corresponding variation in the wave function, one can write

$$\hat{R}_z(\varphi)\psi(x, y, z) = \psi(x + \varphi y, -\varphi x + y, z) = \left(1 - i\varphi \frac{\hat{S}_z}{\hbar}\right) \psi(x, y, z) \quad (2.4.1)$$

In the same manner, for rotations around the other coordinate axes

$$\begin{aligned} \hat{R}_x(\varphi)\psi(x, y, z) &= \left(1 - i\varphi \frac{\hat{S}_x}{\hbar}\right) \psi(x, y, z) \\ \hat{R}_y(\varphi)\psi(x, y, z) &= \left(1 - i\varphi \frac{\hat{S}_y}{\hbar}\right) \psi(x, y, z) \end{aligned} \quad (2.4.2)$$

Generalizing, the rotation operator around any axis u may be written as

$$\hat{R}_u(\varphi) = 1 - i\varphi \frac{\hat{S}_u}{\hbar} \quad (2.4.3)$$

If, in the latter expression, we take into consideration the initial condition $\hat{R}_u(0) = 1$ and the relation

$$\hat{R}_u(\varphi + \Delta\varphi) = \hat{R}_u(\varphi)\hat{R}_u(\Delta\varphi) = \hat{R}_u(\varphi)\left(1 - i\Delta\varphi \frac{\hat{S}_u}{\hbar}\right) \quad (2.4.4)$$

then the rotation operator can be represented in the form

$$\hat{R}_u(\varphi) = e^{-i\varphi \hat{S}_u/\hbar} \quad (2.4.5)$$

Let us consider two systems of reference, (x, y, z) and (X, Y, Z), having the same origin at the point zero, $x = y = z = X = Y = Z = 0$. The Euler angles α, β, γ will characterize the position of one system with respect to the other (Figure 2).

In this case the orientation of reference frame (X, Y, Z) is the result of three consecutive rotations of system (x, y, z): (1) a rotation of angle α around the oz axis transforms the axis oy into the axis u, (2) a rotation of angle β around the axis u transforms the oz axis into the axis OZ, and (3) a rotation of angle γ around

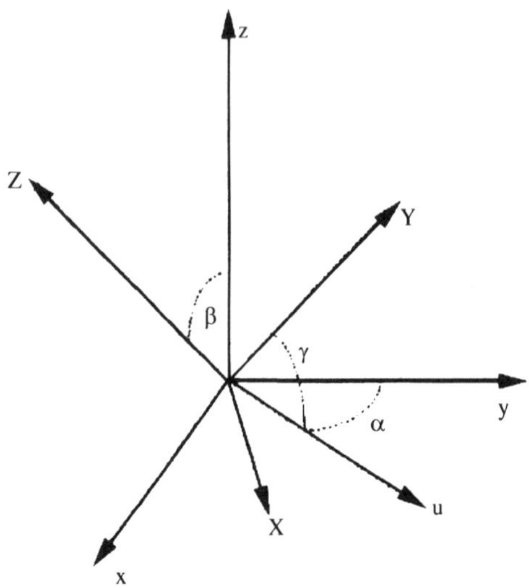

Figure 2. Euler angles of rotation.

the axis OZ. These transformations can be expressed mathematically as an operator:

$$\hat{R}(\alpha, \beta, \gamma) = \hat{R}_Z(\gamma)\hat{R}_u(\beta)\hat{R}_z(\alpha) \tag{2.4.6}$$

In line with the rotation operations, we may write

$$\hat{S}_u = \hat{R}_z(\alpha)\hat{S}_y\hat{R}_z^\dagger(\alpha) \qquad \text{and} \qquad \hat{S}_Z = \hat{R}_u(\beta)\hat{R}_z(\alpha)\hat{S}_z\hat{R}_z(\alpha)\hat{R}_u(\beta) \tag{2.4.7}$$

while expression (2.4.6) finally assumes the form

$$\hat{R}(\alpha, \beta, \gamma) = e^{-i\alpha\hat{S}_z/\hbar}e^{-i\beta\hat{S}_y/\hbar}e^{-i\gamma\hat{S}_z/\hbar} \tag{2.4.8}$$

In terms of the eigenfunctions of the operators of the square of the momentum and its projection $|Sm\rangle$, the following matrix corresponds to the rotation operator:

$$\hat{R}|Sm\rangle = \sum_{m'}|Sm'\rangle\langle Sm'|\hat{R}|Sm\rangle = \sum_{m'} D^{(S)}_{m'm}(\alpha\beta\gamma)|Sm'\rangle \tag{2.4.9}$$

As the operator $\hat{S}_z$ is diagonal within the basis of the eigenfunctions $|Sm\rangle$, it is evident from Eq. (2.4.8) that

$$D^{(S)}_{m'm}(\alpha, \beta, \gamma) = e^{-i(\alpha m' + \gamma m)} d^{(S)}_{m'm}(\beta) \tag{2.4.10}$$

where[32], we have

$$\begin{aligned} d^{(S)}_{m'm}(\beta) &= \langle Sm'| e^{-i\beta \hat{S}_y / \hbar r} |Sm\rangle \\ &= \sum_{t \geq 0} (-1)^t \frac{[(S+m')!(S-m')!(S+m)!(S-m)!]^{1/2}}{t!(S+m'+t)!(S-m-t)!(t+m-m')!} \\ &\quad \cdot \left(\cos\tfrac{1}{2}\beta\right)^{2S+m'-m-2t} \left(\sin\tfrac{1}{2}\beta\right)^{2t+m-m'} \end{aligned} \tag{2.4.11}$$

2.5. Spinors

If $|\frac{1}{2}m\rangle$ are the eignfunctions of the square of the spin S^2 ($S = \frac{1}{2}$) and of its projection, it is possible to show[27], that

$$\begin{aligned} \langle \tfrac{1}{2}m+1|\hat{S}_x|\tfrac{1}{2}m\rangle &= \frac{\hbar}{2}\sqrt{\left(\tfrac{1}{2}-m\right)\left(\tfrac{3}{2}+m\right)} \\ \langle \tfrac{1}{2}m-1|\hat{S}_x|\tfrac{1}{2}m\rangle &= \frac{\hbar}{2}\sqrt{\left(\tfrac{1}{2}+m\right)\left(\tfrac{3}{2}-m\right)} \\ \langle \tfrac{1}{2}m+1|\hat{S}_y|\tfrac{1}{2}m\rangle &= -i\frac{\hbar}{2}\sqrt{\left(\tfrac{1}{2}-m\right)\left(\tfrac{3}{2}+m\right)} \\ \langle \tfrac{1}{2}m-1|\hat{S}_y|\tfrac{1}{2}m\rangle &= i\frac{\hbar}{2}\sqrt{\left(\tfrac{1}{2}+m\right)\left(\tfrac{3}{2}-m\right)} \\ \left\langle \tfrac{1}{2}m|\hat{S}_z|\tfrac{1}{2}m\right\rangle &= \hbar m \end{aligned} \tag{2.5.1}$$

On introducing two-dimensional space formed by the functions

$$|\tfrac{1}{2}, \tfrac{1}{2}\rangle, \qquad |\tfrac{1}{2}, -\tfrac{1}{2}\rangle$$

the operators $\hat{S}_y$, $\hat{S}_x$, and $\hat{S}_z$ will be related associated to the following matrices:

$$S_x = \frac{\hbar}{2}\begin{pmatrix} 0 & 1 \\ 1 & 0 \end{pmatrix}, \qquad S_y = \frac{\hbar}{2}\begin{pmatrix} 0 & -i \\ i & 0 \end{pmatrix}, \qquad S_z = \frac{\hbar}{2}\begin{pmatrix} 1 & 0 \\ 0 & -1 \end{pmatrix} \tag{2.5.2}$$

which are connected to the Pauli $\sigma_x, \sigma_y, \sigma_z$ matrixes[33] by the relations

$$S_x = \frac{\hbar}{2}\sigma_x, \qquad S_y = \frac{\hbar}{2}\sigma_y, \qquad S_z = \frac{\hbar}{2}\sigma_z \tag{2.5.3}$$

A basic property of the Pauli matrices is

$$\sigma_j^2 = I \qquad (j = x, y, z)$$

so it is evident that

$$\sigma_j^{2\kappa} = I, \qquad \sigma_j^{2\kappa+1} = \sigma_j \qquad (j = x, y, z) \tag{2.5.4}$$

Therefore, the exponential representation of the rotation operator

$$\hat{R}_u(\varphi) = e^{-i\varphi\hat{S}_u/\hbar} = e^{-i\varphi\hat{\sigma}_u/2} \tag{2.5.5}$$

enables the exponent to be expanded as a series in such a way, that we finally have

$$\hat{R}_u(\varphi) = \hat{I}\cos\frac{\varphi}{2} - i\hat{\sigma}_u \sin\frac{\varphi}{2} \tag{2.5.6}$$

With the aid of the expression for operator $\hat{S}_y$ and allowance for Eqs. (2.4.10) and (2.4.11), one obtains the following matrix presentation:

$$D_{mm'}^{(1/2)}(\alpha, \beta, \gamma) = e^{-i(\alpha m' + \gamma m)} \begin{pmatrix} \cos\beta/2 & -\sin\beta/2 \\ \sin\beta/2 & \cos\beta/2 \end{pmatrix} \tag{2.5.7}$$

It is now possible to define a two-component function transformed under spatial rotations with the help of matrix $D_{m,m'}^{(1/2)}$; it is called a spinor. Spinors are very important objects in the theory of angular momentum, because spinor spaces are used for the theoretical description of many atomic processes. Spinor spaces of different dimensions could be created, using the rules of spinor algebra. With the help of spinors

$$|\tfrac{1}{2}, m\rangle, \qquad (m = \pm\tfrac{1}{2})$$

or of spinors of the first rank, one can employ the general rules for Clebsch–Gordan coefficients to constrruct spinors of higher ranks. In such a way, a variety of products of the type

$$|\tfrac{1}{2}, m\rangle|\tfrac{1}{2}, m'\rangle \tag{2.5.8}$$

form a spinor of second rank, for which direct multiplication of representations $D^{(1/2)} \times D^{(1/2)}$ is implemented.

Taking the basis of eigenfunctions $|1, m\rangle \quad (m = 0, \pm 1)$:

$$
\begin{aligned}
|1, 1\rangle &= \sum_{mm'} \begin{bmatrix} \frac{1}{2} & \frac{1}{2} & 1 \\ m & m' & 1 \end{bmatrix} |\tfrac{1}{2}, m\rangle |\tfrac{1}{2}, m'\rangle = |\tfrac{1}{2}, \tfrac{1}{2}\rangle |\tfrac{1}{2}, \tfrac{1}{2}\rangle \\
|1, 0\rangle &= \sum_{m} \begin{bmatrix} \frac{1}{2} & \frac{1}{2} & 1 \\ m & -m & 0 \end{bmatrix} |\tfrac{1}{2}, m\rangle |\tfrac{1}{2}, m\rangle = \sqrt{2} |\tfrac{1}{2}, \tfrac{1}{2}\rangle |\tfrac{1}{2}, -\tfrac{1}{2}\rangle \\
|1, -1\rangle &= \sum_{mm'} \begin{bmatrix} \frac{1}{2} & \frac{1}{2} & 1 \\ m & m' & -1 \end{bmatrix} |\tfrac{1}{2}, m\rangle |\tfrac{1}{2}, m'\rangle = |\tfrac{1}{2}, -\tfrac{1}{2}\rangle |\tfrac{1}{2}, -\tfrac{1}{2}\rangle,
\end{aligned}
\tag{2.5.9}
$$

this representation can be split into irreducible forms by taking into consideration that

$$
|1, 1\rangle = i\sqrt{\frac{3}{4\pi}} \cos\theta \qquad \text{and} \qquad |1, \pm 1\rangle = \mp i \sqrt{\frac{3}{4\pi}} \sin\theta \exp(\pm i\varphi).
$$

It is easy to see that, under space transformations, the spinors

$$
|\tfrac{1}{2}, \tfrac{1}{2}\rangle |\tfrac{1}{2}, \tfrac{1}{2}\rangle, \qquad |\tfrac{1}{2}, \tfrac{1}{2}\rangle |\tfrac{1}{2}, -\tfrac{1}{2}\rangle, \qquad |\tfrac{1}{2}, -\tfrac{1}{2}\rangle |\tfrac{1}{2}, -\tfrac{1}{2}\rangle
$$

are transformed as vectors

$$
\mathbf{e}_1 = \frac{\mathbf{e}_y - i\mathbf{e}_x}{\sqrt{2}}, \qquad \mathbf{e}_0 = i\mathbf{e}_z, \qquad \mathbf{e}_{-1} = \frac{\mathbf{e}_y + i\mathbf{e}_x}{\sqrt{2}}
$$

respectively.

Therefore, the unit vectors of the orthogonal frame of reference correspond to the components of a spinor of second rank:

$$
\mathbf{e}_z \leftrightarrow |\tfrac{1}{2}, \tfrac{1}{2}\rangle |\tfrac{1}{2}, -\tfrac{1}{2}\rangle, \qquad \mathbf{e}_1 \leftrightarrow |\tfrac{1}{2}, \tfrac{1}{2}\rangle |\tfrac{1}{2}, \tfrac{1}{2}\rangle, \qquad \mathbf{e}_{-1} \leftrightarrow |\tfrac{1}{2}, -\tfrac{1}{2}\rangle |\tfrac{1}{2}, -\tfrac{1}{2}\rangle
\tag{2.5.10}
$$

2.6. Relationship between the Polarization Direction and Characteristics of a Photon

In Section (2.2) we introduced the direction of polarization as a vector lying in the plane perpendicular to the direction of propagation of the electromagnetic wave, specified by wave vector **k**. Following Yutsys and Banzaitis[(27)] it is convenient to expand this vector in terms of two orthogonal unit vectors determined by the relations

$$\mathbf{e}_1 = \frac{e_y - ie_x}{\sqrt{2}} \qquad \text{and} \qquad \mathbf{e}_{-1} = \frac{e_y + ie_x}{\sqrt{2}}$$

as follows:

$$\mathbf{e} = c_1\mathbf{e}_1 + c_2\mathbf{e}_2 \tag{2.6.1}$$

On the other hand, it is clear that the direction of polarization of a photon should be determined by its internal characteristics. We noted earlier that the spin of a photon is the only internal characteristic of a photon and is equal to unity. In order to define a relationship between the spin projections and polarization direction it should be mentioned that, as there exists no system of reference in which a photon is not moving, the direction of photon propagation is the only preferential axis. Therefore it is possible to characterize the state of a particle by the projection of its spin on this axis, or the *spirality* λ. As the spin of a photon is equal to unity, its spirality may take only three values:

$$\lambda = 0, \pm 1.$$

Referring to expressions (2.5.10), it is clear that the spinor $|\frac{1}{2},\frac{1}{2}\rangle|\frac{1}{2},\frac{1}{2}\rangle$, corresponds to the spirality with $\lambda = 1$. It is transformed under rotations of the frame of reference as the vector $\mathbf{e}_1$. This means that only right circular polarization corresponds to a photon with such a spirality. In the same manner, the spinor $|\frac{1}{2},-\frac{1}{2}\rangle|\frac{1}{2},-\frac{1}{2}\rangle$ corresponds to a photon with spirality $\lambda = -1$ and to polarization vector $\mathbf{e}_{-1}$, which implies that this photon is left circularly polarized. The spinor $|\frac{1}{2},\frac{1}{2}\rangle|\frac{1}{2},-\frac{1}{2}\rangle$ corresponds to a photon with $\lambda = 0$, which is transformed as vector $\mathbf{e}_z$. Such a photon with zero spirality should be polarized along the axis $\mathbf{k}$, which is impossible due to the very strict transversality condition for the electromagnetic wave.

2.7. *Relationship between Stokes Parameters and the Characteristics of an Optical Field*

In order to fully describe the polarization characteristics of the electromagnetic radiation, it is necessary to determine four parameters or conduct four independent measurements. If OZ is the axis of the light beam propagation, these parameters could result from the following measurements:

1. I,

the total radiation intensity within the solid angle $d\Omega$ and the polarizations

$$2.\quad Q=\frac{I_x - I_y}{I} \qquad 3.\quad U=\frac{I_{xy^+} - I_{xy^-}}{I} \qquad 4.\quad V=\frac{I_+ - I_-}{I} \tag{2.7.1}$$

where I_x and I_y are the intensities of light polarized along the x and y axes, respectively, I_{xy^+} and I_{xy^-} are the intensities of light polarized along the diagonals of the first and second quarters (orths e_{xy^+} and e_{xy^-} in Figure 3), while I_+ and I_- are the intensities of the right and left circularly polarized light.

The experimental phenomenological values determined by relations (2.7.1) are called the Stokes parameters. Following Blum[(102)] and Bommier,[138] the Stokes parameters can be expressed in terms of the projections of the electric field strength of the electromagnetic wave $E_{x,y}$:

$$\begin{aligned} I &= \frac{c}{4\pi}\overline{E_x^2(\mathbf{r},t) + E_y^2(\mathbf{r},t)}, & Q &= \frac{c}{4\pi}\overline{E_x^2(\mathbf{r},t) - E_y^2(\mathbf{r},t)} \\ U &= \frac{c}{2\pi}\overline{E_x(\mathbf{r},t)E_y(\mathbf{r},t)}, & V &= \frac{c}{2\pi}\overline{E_x(\mathbf{r},t)E_y(\mathbf{r},t+\tfrac{1}{2}T)} \end{aligned} \tag{2.7.2}$$

where T is the period of oscillations of the electromagnetic wave and the bar denotes temporal averaging.

If the electric field vector is represented by

$$\mathbf{E}(\mathbf{r}) = \frac{1}{\sqrt{2}}[\mathscr{E}(\mathbf{r},t)e^{i\omega t} + \mathscr{E}^*(\mathbf{r},t)e^{-i\omega t}] \tag{2.7.3}$$

where $\mathscr{E}(\mathbf{r},t)$ is a slowly varying function of time that can be expanded in terms of the circular orths, then relations (2.7.2) can be expressed in the following form:

$$\begin{aligned} I &= \frac{c}{4\pi}[\overline{\mathscr{E}_{\infty}\mathscr{E}^*_{\infty}} + \overline{\mathscr{E}_{-\infty}\mathscr{E}^*_{-\infty}}], & Q &= \frac{c}{4\pi}[\overline{\mathscr{E}_{\infty}\mathscr{E}^*_{-\infty}} + \overline{\mathscr{E}_{-\infty}\mathscr{E}^*_{\infty}}] \\ U &= -\frac{c}{4\pi}i[\overline{\mathscr{E}_{\infty}\mathscr{E}^*_{-\infty}} - \overline{\mathscr{E}_{-\infty}\mathscr{E}^*_{\infty}}], & V &= \frac{c}{4\pi}[\overline{\mathscr{E}_{\infty}\mathscr{E}^*_{\infty}} - \overline{\mathscr{E}_{-\infty}\mathscr{E}^*_{-\infty}}] \end{aligned} \tag{2.7.4}$$

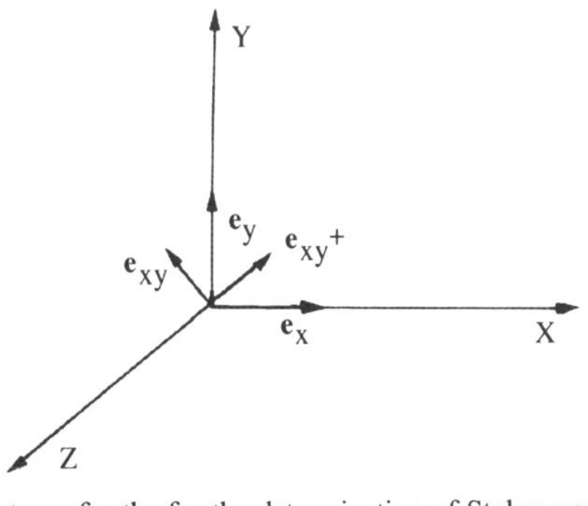

Figure 3. System of orths for the determination of Stokes parameters.

By transforming expression (2.2.9) for the field localized within a volume V to the electromagnetic field, spread over the whole space, and considering expression (2.2.6) and the temporal dependencies, the electric field operator can be written as

$$\mathbf{E}(\mathbf{r}, t) = \int d\mathbf{k} \sum_{\lambda} \sqrt{\frac{\hbar\omega}{(2\pi)^3}}\, i[a_{k\lambda}\mathbf{e}_{\lambda}(k)e^{i[(\mathbf{k},\mathbf{r})+\omega t]} + a^{\dagger}_{k\lambda}\mathbf{e}^{*}_{\lambda}(k)e^{-i[(\mathbf{k},\mathbf{r})+\omega t]}] \tag{2.7.5}$$

If the latter expression is compared with (2.7.3) and the relation

$$d\mathbf{k} = \frac{8\pi^3 \nu^2}{c^3} d\nu d\Omega_k$$

is employed, then we easily derive

$$\frac{c}{4\pi}\mathscr{E}_{\lambda}\mathscr{E}^{*}_{\lambda'} = 2\pi\frac{\hbar\nu^3}{c^2}a^{\dagger}_{k\lambda}a_{k\lambda'} \tag{2.7.6}$$

The polarization matrix with elements determined by the relation

$$I_{\lambda\lambda'}(\nu, \Omega_{\kappa}) = 2\pi\frac{\hbar\nu^3}{c^2}a^{\dagger}_{k\lambda}a_{k\lambda'} \tag{2.7.7}$$

enables expressions (2.7.4) to be transformed as follows:

$$\begin{aligned} I &= \langle I_{11}\rangle + \langle I_{-1-1}\rangle, & Q &= \langle I_{-1-1}\rangle + \langle I_{-11}\rangle \\ U &= -i(\langle I_{1-1}\rangle - \langle I_{-11}\rangle), & V &= \langle I_{11}\rangle - \langle I_{1-1}\rangle \end{aligned} \tag{2.7.8}$$

where the angle brackets denote averaging over the photon states.

From the latter formulas it is clear that the polarization matrix completely determines the polarization properties and anisotropy of the electromagnetic radiation.

2.8. Concluding Remarks

This chapter is of an introductory nature and shows that polarization is one of the principal characteristics of an electromagnetic field, and is directly related to the basic quantum electrodynamic parameters of a photon and general symmetry considerations. One very important result, closely linked to the general aim of this practically oriented book, is that the spectral Stokes parameters incorporated in every remote spectral sensing problem represent in themselves a full set of basic spectroscopic observables related to the fundamental characteristics of an optical field.

3

Theory of Collisional Spectropolarimetric Effects

3.1. Introduction

A principal physical assumption underlying the mathematical formalism used to describe polarization effects is the concept of an ensemble of atomic particles (atoms, ions, molecules) in mixed states. These states are described by linear combinations of the wave functions, corresponding to the stationary states. A complete system could be represented mathematically in this case in terms of the density matrix, which is obtained by means of statistical averaging of the states of particular atoms. The density matrix formalism is widely used for the theoretical analysis of interference phenomena of atomic states.(33,36,102) In order to account in the most adequate way for the symmetry of a system, which could be specified within the framework of many typical problems related to the polarization of an atomic ensemble and the spectropolarimetric effects of its radiation, and to model many real physical situations, it is convenient to consider the density operator in the basis within which the irreducible representations of the rotational group are realized. The matrix elements in this representation are known as polarization moments.

Historically, the polarization moments of an ensemble were first introduced by Fano(28) as the characteristics of the ordering of momenta of atomic particles. Later, Omont(31) and Diakonov and Perel(29,30) related atomic polarization moments to the spectroscopic observables of the polarized radiation measured in the experiment. They have also proposed the basic equations, representing the evolution of polarization moments under collisions. The theory of the density matrix has now been well elaborated and presented in a number of reviews and monographs (see, for example, Baas and Castel(32) and Blum(102)). The general structure of the basic equations for the polarization moments is determined by the nature of the excitation process of an atomic ensemble under study.(93,100) As different types of collisions form the basic excitation mechanisms in ionized media, it is worthwhile studying the principal possibilities of collisional ordering of an atomic ensemble resulting in the emission of polarized radiation. The collisional polarization effects can be divided into two groups regarding the magnitude of the relative velocity of the colliding partners.

The first group comprises processes of excitation related to collisions of slow particles with relative velocities not high enough for the direct excitation of any partner. In this case, the ordering of momenta of initially excited atomic particles is induced only via the collisional relaxation within an array of the fine or hyperfine structure substates or the zeeman sublevels, or, in other words, by transitions between these sublevels, induced by slow collisions within an ensemble of atomic particles. Anisotropies in the velocity space of the colliding partners associated with such an elastic collisional relaxation could provoke complicated mutual kinetics, the transformation of polarization moments with different ranks, and the creation of polarization moments of a particular order from the others. When anisotropies introduced by the slow collisions play a constructive role in the organization of excited atomic ensembles, and thereby in the polarization of optical emission, such a situation is typical for most natural and artificial ionized media, which are sustained due to energy transport in any nonthermal form by static or alternative fields and gradients of parameters. This is valid even for a simple laboratory gas discharge plasma, where excitation of atoms and ions is implemented mostly by the fast isotropically distributed electrons. However, the transitions within the zeeman, fine and hyperfine structure arrays could be strongly affected by the slow drift motion of the atomic particles along the axis of the discharge in the middle of a cylindrical plasma and radially near the wall of the discharge tube.(58,59)

Collisions of fast particles, resulting in the direct excitation of one of the colliding partners, belong to the second group of impact polarization phenomena. Beams of the charged accelerated particles can also be regarded as a general case of the creation of ionized media under laboratory and natural conditions. For example, many extraterrestrial ionized gas entities are formed by accelerated electrons or protons, precipitating to the outer atmosphere of the Earth. Many ionized gas objects in the solar atmosphere are also formed by precipitating beams of fast particles and will be analysed in this book.

The method used to record polarization effects could interfere with the detected spectropolarimetric observables. For direct recording of the spectropolarimetric parameters of radiation(91,92) the collisional process is characterized by axial symmetry specified by an external beam or internal drift motion. In this case only the collisional creation of alignment or quadrupole ordering of angular momenta of excited particles and observation of the linear polarization of optical emission is possible.(10) For more sophisticated techniques in the spectropolarimetry of an excited atomic ensemble, such as when an emitted photon is detected in correlation with an inelastically scattered particle, axial symmetry is violated and different polarization effects can be observed.(86,104–109)

Of utmost importance is the theoretical formalism for a description of the spectropolarimetric effects as a basis for different diagnostic techniques. In the first six sections of this chapter the quantum statistical density operator will be introduced together with the polarization moments as the matrix of this operator within the basis, in which the irreducible representations of the rotational group are realized. It will be shown that, on the one hand, the polarization moments characterize an internal ordering of an excited ensemble while, on the other hand, they are directly related to the spectroscopic observables. In the seventh section, following the results of Bommier,[(138)] the main master kinetic equation will be obtained to describe the evolution of a subsystem caused by interaction with a thermostate. A collisional variation in the density matrix of an atomic ensemble, caused by interaction with different systems (thermostates), will be examined in the following two sections. In the eighth section, which is closer to a real situation in many natural and laboratory ionized media, the role of the thermostate is played by a beam of foreign precipitating particles. Two different cases are analyzed with regard to the velocity of the beam particles. In the case of low relative velocities, this beam is only mixing populations of quantum states of the previously excited atomic system. In the case of high relative velocities, direct impact excitation creates the impact ordering of angular momenta of the atomic ensemble.[(58,59)] Making use of this general approach and studying the interaction of an atomic ensemble with electromagnetic radiation, the role of the thermostate in the ninth section will be played by electromagnetic radiation, affecting an atomic system. Matrix elements of operators from the basic kinetic equation, responsible for the processes of spontaneous, induced radiation and absorption will be obtained there as well.[(137)]

The last three sections can be regarded as an illustration and application for the spectropolarimetric sensing problems of the theoretical apparatus, developed in the first nine sections. Physical aspects of impact alignment creation, the spectropolarimetric features of H_α line radiation, as well as ways to determine the atomic parameters responsible for the description of the polarization of emission of atomic hydrogen will be reviewed in the tenth section. Theoretical aspects of the collisional relaxation of electron polarization moments of an ensemble of hydrogen atoms in the state $n = 2$ under collisions with a low energy proton beam are presented in the eleventh section by employing results of original work.[(94)] This problem is of particular significance from the standpoint of astrophysical applications, because the $2s$ hydrogen state is a metastable one, with the lifetime of 0.12 s, and the excited atoms could be stored in this state under conditions of the interstellar medium. The L_α radiation from the astrophysical objects could provide much valuable information on the process of collisional alignment subject to low relative velocities, observed earlier in laboratory low temperature discharge plasma.[(58,59)] The spectropolartimetric effects on the hydrogen L_α line are also analyzed in more detail.

Recordings of the optical emission of only a proportion of excited particles scattered at some angle specify a particular group of experiments in atomic collisional spectroscopy, known as angular correlation experiments. Because of the lower symmetry of the collisional process, the creation of not only aligned but also oriented states, characterized by dipole ordering of momenta of excited particles, is possible under angular correlation experimental conditions.[86,104–109] Theoretical aspects of experiments under selective recording of radiation from particles scattered at a particular angle,[86,104–109,160,161] as well as atomic parameters inferred from these experiments and possessing fundamental significance for the implementation of impact spectropolarimetric sensing, are analyzed in detail in the final section of this chapter.

3.2. *Density Matrix of an Atomic Ensemble*

If W_n is the probability of an atom to be in a pure state $|n\rangle$ with wave function $\Psi_n(x, t)$ (where x denotes the complete set of spin and spatial variables), one can define the density operator by

$$\hat{\rho}(x, x', t) = \sum_n W_n \Psi_n(x, t) \Psi_n(x', t) \tag{3.2.1}$$

The stationary states of free atoms are described by the full set of functions $\psi_{\alpha jm}(x)$, which are the eigenfunctions of operators $\hat{j}^2$ and $\hat{j}_z$ (the full moment and its projection), with eigenvalues $\hbar j(j+1)$ and $\hbar m$, respectively

$$\hat{j}^2 \psi_{\alpha,j,m}(x) = \hbar^2 j(j+1) \psi_{\alpha,j,m}(x) \qquad \text{and} \qquad j_z \psi_{\alpha,j,m}(x) = \hbar m \psi_{\alpha,j,m}(x) \tag{3.2.2}$$

where all the other quantum numbers relevant to an atomic system, but not connected to the momentum operator, are included in α.

The functions $\psi_{\alpha,j,m}$ as any other wave functions are determined using a particular phase selection. Two methods of phase selection are utilized as a rule. The first is known as standard phase selection[27]

$$\psi^*_{\alpha jm}(x) = (-1)^{j-m} \psi_{\alpha j-m}(x) \tag{3.2.3a}$$

and the second is known as natural phase selection[32]

$$\psi^*_{\alpha jm}(x) = (-1)^m \psi_{\alpha j-m}(x) \tag{3.2.3b}$$

As a rule we shall use here natural phase selection of the wave function. A note will be made, when standard phase selection is used.

If the set of eigenfunctions is complete, a wave function $\Psi_n(x,t)$ may be represented as an expansion

$$\Psi_n(x,t) = \sum_{\alpha jm} C^{(n)}_{\alpha jm}(t)\psi_{\alpha jm}(x) \tag{3.2.4}$$

Substitution of Eq. (3.2.4) in (3.2.1) yields

$$\hat{\rho}(x,x',t) = \sum_{\alpha jm\alpha' j' m'} \rho_{\alpha jm,\alpha' j' m'}(t)\psi_{\alpha jm}(x)\psi^*_{\alpha' j' m'}(x') \tag{3.2.5}$$

Hence

$$\rho_{\alpha jm,\alpha' j' m'}(t) = \langle \alpha jm|\hat{\rho}|\alpha' j' m'\rangle = \sum_n W_n C^{(n)}_{\alpha jm}(t) C^{(n)*}_{\alpha' j' m'}(t) \tag{3.2.6}$$

is the density matrix element in *jm* representation, while W_n is the probability of a particular atom to be at some state with wave function $\Psi_n(x,t)$.

As the density operator $\hat{\rho}$ is Hermitian, we have the following symmetry properties:

$$\rho_{\alpha jm,\alpha' j' m'} = \rho^*_{\alpha' j' m' \alpha jm} \tag{3.2.7}$$

If $|C^{(n)}_{\alpha jm}|^2$ is the probability of an atom to be at a stationary state with wave function $\psi_{\alpha jm}(x)$, the diagonal elements of the density matrix

$$\rho_{\alpha jm,\alpha jm}(t) = \sum W_n |C^{(n)}_{\alpha jm}(t)|^2 \tag{3.2.8}$$

will define the probability of an atom, interacting with the system, to be at some stationary state with wave function $\psi_{\alpha jm}(x)$.

The density matrix is a universal mean to describe pure quantum mechanical effects in ensembles. One such effect, referring to the wave nature of quantum states, is the interference of states. This effect could have a classical analogy only as the interference of electromagnetic fields, and is related to the concept of coherence of electromagnetic waves. There exist a group of interference effects in spectroscopy, mostly in polarization spectroscopy, which arise only if coherence between different states of atomic systems has been created.[2,102,134]

Following analogies with electromagnetic wave phenomena and the definition of the density matrix, one can say that when the density matrix is nondiagonal in the *jm* representation, there exists a coherent superposition of atomic states. The coherence is named optical if the energy gap between the states in the frequency scale belongs to the optical range. If this value belongs to the radiofrequency range, it is called Hertzian coherence. Nondiagonal elements of the density matrix characterize the degree of mixing of quantum states.

Mixing and interference of atomic states lead to polarization of atomic ensembles. Phenomenologically, polarization of atomic ensembles is connected to different interference and polarization effects, observed in optical emissions of atomic systems under natural conditions, or detected in specially organized laboratory experiments.[1,2]

The mean value of a quantum mechanical operator corresponding to an observable quantity is defined in terms of the density operator by

$$\langle \hat{L} \rangle = Sp(\hat{\rho}\hat{L}) = \sum_n W_n \int \Psi_n^*(x,t)\hat{L}\Psi_n(x,t)dx = + \sum_{\alpha jm\alpha' j'm'} \rho_{\alpha jm,\alpha' j'm'} \langle \alpha' j'm|\hat{L}|\alpha jm\rangle \tag{3.2.9}$$

It follows from definitions (3.2.1)–(3.2.3) that the density operator is normalized to unity, but sometimes it may be normalized to the atomic number density. In the latter case, the values W_n are considered to be the number of particles at the $|n\rangle$ state.

The equation describing the variation in the density matrix can be written as

$$\frac{\partial}{\partial t}\hat{\rho} = \left(\frac{\partial \hat{\rho}}{\partial t}\right)_{int} - \hat{\Gamma}\hat{\rho} + \frac{i}{\hbar}\left[\hat{H}_0, \rho\right] \tag{3.2.10}$$

where the first term describes the variation in the density matrix due to interaction with external systems, the second term denotes radiation damping, and the third term comprises the Hamiltonian of the free atom H_0.

The kinetics of the density operator will be examined in more detail in Section 3.7. By considering some variation in the density operator provided by the third term of Eq. (3.2.10), this equation can be expressed in the following matrix form

$$\frac{\partial}{\partial t}\rho_{\alpha jm,\alpha' j'm'} = \frac{i}{\hbar}(E_{\alpha j} - E_{\alpha' j'})\rho_{\alpha jm,\alpha' j'm'} + N_{\alpha jm,\alpha' j'm'} \tag{3.2.11}$$

where matrix $N_{\alpha jm,\alpha' j'm'}$ denotes the excitation process of an atomic ensemble.

From this formula one can see that the nondiagonal *jm* density matrix elements are small for a large energy gap $E_{\alpha j} - E_{\alpha j'}$. An approximation, used in many problems of atomic physics, takes into account only diagonal terms with respect to *jm* in Eq. (3.2.11) and is known as circular. It is evident that the circular approximation is not valid in the case of closely spaced levels in the energy scale and, in particular, for an occasional degeneration.

3.3. Polarization Density Matrix

The state of a photon can be characterized by the projection of the spin moment on the axis, specified by vector **k**. Due to the transversality condition, this projection (called the spirality) can only assume the two values $\lambda = \pm 1$ (see Section 2.5). Hence a photon with given impulse and spirality is in a pure state and can be characterized by the polarization: right circular polarization, $\lambda = 1$; left circular polarization, $\lambda = -1$.

Let us now analyse an ensemble of photons in a light beam. If W_α is the number of photons of a beam that are at a pure state $|\alpha\rangle$, we can construct the polarization density matrix in the spirality representation using relation (3.2.5), as follows

$$\rho_{\lambda\lambda'} = \sum_\alpha W_\alpha C_\lambda^{(\alpha)} C_{\lambda'}^{(\alpha)} \tag{3.3.1}$$

where $\lambda = \pm 1$ and $C_\lambda^{(\alpha)}$ denotes the probability for a photon in a pure state $|\alpha\rangle$ to have a given spirality.

The matrix $\rho_{\lambda\lambda'}$, complex of second order, is generally determined by eight independent parameters. As this matrix is hermitian [see Eq. (3.1.7)], the number of parameters required for a complete determination of this matrix is reduced to four. In this case one of these parameters can be taken as the trace of the matrix ρ, which determines the full intensity of the beam

$$Sp(\rho) = \rho_{11} + \rho_{-1-1} = \sum_\alpha W_\alpha = I \tag{3.3.2}$$

while the Stokes parameters Q, U, and V, given by relation (2.7.1), can be chosen as the other three parameters.

In order to represent the elements of the polarization density matrix in terms of these selected parameters, let us write the density matrix (3.3.1) in the basis $\mathbf{e}_x$, $\mathbf{e}_y$, taking into consideration relations linking the circular orths $\mathbf{e}_{\pm 1}$ to the Cartesian orths $e_{x,y}$

$$\mathbf{e}_y = \frac{\mathbf{e}_1 + \mathbf{e}_{-1}}{\sqrt{2}} \qquad \text{and} \qquad \mathbf{e}_x = \frac{i(\mathbf{e}_1 - \mathbf{e}_{-1})}{\sqrt{2}}$$

The density matrix will then have the form

$$\rho = \begin{pmatrix} \mathrm{Re}(\rho_{11} - \rho_{-11}) & -\mathrm{Im}(\rho_{-11} - \rho_{11}) \\ -\mathrm{Im}(\rho_{11} + \rho_{-11}) & \mathrm{Re}(\rho_{11} + \rho_{-11}) \end{pmatrix} \tag{3.3.3}$$

The intensity of the light beam, polarized along the orth **e**, can be obtained with the aid of the general expression

$$I_{\mathbf{e}} = (\mathbf{e}, \rho\mathbf{e}), \tag{3.3.4}$$

using the definition of the Stokes parameters (2.7.1) and relation (3.3.3). Hence we easily obtain

$$\begin{aligned} I &= I_x + I_y = Sp(\rho) = 2\,\mathrm{Re}(\rho_{11}) \\ IQ &= I_x - I_y = \rho_{xx} - \rho_{yy} = -2\,\mathrm{Re}(\rho_{-11}) \\ IU &= I_{xy^+} - I_{xy^-} = \rho_{xy} + \rho_{yx} = 2\,\mathrm{Im}\,\rho_{-11} \\ IV &= I_+ - I_- = \rho_{11} - \rho_{-1-1} = 2i\,\mathrm{Im}\,\rho_{11} \end{aligned} \tag{3.3.5}$$

Finally, the matrix (3.3.3) in terms of the Stokes parameters will assume the form

$$\rho = \frac{I}{2}\begin{pmatrix} 1+Q & U+iV \\ U-iV & 1-Q \end{pmatrix} \tag{3.3.6}$$

This means that four parameters (the total intensity of radiation and three Stokes parameters) fully determine the polarization density matrix.

If λ_i are eigenvalues of the density matrix, the following relations are valid for the representation in which the density matrix is diagonal

$$Sp\rho = \sum_i \lambda_i = I \qquad \text{and} \qquad Sp\rho^2 = \sum_i \lambda_i^2 \tag{3.3.7}$$

The quantity λ_i/I can be regarded as a probability, so we can set

$$0 \le \frac{\lambda_i}{I} \le 1$$

Hence it is clear that

$$Sp\rho^2 \le (Sp\rho)^2 \le I^2 \tag{3.3.8}$$

Equality occurs only if the system is at a pure state in which the density matrix is diagonal and only one of the diagonal elements is nonzero.

By substituting Eq. (3.3.6) into (3.3.8) it is easy to see that

$$Sp\rho^2 = \frac{I^2}{2}(1 + U^2 + V^2 + Q^2) \le I^2 \tag{3.3.9}$$

and the sum of squares of the Stokes parameters satisfies

$$U^2 + V^2 + Q^2 \leq 1 \tag{3.3.10}$$

with equality again only if the system of photons is at a pure state. Therefore, the quantity

$$P = \sqrt{U^2 + V^2 + Q^2} \tag{3.3.11}$$

can be taken as a parameter specifying the degree of polarization. The photon beam is completely polarized if $P = 1$, nonpolarized if $P = 0$, and partially polarized, if $0 < P < 1$.

In order to determine experimentally three Stokes parameters of an optical field (2.7.1) that consist of the total intensity and combination of polarizations, one should carry out more than one measurement; for example, one should measure the three linearly polarized intensities I_x, I_y, and I_z. In practical problems of polarization spectroscopy, the degree of linear polarization of a light beam is characterized by

$$P = \frac{I_1 - I_2}{I_1 + I_2} \tag{3.3.12}$$

where I_1 and I_2 are the intensities of light with orthogonally linear polarizations. In the case of circular polarization, these values are intensities with right and left circular polarizations.

3.4. *Polarization Moments of the Density Matrix*

The density matrix in the *jm* representation is not the most convenient mathematical object, because for many problems of atomic physics it leads to cumbersome computations and the solution of a system of equations of high order. The main reason is that the *jm* representation does not account for the symmetry properties of the operators on the right hand side of Eq. (3.2.10). Hence for practical problems directly connected to the spectroscopic observables of many particular atomic systems emitting optical radiation, and for elaboration of the spectropolarimetric sensing techniques based on quantitative measurements of these observables, we shall employ polarization moments, which are sets of density matrix elements in a representation known in optical and nuclear spectroscopy as *kq* representation.

The advantages of the polarization moment formalism and the representation of the atomic density matrix in terms of polarization moments, introduced by

Fano[28] and subsequently used in many problems of atomic physics and spectroscopy,[29–31] are straightforward. First, it provides a clear physical meaning for the polarization moments and a direct connection to the spectroscopic observables. Second, this formalism brings much mathematical simplicity because the equation governing the density matrix evolution in the kq representation assumes the simplest form. Third, this formalism provides the most elegant and physically clear link between the symmetry of excitation of atomic ensembles under different natural or artificial conditions, the types of ordering of momenta of excited atoms, and the polarization properties of the emission.

An introduction to polarization moment formalism provided a breakthrough for many problems of atomic and collisional physics and plasma spectroscopy.[1,102,134,135] On the one hand, this formalism provided a better understanding of the internal mechanisms of collisions, the structure of electron shells of atomic particles at the post-collision phase, and the collisional spectroscopic effects in the optical emission of an atomic ensemble. On the other hand, the polarization moment formalism enabled one to understand and describe clearly the transformation of orderings and anisotropies in the velocity space of particles exciting the optical emission of a plasma into the macroscopic orientation of the electron shells of the excited atomic particles, responsible for an optical emission. As a result, spectropolarimetry could be transformed into a quantitative sensing technique.[1,2,10,11,134]

A mathematical introduction to this formalism is straightforward. Instead of a set of products $\psi_{\alpha jm}(x)\psi^*_{\alpha' j' m'}(x')$ for which the representation of the $D^{(j)} \otimes D^{(j')}$ rotational group is realized, we use a set of operators $\hat{T}^k_q(x, x'; \alpha j, \alpha' j)$ related to the first set by

$$\psi_{\alpha jm}(x)\psi^*_{\alpha' j' m'}(x') = \sum_{k,q} (-1)^{j'-m'} \begin{bmatrix} j & j' & k \\ m & -m' & q \end{bmatrix} \hat{T}^k_q(x, x'; \alpha j, \alpha' j') \tag{3.4.1}$$

For the second set of operators a direct product of representations could be split into irreducible sets. The transformation of these quantities under space rotation is given by

$$\hat{T}^k_q(x, x'; \alpha j, \alpha j') = \sum_{q'} D^{(k)}_{q'q}(\alpha, \beta, \gamma)\hat{T}^k_{q'}(x, x'; \alpha j, \alpha' j') \tag{3.4.2}$$

where (α, β, γ) are Euler angles. By these means, the set $\hat{T}^k_{q'}(x, x'; \alpha j, \alpha' j')$ became irreducible tensor operators.

The orthogonality of the Clebsh–Gordan coefficients yields an inverse relation with respect to (3.4.1) in matrix form

$$\langle \alpha jm | \hat{T}^k_q(x, x'; \alpha j, \alpha' j') | \alpha' j' m' \rangle = (-1)^{j'-m'} \begin{bmatrix} j & j' & k \\ m & -m' & q \end{bmatrix} \tag{3.4.3}$$

On the other hand, using the Wigner–Eckart theorem for irreducible tensor operators,[27] one can write

$$\langle \alpha jm | \hat{T}_q^k(x, x'; \alpha j, \alpha' j') | \alpha' j' m' \rangle = \langle \alpha j \| \hat{T}^k(x, x') \| \alpha' j' \rangle \begin{bmatrix} j' & k & j \\ m' & q & m \end{bmatrix} \tag{3.4.4}$$

where $\langle \alpha j \| \hat{T}^k(x, x') \| \alpha' j' \rangle$ is the reduced matrix element.

By comparing Eqs. (3.4.3) and (3.4.4) one can see that

$$\langle \alpha j \| \hat{T}^k(x, x') \| \alpha' j' \rangle = \sqrt{(2k+1)/(2j+1)} \tag{3.4.5}$$

Let us denote $\hat{T}_q^{k+}(x, x'; \alpha j, \alpha' j')$ the adjoint operator with respect to $\hat{T}_q^k(x, x'; \alpha j, \alpha' j')$ for the matrix elements we can write

$$\langle \alpha jm | \hat{T}_q^{k+}(x, x'; \alpha j, \alpha' j') | \alpha' j' m' \rangle = \langle \alpha' j' m' | \hat{T}_q^k(x, x'; \alpha j, \alpha' j') | \alpha jm \rangle \tag{3.4.6}$$

Subsequently, the symmetry condition takes the form

$$\hat{T}_q^{k+}(x, x'; \alpha j, \alpha' j') = (-1)^{j-j'-q} \hat{T}_{-q}^k(x, x'; \alpha j; \alpha' j') \tag{3.4.7}$$

and the orthogonality relation can be expressed as follows

$$Sp\Big[\hat{T}_q^{k+}(x, x'; \alpha j, \alpha' j') \cdot \hat{T}_{q'}^{k'}(x, x'; \alpha j, \alpha' j')\Big] = \delta_{kk'} \delta_{qq'} \tag{3.4.8}$$

It is clear from the latter formula, that the trace of the operator with rank greater than zero, is zero; or

$$Sp \hat{T}_q^k(x, x'; \alpha j, \alpha' j') = \sqrt{2k+1}\, \delta_{k0} \delta_{q0} \tag{3.4.9}$$

Now, formula (3.2.4) with the help of relation (3.4.1) may be written in the following way:

$$\hat{\rho}(x, x'; t) = \sum_{kq} \rho_q^k(\alpha j, \alpha' j'; t) \hat{T}_q^k(x, x'; \alpha j, \alpha' j') \tag{3.4.10}$$

where

$$\rho_q^k(\alpha j, \alpha' j'; t) = \sum_{mm'} (-1)^{j'-m'} \begin{bmatrix} j & j' & k \\ m & -m' & q \end{bmatrix} \rho_{\alpha jm, \alpha' j' m'}(t) \tag{3.4.11}$$

are known as polarization moments.

The character of the polarization moments is defined by the physical nature of the angular momentum j. If we examine the orbital momentum $j = l$, one can speak about the orbital polarization moments. If j is the total electron moment, or the sum of the orbital and spin moments, then the quantities expressed by the general definition (3.4.11) are called the electron polarization moments. In the case of nonzero nuclear spin I, the quantum atomic stats are characterized by the total momentum as the sum of the electronic and nuclear momenta $F = I + j$, and then the corresponding polarization moments, given by (3.4.11), are called the electron-nuclear polarization moments.

Formally speaking, formula (3.4.11) may be considered as an expansion of density operator in terms of irreducible tensor operators. Making use of this approach, the polarization moments are effectively the coefficients of this expansion. The relation for the polarization moments, similar to Eq. (3.2.7), can be obtained by utilizing (3.4.11) as the formal definition of the polarization moments

$$\rho_q^{k^*}(\alpha j, \alpha' j'; t) = (-1)^{j-j'+q} \rho_{-q}^k(\alpha j, \alpha' j'; t) \tag{3.4.12}$$

3.5. Physical Meaning of Polarization Moments

As already mentioned, the density matrix elements at the kq representation or the polarization moments have a deeper physical meaning than that at the jm represention; they describe the different types of ordering of an atomic ensemble. This is easily illustrated, because the result of a statistical averaging of irreducible tensor operators is related to the polarization moments.

The diagonal over quantum numbers α and j for a polarization moment of zero rank is proportional to the population of state $n_{\alpha j}$

$$\rho_0^0(\alpha j, \alpha j; t) = \frac{\sum_m \rho_{\alpha j m \alpha j m}(t)}{\sqrt{2j+1}} = \frac{n_{\alpha j}}{\sqrt{2j+1}} \tag{3.5.1}$$

In order to realize the physical meaning of a polarization moment of rank higher than zero, let us perform a statistical averaging of the q-component of an irreducible tensor operator of rank k

$$\begin{aligned} \langle \hat{T}_q^k \rangle &= Sp(\hat{\rho} \hat{T}_q^k) \\ &= (-1)^q \sum_{\alpha j \alpha' j'} \langle \alpha j' \| T^k \| \alpha j \rangle \sqrt{\frac{2j+1}{2k+1}} (-1)^{j-j'} \rho_{-q}^k(\alpha j, \alpha' j'; t) \end{aligned} \tag{3.5.2}$$

It follows from Eq. (3.5.2) that the q-component of the irreducible tensor operator of rank k, averaged over the ensemble, is a linear combination of polarization moments of the same rank. The averaged values of the irreducible tensor operators, constructed from the components of the angular moment and the dipole moment of an allowed optical transition, are of most interest in spectroscopic applications.

The formalism of irreducible tensor operators and polarization moments is not limited only to an optimal description of the optical spectroscopic effects associated with the dipole moment of an optically allowed atomic transition. The same methodology can be applied to the magnetic dipole and the electric quadrupole moments, which are related to many effects of coherent resonance optics, interaction of atomic systems with fields, collisional physics, and nonlinear optical processes. The magnetic dipole moment is proportional to the mean value of the angular moment

$$\boldsymbol{\mu} = -g_j \mu_0 \langle j \rangle \tag{3.5.3}$$

where g_j and μ_0 are the Landé factor and Bohr magneton.

The electric quadrupole moment is the mean value of the symmetrical second rank tensor, constructed from angular moment components.

In order to provide a link between polarization moments and the physical characteristics of an atomic ensemble, let us consider an operator $\hat{J}_q^k$ or the q-component of the irreducible tensor operator of rank k, constructed from components of the total angular moment. As the Hamiltonian of an atom and $\hat{j}^2$ operator commute with operators $\hat{j}_x, \hat{j}_y, \hat{j}_z$, the nondiagonal over matrix elements α and j will be zero. Therefore, Eq. (3.5.2) yields

$$\langle \hat{J}_q^k \rangle = (-1)^q \sum_{\alpha j} \langle \alpha j \| \hat{J}^k \| \alpha j \rangle \sqrt{\frac{2j+1}{2k+1}} \rho_{-q}^k(\alpha j; t) \tag{3.5.4}$$

Consequently, the mean value of $\hat{J}_q^k$ is a linear combination of polarization moments of identical rank. Hence the mean value of $\hat{J}_q^1$ is a linear combination of the quantities ρ_q^1, $q = 0, \pm 1$, which are transformed as a vector under space rotations. Three components of ρ_q^1 form the orientation vector. According to Eq. (3.5.3) the components of this vector represent the magnetic dipole moment of an atomic ensemble. The five elements of ρ_q^2, where $q = 0, \pm 1, \pm 2$, are called the alignment tensor. The components of this tensor define the mean value of the quadrupole moment of a system.

Quantities $\rho_q^1(\alpha j; \alpha j; t)$ and $\rho_q^2(\alpha j; \alpha j; t)$ are usually known as the orientation and alignment of an αj level. The orientation and alignment of an atomic system are the linear combinations of the corresponding values of the momentum projections for a level specifying an energy state of the system (3.5.2). The following expressions link the orientation and alignment with the mean values of tensor $\hat{j}_q^k$

$$\begin{aligned}
\langle j_z \rangle &= \sqrt{2j(2j+1)(2j+2)/12}\,\rho_0^1(\alpha j, \alpha j; t) \\
\langle j_\pm \rangle &= \pm\sqrt{2j(2j+1)(2j+2)/6}\,\rho_{\mp 1}^1(\alpha j, \alpha j; t) \\
\langle 3j_z^2 - j^2 \rangle &= \sqrt{(2j-1)2j(2j+1)(2j+2)(2j+3)/20}\,\rho_0^2(\alpha j, \alpha j; t) \qquad (3.5.5) \\
\langle j_z j_\pm + j_\pm j_z \rangle &= \mp\sqrt{(2j-1)2j(2j+1)(2j+2)(2j+3)/60}\,\rho_{\mp 1}^2(\alpha j, \alpha j; t) \\
\langle j_\pm^2 \rangle &= \sqrt{(2j-1)2j(2j+1)(2j+2)(2j+3)/30}\,\rho_{\pm 2}^2(\alpha j, \alpha j; t)
\end{aligned}$$

The third formula in (3.5.5) is the most illustrative and is presented in Figure 4 in graphical form. This picture shows the angular dependence of the quantity

$$\frac{3\cos^2\theta - 1}{2} = \frac{\langle 3j_z^2 - j^2\rangle}{2}\Big|_{j=1}$$

which demonstrates the longitudinal alignment $\rho_0^2(j)$ of an atomic ensemble with $j = 1$ and the axially symmetric distribution of momenta with respect to the OZ axis.

If all the moments are parallel to the OZ axis ($\langle\cos\theta\rangle = 1$), the degree of alignment is maximal and equal to 1; if all the momenta are perpendicular to this axis ($\langle\cos\theta\rangle = 0$), it is minimal and equal to $-\frac{1}{2}$. For $\langle\cos\theta\rangle = 1/\sqrt{3}$ ($\theta \approx 54.73°$), the longitudinal alignment is absent while the distribution of momenta is ordered. This angle is often called the magic angle. For an axially symmetric system invariant with respect to reflections in the XY plane, the orientation is also zero. Hence for such a system at the magic angle only the population, the zero-order component of alignment and probably the zero-order components of the higher-rank polarization moments are not zero. This means that for an axially symmetric ensemble of particles with angular moments $j = 1$ at the magic angle, only the population is present, although the momenta are

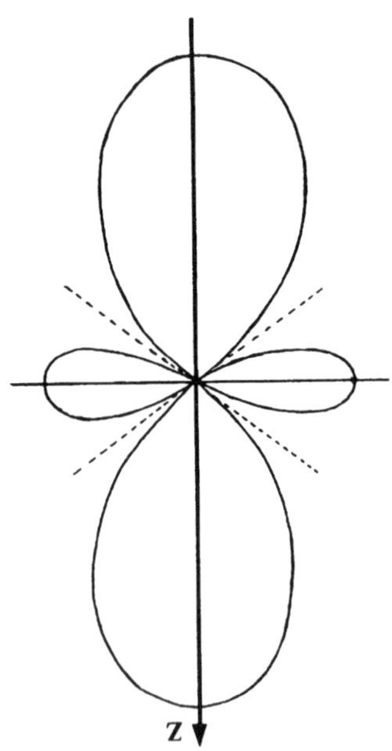

Figure 4. Angular dependence of the longitudinal alignment.

ordered ($\langle\cos\theta\rangle \neq 0$). This phenomenon may be used to create nonaligned ensembles under typical experimental conditions with axial symmetry of the excitation process.[141]

As the Hamiltonian operator does not commute with the dipole moment components, the mean values of operators comprising these components are represented by a formula analogous to Eq. (3.5.2). In particular, the mean values of the second rank-tensor components are formally related to the components of the quadrupole moment

$$\langle\hat{Q}_q\rangle = \sum_{\alpha j\alpha' j'} (-1)^{j-j'+q} \langle\alpha j\|2\|\alpha' j'\rangle \sqrt{\frac{2j+1}{5}} \rho^2_{-q}(\alpha j, \alpha' j'; t) \tag{3.5.6}$$

Therefore, speaking in terms of the polarization moments of an ensemble, the five components of the alignment tensor define the mean value of the quadrupole moment of this ensemble.

Generally, the polarization moments are connected to the symmetry of the system. For spherical symmetry, all polarization moments with $k > 0$ are zero. For an axially symmetric system, all polarization moments with $q \neq 0$ are also zero. The last two considerations follow directly from the requirement of invariance of the system with respect to rotations around the center of symmetry and the axis of symmetry. If a system is spherically symmetric, the polarization moment should be invariant under rotations which, according to Eq. (3.4.2), is possible only if $k = 0$. The polarization moment ρ^k_q for the axially symmetric system with respect to the OZ axis, after rotation at angle α, acquires the phase factor $\exp(i\alpha q)$. Therefore, the invariance of the polarization moments under rotations with respect to the symmetry axis will be realized only if $q = 0$. It is illustrated in Figure 5 where arrows indicate the momenta of particles for the system under study.

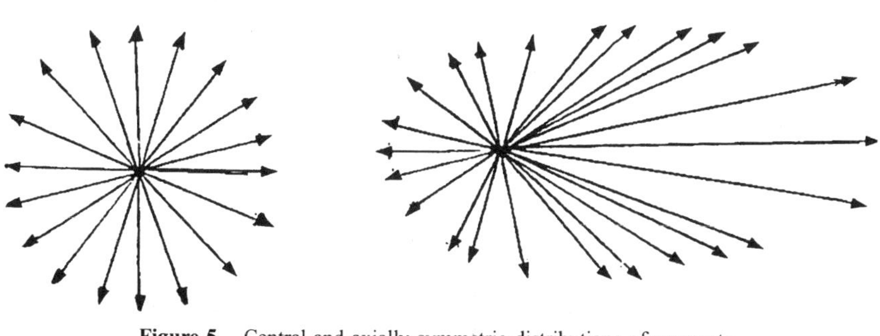

Figure 5. Central and axially symmetric distributions of momenta.

3.6. Polarization Moments and Spectral Characteristics of an Atomic Ensemble

The polarization moments represent the ordering of an ensemble, so this formalism is especially convenient to describe the polarization phenomena of dipole radiation. Suppose an atomic ensemble is excited to a state that is a coherent superposition of atomic states $|\alpha j\rangle$. The intensity of the polarized radiation, emitted on transition to the ground state $|\alpha_0 j_0\rangle$, is expressed with the help of the polarization matrix ρ [see Eq. (3.3.1)] as follows

$$I_\lambda = \langle e_\lambda|\rho|e_\lambda\rangle = \sum_{j_0 m_0} \langle j_0 m_0 e_\lambda|\rho|j_0 m_0 e_\lambda\rangle \tag{3.6.1}$$

where $\mathbf{e}_\lambda$ is the polarization vector.

The dipole approximation enables this formula to be written in the form

$$I_\lambda = I_0 \sum_{j_0 m_0 m m_1} \langle j_0 m_0|(\mathbf{e}_\lambda, \mathbf{d})|jm\rangle \langle jm_1|(\mathbf{e}_\lambda, \mathbf{d})^*|\, j_0 m_0\rangle \rho_{jm,jm_1} \tag{3.6.2}$$

where $\mathbf{d}$ is the dipole moment.

The polarization orth $\mathbf{e}_\lambda$ can be expressed in terms of the circular unit vectors

$$\mathbf{e}_\lambda = \sum_q a_q^{(\lambda)} \mathbf{e}_q$$

with the help of the orthogonality of the Clebsh–Gordan coefficients. Hence

$$I_\lambda = I_0 |\langle \alpha_0 j_0 \| 1 \| \alpha j\rangle|^2 (2j_0 + 1)(-1)^{j+j_0+1} \cdot \sum_{kq} \begin{Bmatrix} 1 & 1 & k \\ j & j & j_0 \end{Bmatrix} \Phi^k_{-q}(\mathbf{e}_\lambda) \rho^k_q(\alpha j, \alpha j; t) \tag{3.6.3}$$

where $\langle \alpha_0 j_0 \| 1 \| \alpha j\rangle$ is the reduced matrix element and

$$\Phi^k_q(\mathbf{e}_\lambda) = \sum_{q'q''} (-1)^{q'} a_{q'}^{(\lambda)} a_{q''}^{(\lambda)*} \begin{bmatrix} 1 & 1 & k \\ q' & q'' & q \end{bmatrix} \tag{3.6.4}$$

is the polarization tensor.

It is evident from Eq. (3.6.4), that

$$\Phi^{k*}_q(\mathbf{e}_\lambda) = (-1)^q \Phi_{k-q}(\mathbf{e}_\lambda) \tag{3.6.5}$$

One can see from relation (3.6.3) that, due to the properties of the Wigner coefficients, the intensity of polarized dipole radiation is a linear combination of polarization moments with ranks less than two. The moments of rank higher than two are called "hidden", because they do not contribute to the dipole radiation. Higher-rank polarization moments may influence the polarization characteristics of radiation, giving rise to the "polarization beats" phenomenon.[34]

For spectral emission polarized in the direction defined by spherical angles ϑ and φ, the elements of the polarization tensor take the form

$$\Phi_0^0 = 1/\sqrt{3}; \qquad \Phi_q^1 = 0, \qquad q = 0, \pm 1$$
$$\Phi_0^2 = (1 - 3\cos^2\vartheta)/\sqrt{6}, \qquad \Phi_{-1}^2 = \sin 2\vartheta e^{i\varphi}/2 \qquad \Phi_{-2}^2 = -\sin^2\vartheta e^{2i\varphi}/2 \tag{3.6.6}$$

Substitution of Eq. (3.6.2) in (3.6.3) for the intensity of linear polarized radiation yields

$$I_\lambda = I_0 |\langle \alpha_0 j_0 \| 1 \| \alpha j \rangle|^2 (2j_0 + 1)(-1)^{j+j_0+1} \left\{ \frac{(-1)^{j+j_0+1}}{3\sqrt{2j+1}} \rho_0^0(\alpha j; t) \right.$$
$$+ \begin{Bmatrix} 1 & 1 & 2 \\ j & j & j_0 \end{Bmatrix} \left[\frac{(1 - 3\cos^2\vartheta)\rho_0^2(\alpha j; t)}{\sqrt{6}} + \mathrm{Re}[\sin 2\vartheta e^{i\varphi} \rho_1^2(\alpha j; t)] \right.$$
$$\left.\left. - \mathrm{Re}[\sin^2\vartheta e^{2i\varphi} \rho_2^2(\alpha j; t)] \right] \right\} \tag{3.6.7}$$

For the intensity of circular polarized radiation propagating in the direction defined by spherical angles ϑ and φ, the elements of the polarization tensor take the form

$$\Phi_0^0 = 1/\sqrt{3}, \qquad \Phi_0^1 = \varepsilon \cos\vartheta/\sqrt{2}, \qquad \Phi_1^1 = \varepsilon e^{i\varphi} \sin\vartheta/2$$
$$\Phi_0^2 = (3\cos^2\vartheta - 1)/2\sqrt{6}, \qquad \Phi_1^2 = -e^{i\varphi}\sin 2\vartheta/4, \qquad \Phi_2^2 = \sin^2\vartheta e^{i2\varphi}/4 \tag{3.6.8}$$

where $\varepsilon = \pm 1$ denotes right and left circular polarizations, respectively.

Equation (3.6.3) enables the following formula to be obtained:

$$I_\lambda = I_0 |\langle \alpha_0 j_0 \| 1 \| \alpha j \rangle|^2 (2j_0 + 1)(-1)^{j+j_0+1} \left\{ \frac{(-1)^{j+j_0+1}}{3\sqrt{2j+1}} \rho_0^0(\alpha j; t). \right.$$
$$- \varepsilon \begin{Bmatrix} 1 & 1 & 1 \\ j & j & j_0 \end{Bmatrix} \left[\frac{\cos\vartheta \rho_0^1(\alpha j; t)}{\sqrt{2}} + \mathrm{Re}[e^{i\varphi} \sin\vartheta \rho_1^2(\alpha j; t)] \right]$$
$$+ \begin{Bmatrix} 1 & 1 & 2 \\ j & j & j_0 \end{Bmatrix} \left[\frac{(3\cos^2\vartheta - 1)\rho_0^2(\alpha j; t)}{2\sqrt{6}} - \frac{1}{2}\mathrm{Re}[e^{i\varphi} \sin 2\vartheta \rho_1^2(\alpha j; t)] \right.$$
$$\left.\left. + \frac{1}{2}\mathrm{Re}[\sin^2\vartheta e^{2i\varphi} \rho_2^2(\alpha j; t)] \right] \right\} \tag{3.6.9}$$

Expressions (3.6.7) and (3.6.9) consist of an isotropic part proportional to the zero-rank polarization moment, and an anisotropic part determined by higher-rank polarization moments. Within the dipole approximation the latter are exclusively the orientation and alignment of an excited state. These relations express the intensity I_λ in terms of polarization moments and thereby define the polarization and spatial properties of the light emission.

On the other hand, if the intensity anisotropy and polarization are known, the general relations (3.6.7) and (3.6.9) enables one to resolve the inverse problem of the determination of the polarization moments and consequently the parameters pertaining to the ordering of an atomic ensemble. These parameters are connected, in turn, to the specification of the excitation process of atomic ensembles, resulting either from the local parameters of the medium or from various external interactions. Therefore, these general relations constitute a principal basis for the spectropolarimetric sensing technique. Using the three values $I_{x,y,z}$ of the linear polarizations one can construct different intensity combinations, which are easily related to the polarization moments. The sum of the linear polarized intensities is invariant under space rotation, being connected to the scalar quantity of the higher level population

$$I_x + I_y + I_z = I_0 |\langle \alpha_0 j_0 \| 1 \| \alpha j \rangle|^2 \frac{2j_0 + 1}{\sqrt{2j + 1}} \rho_0^0(\alpha j; t) \tag{3.6.10}$$

Other combinations of intensities polarized in two mutually orthogonal directions are given by the relations

$$(I_x - I_y)_z = 2I_0 |\langle \alpha_0 j_0 \| 1 \| \alpha j \rangle|^2 (2j_0 + 1)(-1)^{j+j_0+1} \begin{Bmatrix} 1 & 1 & 2 \\ j & j & j_0 \end{Bmatrix} \cdot \mathrm{Re}[\rho_2^2(\alpha j; t)] \tag{3.6.11}$$

$$\begin{aligned} (I_x - I_z)_y = {} & I_0 |\langle \alpha_0 j_0 \| 1 \| \alpha j \rangle|^2 (2j_0 + 1)(-1)^{j+j_0+1} \begin{Bmatrix} 1 & 1 & 2 \\ j & j & j_0 \end{Bmatrix} \\ & \cdot \left[\frac{3\rho_0^2(\alpha j; t)}{\sqrt{6}} - \mathrm{Re}\, \rho_2^2(\alpha j; t) \right] \end{aligned} \tag{3.6.12}$$

$$\begin{aligned} I_{xz^+} - I_{xz^-} = {} & \sqrt{2} I_0 |\langle \alpha_0 j_0 \| 1 \| \alpha j \rangle|^2 (2j_0 + 1)(-1)^{j+j_0+1} \begin{Bmatrix} 1 & 1 & 2 \\ j & j & j_0 \end{Bmatrix} \\ & \cdot \mathrm{Re}\, [\rho_1^2(\alpha j; t)] \end{aligned} \tag{3.6.13}$$

where I_{xz^+} and I_{xz^-} are intensities polarized in directions defined by spherical angles $\vartheta = \pi/4$, $\varphi = 0$ and $\varphi = \pi$, respectively.

The expression for the $\rho_0^2(\alpha j, \alpha j; t)$ moment may be obtained by combining these intensities:

$$(I_x + I_y + I_z)/3 - I_z = \frac{2I_0}{\sqrt{6}} |\langle \alpha_0 j_0 \| 1 \| \alpha j \rangle|^2 (2j_0 + 1) \cdot (-1)^{j+j_0+1} \begin{Bmatrix} 1 & 1 & 2 \\ j & j & j_0 \end{Bmatrix} \rho_0^2(\alpha j; t) \qquad (3.6.14)$$

The difference between the right and left circularly polarizated intensities is proportional to the orientation vector and, in turn, to the mean value of the momentum projections:

$$\begin{aligned} (I_+ - I_-)_z &= K_{j_0 j} \rho_0^1(\alpha j; t) \sim \langle j_z \rangle \\ (I_+ - I_-)_y &= K_{j_0 j} \operatorname{Im} \rho_1^1(\alpha j; t) \sim \langle j_y \rangle \\ (I_+ - I_-)_x &= K_{j_0 j} \operatorname{Re} \rho_1^1(\alpha j; t) \sim \langle j_x \rangle \end{aligned} \qquad (3.6.15)$$

where

$$K_{j_0 j} = (-1)^{j+j_0+1} I_0 |\langle \alpha_0 j_0 \| 1 \| \alpha j \rangle|^2 (2j_0 + 1)$$

Finally, the spatial anisotropy of linear polarized radiation is completely determined by the five alignment tensor components, while three components of the orientation vector define the spatial anisotropy of circularly polarized light.

To conclude this section, it is noteworthy that the above discussion referred to the simple transition of an atomic particle from a state with momentum j to the lower energy state j_0. In case of a hyperfine structure or any occasional degeneration, the basic relation (3.6.3) should be changed. In particular, for the intensity of radiation under transitions of atomic particles from an excited state with principal quantum number n to a state with n_0 in the presense of occasional degeneration, the following relation is valid:

$$I_\lambda = I_0 \sum_{ll'_{kQ}} W^{(k)}(ll') \rho_{-Q}^k(ll') \Phi_Q^k(\mathbf{e}_\lambda) \qquad (3.6.16)$$

where the geometrical factor $\Phi_Q^K(\mathbf{e}_\lambda)$ is determined by the same formula (3.6.4), but the term $W^K(ll')$ is given by

$$W^{(k)}(ll') = \sum_{l_0} \sqrt{(2l+1)(2l_0+1)} i^{l'-l} [n_0 l_0 || 1 || nl] \cdot [nl' || 1 || n_0 l_0] \begin{Bmatrix} 1 & 1 & k \\ l & l' & l_0 \end{Bmatrix}, \qquad (3.6.17)$$

The reduced matrix elements in square brackets are considered as a representation of the standard phase selection.[27]

After integration with respect to the angular variables, the latter expression can be transformed in terms of the radial matrix elements of operator **r**, or r_{nl,n_0l_0}, as follows:

$$W^k(ll') = \sqrt{(2l+1)(2l'+1)} \sum_{l_0} \begin{bmatrix} l & 1 & l_0 \\ 0 & 0 & 0 \end{bmatrix} \begin{bmatrix} l' & 1 & l_0 \\ 0 & 0 & 0 \end{bmatrix} \cdot r_{nln_0l_0} r_{nl'n_0l_0} \begin{Bmatrix} 1 & 1 & k \\ l & l' & l_0 \end{Bmatrix} \tag{3.6.18}$$

3.7. Density Matrix Relaxation Theory

Relaxation phenomena are expressed mathematically in terms of nonstationary perturbation theory. Let us consider an atomic system in which every atom is characterized by a certain nonperturbed Hamiltonian H_0 and interaction of an atom with the environment is described by operator $V(t)$. The Hamiltonian operator of a particular atom can be written as

$$H(t) = H_0 + V(t) \tag{3.7.1}$$

The wave function $\Psi(x, t)$ describing the state of this atom satisfies the nonstationary Schrödinger equation

$$i\hbar \frac{\partial}{\partial t} \Psi(x, t) = [H_0 + V(t)]\Psi(x, t) \tag{3.7.2}$$

with initial condition

$$\Psi(x, 0) = \Psi_0(x) \tag{3.7.3}$$

If the solution of Eq. (3.7.2) is written formally as

$$\Psi(x, t) = U(t)\Psi_0(x) \tag{3.7.4}$$

it is easy to see that the operator $U(t)$ satisfies the equation

$$i\hbar \frac{\partial}{\partial t} U(t) = H(t)U(t) \tag{3.7.5}$$

with

$$U(0) = I \tag{3.7.6}$$

The density operator, which was introduced in Eq. (3.2.1), may be expressed for this system in the form

$$\rho(x, x', t) = \langle \Psi(x, t)\Psi^*(x't) \rangle \tag{3.7.7}$$

where the angle brackets mean ensemble averaging.

The value of the density operator at any time can be related, with the help of (3.7.4) to its value at the initial moment by means of the equation

$$\rho(t) = U(t)\rho(0)U^\dagger(t) \tag{3.7.8}$$

where all other arguments of the density operator are omitted for simplicity.

Differentiation of the latter expression with respect to time and consideration of Eq. (3.7.5) show that the density operator satisfies the following equation:

$$i\hbar \frac{\partial}{\partial t}\rho(t) = i\hbar \frac{\partial U(t)}{\partial t}\rho(0)U^\dagger(t) - U(t)\rho(0)\left[i\hbar \frac{\partial}{\partial t}U(t)\right]^\dagger = [H(t), \rho(t)] \tag{3.7.9}$$

which is called the Liouville equation for the Schrödinger representation.

Equations (3.7.5) and (3.7.9) are not quite convenient for the theoretical description of an interaction process, because the Hamiltonian operator on the right hand side consists of terms referring to the nonperturbed atom. In order to identify a variation in the density matrix caused by the interaction term $V(t)$, it is necessary to transform into the interaction representation.

Within the framework of the interaction representation, the wave function $\tilde{\Psi}(x, t)$ can be obtained from the same quantity for the Schrödinger representation with the aid of the transformation:

$$\tilde{\Psi}(x, t) = e^{iH_0 t/\hbar}\Psi(x, t) \tag{3.7.10}$$

Then, using Eq. (3.7.8), the density operator in this representation is given by

$$\tilde{\rho}(t) = \langle \tilde{\Psi}(x, t)\tilde{\Psi}^*(x', t) \rangle = e^{iH_0 t/\hbar}\rho(t)e^{-iH_0 t/\hbar} \tag{3.7.11}$$

If the latter expression is differentiated with respect to time with allowance for Eq. (3.7.9), the density matrix for the interaction representation is found to satisfy the Liouville equation in a form specific for the interaction representation:

$$i\hbar \frac{\partial}{\partial t}\tilde{\rho}(t) = [\tilde{V}(t), \tilde{\rho}(t)] \tag{3.7.12}$$

where $\tilde{V}(t)$ denotes the perturbation operator in the interaction representation:

$$\tilde{V}(t) = e^{iH_0t/\hbar} V(t) e^{-iH_0t/\hbar} \tag{3.7.13}$$

An expression for the density operator within the interaction representation, analogous to Eq. (3.7.8), can be obtained with the aid of Eqs. (3.7.10) and (3.7.11):

$$\tilde{\rho}(t) = \tilde{U}(t)\tilde{\rho}(0)\tilde{U}^{\dagger}(t) \tag{3.7.14}$$

where it follows from Eq. (3.7.4) that

$$\tilde{U}(t) = e^{iH_0t/\hbar} U(t) \tag{3.7.15}$$

Differentiating this expression with respect to time, yields that the evolution operator in the interaction representation is given by the equation

$$i\hbar \frac{\partial \tilde{U}(t)}{\partial t} = \tilde{V}(t)\tilde{U}(t) \tag{3.7.16}$$

where $\tilde{V}(t)$ is determined by expression (3.7.13).

Equations (3.7.12) and (3.7.16) illustrate the advantages of the interaction representation in that variations in operators $\tilde{\rho}(t)$ and $\tilde{V}(t)$ are formally determined only by an interaction characterized by operator $V(t)$. The solution of Eq. (3.7.16) can be obtained by an iteration technique, taking a zero approximation as an initial condition:

$$\tilde{U}^{(0)}(t) = \tilde{U}(0) = I \tag{3.7.17}$$

All other approximations will then take the form

$$\tilde{U}^{(1)}(t) = -\frac{i}{\hbar}\int_0^t \tilde{V}(t')dt'$$

$$\tilde{U}^{(2)}(t) = \left(-\frac{i}{\hbar}\right)^2 \int_0^t dt' \tilde{V}(t') \int_0^{t'} dt'' \tilde{V}(t'') \tag{3.7.18}$$

We now analyze some physical aspects of the evolution of an atomic system related to the general time reversal transformation. Equations (3.7.12) and (3.7.18), describing the microscopic properties of a system, possess time reversal. On the other hand, it is known that if a macroscopic system is out of balance, it will regain the equilibrium state during some period of time while undergoing a process of relaxation. The relaxation process of a macroscopic system does not possess time reversal. Therefore, one of the principal problems in the evolution of the system, connected to concepts of excitation and relaxation of an atomic ensemble, is to understand the physical reasons and the formulation of conditions under which joint action of the many microscopic time-reversal processes could lead to a macroscopic nonreversal result.

In considering this problem, we introduce an atomic system, denoted by subscript S, interacting with another system, called the thermostate and denoted by subscript T. Concerning the thermostate we suppose initially that the number of degrees of freedom is so high that any perturbation of the thermostate, caused by interaction with the atomic system, decays within the thermostate without affecting the state of the system (no back reaction approximation), and second, that the thermostate is in thermal balance regardless of the influence of the atomic system. The Hamiltonian operator of the full system, which includes an atomic system plus a thermostate, can be written in the form

$$H_{ST} = H_S + H_T + V(t) \tag{3.7.19}$$

where operator $V(t)$ describes the interaction of the system with the thermostate.

The density matrix $\tilde{\rho}_{ST}(t)$ of the full system, comprising the studied atomic system and the thermostate, can be represented at any time as the direct sum of the density matrices of the atomic system and of the thermostate. For the thermostate it is possible to take the value at the initial moment as

$$\tilde{\rho}_{ST}(t) = \tilde{\rho}_S(t) \oplus \tilde{\rho}_T(0) \tag{3.7.20}$$

where the index $\sim$ denotes that the corresponding value is for the interaction representation.

The third supposition, concerning interaction between the atomic system and the thermostate, known also as the impact approximation, is that the time interval between two subsequent interactions ΔT is much more that the interaction duration Δt. We introduce the time interval τ, usually called the time of interest, which is so small that

$$\Delta t \ll \tau \ll \Delta T \tag{3.7.21}$$

For a variation in the density matrix (3.7.14) during the time of interest, one has

$$\begin{aligned}\tilde{\rho}_S(\tau) - \tilde{\rho}_S(0) &= \tau \frac{\partial}{\partial t} \tilde{\rho}_S(0) \\ &= Sp_T[\tilde{U}(\tau)\tilde{\rho}_S(0)\tilde{\rho}_T(0)\tilde{U}^{\dagger}(\tau) - \tilde{\rho}_S(0)\tilde{\rho}_T(0)]\end{aligned} \tag{3.7.22}$$

where Sp_T means that the summation in the right hand side of the equation is implemented over the thermostate variables.

As the interest time τ is much higher than the interaction time, one could substitute τ in the right hand side of Eq. (3.7.22) by infinity. Then, introducing the scattering matrix $\tilde{S} = \tilde{U}(\infty)$, the equation can be written in the form

$$\tau \frac{\partial}{\partial t} \tilde{\rho}_S(0) = Sp_T \left[\tilde{S} \tilde{\rho}_S(0) \tilde{\rho}_T(0) \tilde{S}^\dagger - \tilde{\rho}_S(0) \tilde{\rho}_T(0) \right] \tag{3.7.23}$$

By denoting the eigenfunctions of the thermostate and the atomic system by greek and latin characters, respectively, Eq. (3.7.23) can be written in the matrix form

$$\tau \frac{\partial}{\partial t} \langle i|\tilde{\rho}_S| j\rangle = -\langle i|\tilde{\rho}_S| j\rangle + \sum_{\alpha\beta\gamma\kappa l} \langle i\alpha|\tilde{S}|\kappa\beta\rangle \langle \kappa|\tilde{\rho}_S|l\rangle \langle \beta|\tilde{\rho}_T|\gamma\rangle \langle l\gamma|\tilde{S}^\dagger| j\alpha\rangle \tag{3.7.24}$$

Further simplification of the latter equation is related to an expansion of the scattering matrix into series similar to Eq. (3.7.18) but limited to square terms with respect to the interaction

$$\tilde{S} = I + \frac{i}{\hbar} \tilde{S}^{(1)} - \frac{1}{\hbar^2} \tilde{S}^{(2)} \tag{3.7.25}$$

where, according to Eq. (3.7.18), the interaction terms are expressed as

$$\tilde{S}^{(1)} = \int_{-\infty}^{\infty} dt\ \tilde{V}(t) \qquad \text{and} \qquad \tilde{S}^{(2)} = \int_{-\infty}^{\infty} dt\ \tilde{V}(t) \int_{-\infty}^{+\infty} dt'\ V(t') \tag{3.7.26}$$

Substituting Eq. (3.7.25) into (3.7.24) yields

$$\begin{aligned} \tau \frac{\partial}{\partial t} \langle i|\tilde{\rho}_S|j\rangle = & \frac{1}{\hbar^2} \sum \langle i\alpha|\tilde{S}^{(1)}|\kappa\beta\rangle \langle l\gamma|\tilde{S}^{(1)}|j\alpha\rangle \langle \kappa|\tilde{\rho}_S|l\rangle \langle \beta|\tilde{\rho}_T|\gamma\rangle \\ & - \frac{1}{\hbar^2} \sum \langle \kappa\beta|\tilde{S}^{(2)}|j\alpha\rangle \langle i|\tilde{\rho}_S|\kappa\rangle \langle \alpha|\tilde{\rho}_T|\beta\rangle \\ & - \frac{1}{\hbar^2} \sum \langle \alpha i|\tilde{S}^{(2)}|\kappa\beta\rangle \langle \kappa|\tilde{\rho}_S|j\rangle \langle \beta|\tilde{\rho}_T|\alpha\rangle \end{aligned} \tag{3.7.27}$$

where the summation is undertaken over all indices except i and j and the values of $\tilde{S}^{(1)}$ and $\tilde{S}^{(2)}$ are determined by Eq. (3.7.25).

After rather complicated transformations Eq. (3.7.27) can be expressed in the form[(138)]

$$\frac{\partial}{\partial t}\langle i|\tilde{\rho}_S|j\rangle = \frac{1}{2}\sum_{\kappa l}[\Gamma_{\kappa l,ij} + \Gamma^*_{l\kappa ji}]\langle\kappa|\tilde{\rho}_S|l\rangle$$
$$-\frac{1}{2}\sum_{\kappa l}[\Gamma^*_{i\kappa,ll}\langle\kappa|\tilde{\rho}_S|j\rangle + \Gamma_{j\kappa,ll}\langle i|\tilde{\rho}_S|\kappa\rangle] \qquad (3.7.28)$$

where the matrix element $\Gamma_{kl,ij)}$ can be written as follows:

$$\Gamma_{\kappa l,ij} = \frac{2}{\hbar^2\tau}Sp_T\left[\int_{-\infty}^{\infty}dt\langle i|\tilde{V}(t)|\kappa\rangle\int_{-\infty}^{t}dt'\langle l|\tilde{V}(t')|j\rangle\right]$$
$$= \frac{2}{\hbar^2\tau}\sum_{\alpha\beta\gamma}\int_{-\infty}^{\infty}dt\langle i\alpha|\tilde{V}(t)|\kappa\beta\rangle\int_{-\infty}^{t}dt'\langle l\gamma|\tilde{V}(t)|j\alpha\rangle\langle\beta|\rho_T|\gamma\rangle \qquad (3.7.29)$$

Equation (3.7.28) is called the master kinetic equation. The first term of this equation describes the transfer of population to levels i and j, while the second term denotes the opposite effect, i.e., transfer of populations from levels i and j to level l.

In order to simplify the master equation we use the condition that there are no transitions between the energy states of the thermostate induced by the interaction with the atomic system. This means that $\tilde{\rho}_T$ and H_T operators commutate and that the ρ_T operator is the same in the Schrödinger and interaction representations. For further simplification of Eq. (3.7.29), we write the operators on the right hand side of Eq. (3.7.29) in the Schrödinger representation

$$\Gamma_{\kappa l,ij} = \frac{2}{\hbar^2\tau}\sum_{\alpha\beta\gamma}\int_{-\infty}^{\infty}dt\langle i\alpha|V(t)|\kappa\beta\rangle e^{i(\omega_{i\kappa}+\omega_{\gamma\beta})t}$$
$$\cdot\int_{-\infty}^{t}dt'\langle l\gamma|V(t')|\,j\alpha\rangle e^{i(\omega_{lj}+\omega_{\gamma\alpha})t'}\langle\beta|\rho_T|\gamma\rangle \qquad (3.7.30)$$

where the new variable $t'' = t - t'$ has been introduced in the second integral of this expression. As a result, we have

$$\Gamma_{\kappa l,ij} = \frac{2}{\hbar^2\tau}\sum_{\alpha\beta\gamma}\int_{-\infty}^{\infty}dt\langle i\alpha|V(t)|\kappa\beta\rangle e^{i(\omega_{i\kappa}+\omega_{lj}+\omega_{\gamma\beta})t}$$
$$\cdot\int_{0}^{\infty}dt''\langle l\gamma|V(t-t'')|\,j\alpha\rangle e^{i(\omega_{lj}-\omega_{\gamma\alpha})t''} \qquad (3.7.31)$$

The quantity t in the exponent is of the order of the interest time τ, due to the uncertainty principle $\omega_{ij}\Delta T \sim 1$ and the relation $\tau \ll \Delta T$, so $(\omega_{ik} + \omega_{lj})t$ is small. Moreover, there are no transitions between different energy states of the thermostate, so we can set $\omega_{\gamma\beta} = 0$. Therefore, if the exponent in the integral on the right hand side is equal to unity, we have

$$\Gamma_{\kappa l,ij} = \frac{2}{\hbar^2\tau}\sum_{\alpha\beta\gamma}\int_{-\infty}^{\infty} dt\langle i\alpha|V(t)|\kappa\beta\rangle \cdot \int_0^{\infty} dt'\langle l\gamma|V(t-t')|j\alpha\rangle e^{i(\omega_{lj}-\omega_{\gamma\alpha})t'}\langle\beta|\rho_T|\gamma\rangle \tag{3.7.32}$$

Subsequently, the interaction can be represented in the form:

$$V(t) = \sum_m P_m Q_m(t) \tag{3.7.33}$$

where P and Q are operators connected to the atomic system and the thermostate, respectively. If stationary correlation functions

$$\langle\tilde{Q}_{m'}(t)\tilde{Q}_m(t-t')\rangle = Sp_T\left[\tilde{Q}_{m'}(t)\tilde{Q}_m(t-t')\rho_T\right] \tag{3.7.34}$$

are introduced, then expression (3.7.32) can be transformed such that

$$\Gamma_{\kappa l,ij} = \frac{2}{\hbar^2\tau}\sum_{mm'}\langle i|P|\kappa\rangle\langle l|P|j\rangle\int_{-\infty}^{\infty} dt\int_0^{\infty} dt'\, e^{i\omega_{jl}t'}\langle\tilde{Q}_m(t)\tilde{Q}_{m'}(t-t')\rangle \tag{3.7.35}$$

We make use of the condition, that the thermostate possesses a large number of degrees of freedom and all perturbations caused by interaction with the atomic system decay rapidly. In such a situation, the correlation, functions on the right hand side of Eq. (3.7.35) can be considered nonzero for $t < t_c$, where t_c is the characteristic correlation time of the thermostate.

We now consider two different aspects of a general theoretical concept concerning interaction of an atomic system with a thermostate; both are directly related to the theory of atomic collisions and spectroscopy. In the first case the role of the thermostate is played by structure-free particles, changing the states of the atomic system under collisions. In the second case the thermostate is represented by electromagnetic radiation, interacting with the atomic system.

3.8. Collisional Evolution of Polarization Moments

Two different physical situations should be considered when analyzing the kinetics of polarization moments during atomic collisions. The first consists of slow collisions with a previously excited atomic ensemble. In this case there is an external process of excitation of the atomic ensemble, while the role of slow collisions results only in mixing the populations of the excited states. This situation often takes place in a gas discharge where the excitation of atomic particles in many cases is performed by the isotropically distributed fast electrons of the low temperature plasma and the populations can be mixed simultaneously by slow collisions between atoms or ions. The second case is realized under collisional excitation of the fast projectiles with an ensemble of atomic particles. Here, collisions play the role of an excitation mechanism. In both cases the problem of determining the collisional relaxation rate constants will be solved mathematically by summing all the contributions of collisions with relative velocities lying within the solid angle $d\Omega$ inclined at angle θ with respect to the OZ axis of a laboratory system of reference (Figure 6).

Let us introduce two systems of reference: a laboratory system with (X, Y, Z) axes and a collisional system with $O\tilde{Z}$ axis directed antiparallel to the relative velocity vector and making an angle θ with the OZ axis of the laboratory coordinate system. Expression (3.7.23) can be employed to describe the collisional variation in the density matrix. The precipitating particles are regarded as structure-free, so calculation of the track on the right hand side of this expression is reduced to averaging over the relative velocities and multiplying by the number of collisions, corresponding to the interval $b, b + db$ of the impact parameter (Figure 7), with subsequent integration over the impact parameter b.

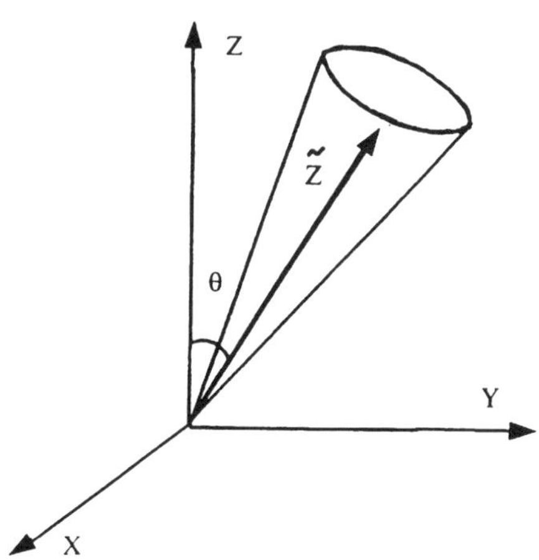

Figure 6. Orientation of the differential solid angle in the laboratory frame of reference for computation of collisional relaxation rate constants.

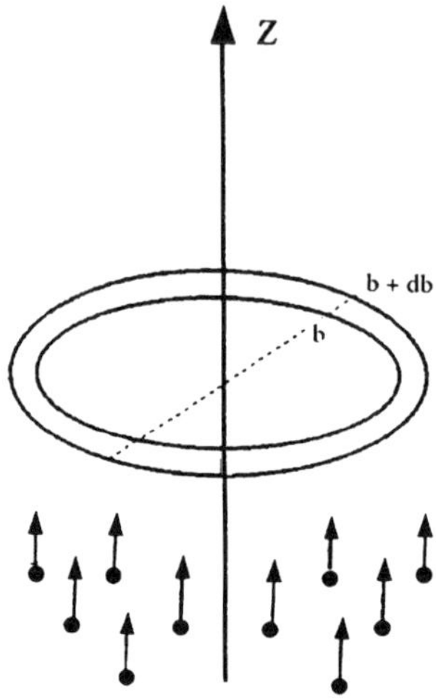

Figure 7. Impact parameter for computing the collisional variation in the density matrix.

With the help of relation (3.7.23), the variation in the density matrix due to collisions with velocity vectors within a solid angle $d\Omega$ can be represented as

$$\left[\frac{\partial \tilde{\hat{\rho}}}{\partial t}\right]_{coll} = n_0|v|f(v)dv \int b\, db[\tilde{\hat{S}}(b, v)\tilde{\hat{\rho}}^{(i)}\tilde{\hat{S}}^{+}(b, v) - \tilde{\hat{\rho}}^{(i)}] \tag{3.8.1}$$

where n_0 is the foreign gas density, $dv = v^2\, dv\, d\Omega$, $f(v)$ is the relative velocity distribution function, $\hat{\rho}$) is the density operator of the atomic system, and the index '$\sim$' indicates that the corresponding values are represented in the collision system of reference.

We assume that the density matrix of an atomic system is the direct sum of the density matrices of the ground and excited states, as there is no coherence between the ground and excited states. Expressing Eq. (3.8.1) in terms of polarization moments, we have

$$\left[\frac{\partial}{\partial t}\tilde{\rho}_q^k(jj')\right]_{coll} = n_0|v|f(v)dv \sum_{k'q'j_1j_1'} \tilde{R}_{qq'}^{kk'}(jj', j_1j_1'|v)\tilde{\rho}_{q'}^{k'}(j_1j_1') \tag{3.8.2}$$

where the elements of matrix R may be written in the form

$$\tilde{R}_{qq'}^{kk'}(jj', j_1j_1'|v) = \sum_{mm'm_1m_1'} (-1)^{m'-m_1'} \begin{bmatrix} j & j' & k \\ m & -m' & q \end{bmatrix} \begin{bmatrix} j_1 & j_1' & k' \\ m_1 & -m_1' & q' \end{bmatrix} \cdot \int_0^\infty b\, db(\tilde{S}_{mm_1}^{jj_1}\tilde{S}_{m'm_1'}^{j'j_1'*} - \delta_{mm_1}\delta_{m'm_1'}\delta_{jj_1}\delta_{j'j_1'}) \tag{3.8.3}$$

Transformation of the polarization moments $\tilde{\rho}_q^k$ into the laboratory frame of reference is implemented by a rotation matrix, as follows

$$\left[\frac{\partial}{\partial t}\rho_q^k(jj')\right]_{coll} = n_0|v|f(v)dv \sum_{k'q_1q_1'}(-1)^{q+q'}e^{-i\alpha(q_1-q)}e^{-i\gamma(q_1'-q')} \cdot d_{-q-q'}^{(k)}(\theta)d_{q_1q_1'}^{(k')}(\theta)\tilde{R}_{q'q_1'}^{kk'}(jj',j_1j_1'|v)\rho_{q'}^{k'}(j_1j_1') \quad (3.8.4)$$

The particular structure of expression (3.8.4) is linked to the characteristics of the process under study. We shall analyze here three basic processes:

(a) Emission of the preliminary excited ensemble of atoms or ions, undergoing collisions with slow foreign atomic particles.
(b) Emission of excited atoms or ions induced by collisions with fast structureless particles.
(c) Emission of an ensemble of atoms or ions excited by impact with a foreign particle which is scattered at a given angle.

In case (a), the relative velocities of the colliding particles are small and the influence of collisions on the polarization characteristics of emission takes place owing to the population mixing of the close excited levels j, j', j_1, and j_1' or magnetic substates of the preliminary excited energy state of atomic particles belonging to the radiating ensemble. Suppose the excitation process does not change a number density of particles at the ground state. In this situation, the collisional variation of the polarization moments within a laboratory frame of reference could be obtained by summing the contributions of individual collisions lying within the solid angle $d\Omega$ (Figure 6). By conducting this summation and accounting for the invariance of the collisional process with respect to rotation around the $O\tilde{Z}$ axis, we have

$$\left[\frac{\partial}{\partial t}\rho_q^k(jj')\right]_{col} = n_0\sum_{k_1q'}\langle v\sigma_{qq'}^{kk'}(jj',j_1j_1')\rangle\rho_{q'}^{k'}(j_1j_1') \quad (3.8.5)$$

where for

$$\tilde{\sigma}_q^{kk'}(jj',j_1j_1'|v) = \tilde{R}_{qq}^{kk'}(jj',j_1j_1'|v)$$

one obtains

$$\langle v\sigma_{qq'}^{kk'}(jj',j_1j_1')\rangle = \sum_{q''}(-1)^{q+q''}\begin{bmatrix} k & k' & l \\ q & q' & m \end{bmatrix}\begin{bmatrix} k & k' & l \\ q'' & -q'' & 0 \end{bmatrix} \cdot \int d(v)|v|D_{m_0}^{(l)}(\alpha,\theta,0)\tilde{\sigma}_{q''}^{kk'}(jj',j_1j_1'|v) \quad (3.8.6)$$

This formula is an expression for the matrix elements of a matrix, which is known as the matrix of rate constants of the collisional relaxation of polarization moments, with allowance for the dependence of the cross section on the velocity.

If the velocity distribution of colliding particles is axially symmetric, we finally have

$$\left[\frac{\partial}{\partial t}\rho_q^k(jj')\right]_{col} = n_0\sum_{k'}\langle v\sigma_q^{kk'}(jj',j_1j_1')\rangle\rho_q^{k'}(j_1j_1') \tag{3.8.7}$$

where

$$\langle v\sigma_q^{kk'}(jj',j_1j_1')\rangle = 2\pi\sum_{q'l}(-1)^{q'+q}\begin{bmatrix} k & k' & l \\ q & -q & 0 \end{bmatrix}\begin{bmatrix} k & k' & l \\ q' & -q' & 0 \end{bmatrix} \cdot \int v^3 f(v)\sin\theta P_l(\cos\theta)\tilde{\sigma}_{q'}^{kk'}(jj',j_1j_1'|v)dv\,d\theta$$

Hence in this case collisions mix polarization moments of different ranks while maintaining invariant their projections on the axis of anisotropy.

The latter expression can be written in a more compact form, by expanding the relative velocity distribution function in terms of Legendre polynomials

$$f(v) = \sum_k f_k(v)P_k(\cos\theta) \tag{3.8.8}$$

where $f_k(v)$ are the multipole moments of the relative velocity distribution function determined by the relation[136]

$$f_k(v) = \frac{2l+1}{4\pi}\int f(v)P_k(\cos\theta)\sin\theta\,d\theta \tag{3.8.9}$$

If we substitute Eq. (3.8.8) into (3.8.7) and assume that the cross section depends only on the absolute values of the relative velocity, we obtain

$$\langle v\sigma_q^{\kappa\kappa'}(jj'j_1j_1')\rangle = \sum_{q'}(-1)^{q+q'}\begin{bmatrix} \kappa & \kappa' & l \\ q & -q & 0 \end{bmatrix}\begin{bmatrix} \kappa & \kappa' & l \\ q' & -q' & 0 \end{bmatrix} \cdot \frac{4\pi}{2l+1}\int_0^\infty v^3 f_l(v)\sigma_{q'}^{\kappa\kappa'}(jj',j_1j_1'|v)dv \tag{3.8.10}$$

In case (b) we suppose that the density of foreign particles is not high. Then one can assume that collisions do not change the population of the ground level and, as a result, the density matrix of the ground state is proportional to the unit matrix. Before the collisions there was no population of the excited state, so relations (3.8.2) and (3.8.3), which describe the collisional variation in the polarization moments of the excited state in this case, will have the form

$$\left[\frac{\partial}{\partial t}\tilde{\rho}_q^k(jj')\right]_{ex} = n_0|v|\tilde{\sigma}_q^k(j,j';j_0;v) \tag{3.8.11}$$

where $\sigma_q^k(j,j';j_0;v)$ is the cross section of the impact ordering of momenta of the excited particles. Equation (3.8.3) can be used to transform the latter quantity, as follows

$$\tilde{\sigma}_q^k(j,j';j_0;v) = f(v)dv \sum_{mm'm_0}(-1)^{j'-m'}\begin{bmatrix} j & j' & k \\ m & -m' & q \end{bmatrix} \cdot \int_0^\infty dbbS_{mm_0}^{jj_0}S_{m'm_0}^{j'j_0^*} \tag{3.8.12}$$

Then, in a laboratory frame of reference, we have

$$\left[\frac{\partial}{\partial t}\rho_q^k(jj')\right]_{ex} = n_0\int dv|v|f(v)\tilde{\sigma}_0^k(j,j';j_0;v)D_{q0}^{k*}(\alpha,\theta,0) \tag{3.8.13}$$

This formula, in the case of an axially symmetric velocity distribution function, can be transformed into

$$\left[\frac{\partial}{\partial t}\rho_0^k(jj')\right]_{ex} = 2\pi n_0\int dv\,d\theta|v|^3 f(v)\sin\theta\,\tilde{\sigma}_0^k(j,j';j_0;v)P_k(\cos\theta) \tag{3.8.14}$$

By making the same transformations as before and substituting Eq. (3.8.8) into the last expression, we have

$$\frac{\partial\rho_0^k(jj')}{\partial t} = n_0\langle v\sigma_0^k(jj';j_0;v)\rangle \tag{3.8.15}$$

where $\langle v\sigma_0^k(jj';j_0;v)\rangle$ is the rate constant of collisional polarization of an excited state.

Suppose again that the cross section of the creation of an ordering of momenta of the excited particles is dependent only on the absolute value of the relative velocity. The right hand side of Eq. (3.8.15) then yields

$$\langle v\sigma_0^k(jj';j_0;v)\rangle = \frac{4\pi}{2k+1}\int dv\,v^3 f_k(v)\sigma_0^k(jj';j_0;v) \tag{3.8.16}$$

Therefore, collisions induce only zero-order components of polarization moments in this case.

Variation of the polarization moments within the laboratory frame of reference in the case (c), when radiation of atoms or ions scattered at a given angle is recorded, is given by the equation

$$\frac{\partial}{\partial t}\rho_q^k(j,j';b) = n_0|v|\sigma_q^k(j,j';j_0;v;b) \tag{3.8.17}$$

where $\sigma_q^k(j,j';j_0;v;b)$ is the cross section of the impact ordering of momenta of an excited atomic ensemble, scattered at angle θ, while b is the impact parameter.

The cross section of the impact ordering of momenta of the excited atomic ensemble is determined by the impact parameter b and this dependency can be calculated, for instance, with the help of the Rutherford formula. It is clear that the right hand side of Eq. (3.8.17) is similar to that of Eq. (3.8.12) only if integration over the impact parameter is excluded.

3.9. Interaction with an Electromagnetic Field

The evolution of the density matrix in this case is given by the master kinetic equation (3.7.28), in which the interaction operator of the atomic system with the electromagnetic field is given within the dipole approximation

$$V = (\mathbf{d}, \mathbf{E}_0) \tag{3.9.1}$$

where $\mathbf{d}$ is the dipole moment operator and $\mathbf{E}_0$ is the electric field strength at the nuclear location coordinate.

Expression (2.5.10) for the electric field strength, and expanding the dipole moment vector in terms of the circular orths:

$$\mathbf{d} = \sum_q d_q \mathbf{e}_\kappa \tag{3.9.2}$$

enables the interaction operator (3.9.1) to be written as

$$V = i\sum_q d_q \sum_\lambda \int d\boldsymbol{\kappa}\sqrt{\frac{\hbar\omega}{(2\pi)^3}}\,[a_{\kappa\lambda}(\mathbf{e}_q, \mathbf{e}_\lambda) - a_{\kappa\lambda}^\dagger(\mathbf{e_q}, \mathbf{e}_\lambda^*)] \tag{3.9.3}$$

In this expression we note that two orths, e_λ $(\lambda = \pm 1)$, lie in the plane perpendicular to the direction of propagation and three orths, e_q $(q = 0, \pm 1)$ and chosen in such a way that orth e_0 is oriented along the magnetic field strength and determines the quantization axis. In line with Eq. (3.9.2), quantities $(\mathbf{e}_q, \mathbf{e}_\lambda)$ are the coefficients of expansion of vector $\mathbf{e}_\lambda$ in terms of orths $\mathbf{e}_q$. By setting $(\mathbf{e}_q, \mathbf{e}_\lambda) \equiv (\mathbf{e}_\lambda)_q$, we have

$$(\mathbf{e_q}, \mathbf{e}_\lambda^*) = (-\mathbf{1})^{\mathbf{q}}(\mathbf{e}_{-\mathbf{q}}^*, \mathbf{e}_\lambda)^* = (-\mathbf{1})^{\mathbf{q}}(\mathbf{e}_\lambda)_{-\mathbf{q}} \tag{3.9.4}$$

If Eq. (3.9.3) is split into two terms and the sign of q changed in the second term, then the equation can be transformed to

$$V = i\sum_q d_q \sum_\lambda \int d\boldsymbol{\kappa} \sqrt{\frac{\hbar\omega}{(2\pi)^3}} a_{\kappa\lambda}(\mathbf{e}_\lambda)_q - i\sum_q d_q^* \sum_\lambda \int d\boldsymbol{\kappa} \sqrt{\frac{\hbar\omega}{(2\pi)^3}} a^\dagger_{\kappa\lambda}(\mathbf{e}_\lambda)^*_q \tag{3.9.5}$$

The latter expression consists of two contributions, each of which is in turn the sum of products of a pair of operators, the first of which acts only on the atomic variables and the second on the variables of the electromagnetic field. Therefore, the operator (3.9.5) is similar to the operator (3.7.33). On substituting the last expression into Eq. (3.7.29) we take into consideration that there exist no transitions with different population numbers for the electromagnetic field, contrary to transitions between the states of a thermostate with different energies. The latter feature can be expressed mathematically as

$$\langle a_{\kappa\lambda} a_{\kappa'\lambda'} \rangle_T = \langle a^\dagger_{\kappa\lambda} a^\dagger_{\kappa'\lambda'} \rangle = 0 \tag{3.9.6}$$

Hence, the matrix element of the master kinetic equation takes the form

$$\begin{aligned} \Gamma_{\kappa l, ij} = \frac{2}{\hbar^2 \tau} \sum_{qq'\lambda\lambda'} \int \frac{\hbar\omega}{(2\pi)^3} \int_{-\infty}^{\infty} dt \int_0^\infty dt' e^{i\omega_{jl}t'} [\langle i|d_q|\kappa\rangle \\ \cdot \langle l|d^*_{q'}|j\rangle \langle a_{\kappa\lambda'} a^\dagger_{\kappa\lambda}\rangle_T e^{i\omega t'} (\mathbf{e}_\lambda)_q (\mathbf{e}_{\lambda'})^*_{q'} \\ + \langle i|d_{q'}|\kappa\rangle \langle l|d^*_q|j\rangle \langle a^\dagger_{\kappa\lambda} a_{\kappa\lambda}\rangle_T e^{-i\omega t'} (\mathbf{e}_{\lambda'})_{q'} (\mathbf{e}_\lambda)^*_q] \end{aligned} \tag{3.9.7}$$

The first term of this expression is responsible for radiation because the rising operator, reflecting the creation of a photon, is on the right side of the operator product. The second term is responsible for absorbtion because the lowering operator, expressing the decay of a photon, is on the right side of the operator product.

In order to simplify expression (3.9.7) we note that the integral $\int_{-\infty}^{\infty}$ equals a value of the same order of magnitude as the interest time τ. By introducing the polarization tensor

$$\langle I_{\lambda\lambda'} \rangle = \frac{\hbar v^3}{c^2} \delta(\boldsymbol{\kappa} - \boldsymbol{\kappa}') \langle a^\dagger_{\kappa\lambda} a_{\kappa\lambda'} \rangle \tag{3.9.8}$$

similarly to Eq. (2.7.7) such that the components of combinations of this tensor, averaged over the ensemble, determine the Stokes parameters, and using the fact that the expansion coefficients of vector $\mathbf{e}_\lambda$ in terms of the $\mathbf{e}_q$ basis are elements of the rotation matrix, the absorption part of expression (3.9.7) can be expressed in the form

$$\Gamma^{(abs)}_{\kappa l,ij} = \frac{2}{\hbar^2}\sum_{qq'}\langle i|d_q|\kappa\rangle\langle l|d^*_{q'}|j\rangle \int_0^\infty dt \int d\nu\, d\Omega_\kappa e^{-i(\omega-\omega_0)t} \cdot \sum_{\lambda\lambda'}\langle I_{\lambda\lambda'}\rangle D^{(1)}_{q'\lambda'}(\alpha\beta\gamma)D^{(1)}_{q\lambda}(\alpha\beta\gamma) \tag{3.9.9}$$

The commutation relations for the field operators

$$a_{\kappa\lambda'}a^\dagger_{\kappa\lambda} - a^\dagger_{\kappa\lambda}a_{\kappa\lambda'} = \delta_{\lambda\lambda'} \tag{3.9.10}$$

are used together with the notation

$$d\mathbf{k} = k^2\, dk\, d\Omega_k \omega_{ij} = \omega_0$$

to obtain an expression for the emission part, similar to Eq. (3.9.9):

$$\Gamma^{(em)}_{\kappa l,ij} = \frac{2}{\hbar^2}\sum_{qq'}\langle i|d_q|\kappa\rangle\langle l|d^*_{q'}|j\rangle \int_0^\infty dt \int d\nu\, d\Omega_\kappa \frac{2\pi}{c} e^{i(\omega-\omega_0)t} \times \sum_{\lambda\lambda'}\left[\delta_{\lambda\lambda'}\frac{\hbar\nu^3}{c^2} + \langle I_{\lambda\lambda'}\rangle\right]D^{(1)}_{q'\lambda'}(\alpha\beta\gamma)D^{(1)*}_{q\lambda}(\alpha\beta\gamma) \tag{3.9.11}$$

Two important results follow from the last expression. By choosing three Euler angles in such a way that the first one is $\alpha = 0$, it is possible to integrate over the angle γ with allowance for the γ dependency of the differential of the solid angle:

$$d\Omega_k = \sin\beta\, d\beta\, d\gamma$$

Then for spontaneous emission we have

$$\Gamma^{(sp)}_{\kappa l,ij} = \frac{2}{\hbar^2}\sum_{q}\langle i|d_q|\kappa\rangle\langle l|d^*_{q}|j\rangle \int_0^\infty dt \frac{8\pi\hbar\nu^3}{3c^3} e^{i(\omega-\omega_0)t} \tag{3.9.12}$$

and for stimulated emission

$$\Gamma^{(ind)}_{\kappa l,ij} = \frac{4\pi}{c\hbar^2}\sum_{qq'}\langle i|d_q|\kappa\rangle\langle l|d_{q'}|j\rangle \int_0^\infty dt \int d\nu\, d\Omega_\kappa e^{i(\omega-\omega_0)t} \cdot \sum_{\lambda\lambda'}\langle I_{\lambda\lambda'}\rangle D^{(1)}_{q'\lambda}(\alpha\beta\gamma)D^{(1)*}_{q\lambda'}(\alpha\beta\gamma) \tag{3.9.13}$$

Formulas (3.9.11)–(3.9.13) describe the transfer of population between atomic levels. The efficiency of this process is determined by the atomic characteristics via the matrix elements of the dipole moment and by the field characteristics incorporated in the polarization tensor. The latter, in turn, is connected to the Stokes parameters of the electromagnetic wave. As a result, we obtain a very important formula relating the transition probability for populations between atomic levels to the polarization characteristics of an incident electromagnetic wave.

3.10. Impact Alignment Cross Section

The dipole optical radiation emitted by an atomic ensemble excited by a unidirectional particle beam has a nonuniform angular distribution. A spatial distribution of emission is related simply to the polarization degree P^{90° of radiation propagating within the solid angle, with an axis oriented perpendicular to the beam of projectiles. Assuming a two-step process of excitation of an upper quantum state of the atom and subsequent spontaneous emission of a polarized photon, it is easy to express the polarization of radiation in terms of populations of the excited magnetic sublevels.[140] In the case of axially symmetric excitation, it is interesting to exploit the symmetry of the system and introduce the state multipole ρ_q^κ of the density matrix ρ.

The polarization moments of the density matrix of the upper atomic state created via impact excitation of atoms in a monoenergetic electron or proton beam may be expressed by employing the elements of the density matrix in the representation of magnetic quantum numbers $\rho_{Jm,Jm'}$ [see Eq. (3.4.11)], as follows:

$$\rho_q^k(J) = \sum_{mm'}(-1)^{J-m'}\begin{bmatrix} J & J & k \\ m & -m' & q \end{bmatrix}\rho_{Jm,Jm'} \tag{3.10.1}$$

If the initial state is spherically symmetric, the excited state will be axially symmetric and the nonzero polarization moments will have even κ-numbers and only zero q-numbers.[102,141] Therefore, in a frame of reference with axis of quantization parallel to the beam, the degree of linear polarization $P(\mathbf{r})$, of the optical emission of a homogeneous plasma volume defined by coordinate $\mathbf{r}$ is given by[1,134]

$$P(\mathbf{r}) = \left\{\left[(-1)^{J+J_1}\left(\frac{3}{2}\right)^{3/2}\sqrt{2J+1}\begin{Bmatrix} 1 & 1 & 2 \\ J & J & J_1 \end{Bmatrix}\frac{\rho^{(2)}(\mathbf{r})}{\rho^{(0)}(\mathbf{r})}\right]^{-1} + \frac{1}{3}\right\}^{-1} \tag{3.10.2}$$

Here, J and J_1 are the angular moments of the upper and lower levels of the optical transition; $\rho^{(0)}(\mathbf{r})$ and $\rho^{(2)}(\mathbf{r})$ are the population and alignment of the upper level and are expressed in terms of the impact excitation cross section of the upper level $\sigma^{(0)}(v)$ and the impact alignment cross section $\sigma^{(2)}(v)$ as follows

$$\rho^{(0)}(\mathbf{r}) = 4\pi N_0 N_p \Gamma_0^{-1} \int_{v_t}^{\infty} dv\, v^3 f_0(v, \mathbf{r}) \sigma^{(0)}(v) \tag{3.10.3}$$

$$\rho^{(2)}(\mathbf{r}) = 4\pi 5^{-1/2} N_0 N_p \Gamma_2^{-1} \int_{v_t}^{\infty} dv v^3 f_2(v, \mathbf{r}) \sigma^{(2)}(v) \tag{3.10.4}$$

where v_t is the threshold velocity of the projectile for impact excitation, Γ_0 and Γ_2 are relational constants for population and alignment, N_0 and N_p are concentrations of the ground state atoms and perturbers, while $f_0(v, \mathbf{r})$ and $f_2(v, \mathbf{r})$ are zero-order and quadrupole moments of the relative velocity distribution function defined by relation (3.8.9).

The cross sections $\sigma^{(0)}(v)$ and $\sigma^{(2)}(v)$ can be expressed in terms of the cross sections σ_m for excitation of sublevels Jm

$$\sigma^{(0)} = (2J+1)^{-1/2} \sum_m \sigma_m \tag{3.10.5}$$

$$\sigma^{(2)}(v) = \sum_m (-1)^{J-m} \begin{bmatrix} J & J & 2 \\ m & -m & 0 \end{bmatrix} \sigma_m \tag{3.10.6}$$

In general, the simplest way to determine the impact-alignment cross section is based on spectropolarimetric measurements of the linear polarization of atoms excited by a monoenergetic particle beam. In this case, $\sigma^{(2)}(v)$ may be expressed using two quantities: the energy dependence of $P_{90}(v)$ or the degree of polarization measured at right angles with respect to the beam direction and the velocity dependence of the excitation cross section $\sigma^{(0)}(v)$[(142)]

$$\sigma^{(2)}(v) = \frac{\sqrt{8}}{27} \frac{(-1)^{J+J_1}}{\sqrt{2J+1}} \begin{Bmatrix} 1 & 1 & 2 \\ J & J & J_1 \end{Bmatrix}^{-1} \frac{\Gamma_2}{\Gamma_0} \sigma^0(v) \frac{P_{90}(v)}{1 - \frac{1}{3} P_{90}(v)} \tag{3.10.7}$$

Another possibility is to make use of different polarization spectroscopic parameters. A significant number of versions of electron–photon angular correlation experiments, in which the correlation between the exciting particle and the subsequently emitted photon is measured, have been performed in atomic physics. Some results of these experiments, which could be utilised for computations of the impact-alignment cross sections, have been reviewed elsewhere.[(102,109,149)]

The impact-alignment cross section for atomic states by fast electrons or protons can be computed subject to the Born limit approximation. The expression of $\sigma^{(2)}(v)$ in terms of the differential cross section of the impact excitation of the atomic state $\sigma(\theta, v)$ has been obtained,[142] using specific symmetry considerations of the electron impact excitation operator.[141] Amplitudes of the zeeman sublevels of atomic states C_{J_m} are proportional to the matrix elements of the electron impact excitation operator, $\sum_e \exp(-i\mathbf{k}'\hat{\mathbf{r}}_e)$, where $\hat{\mathbf{r}}_e$ are atomic electron coordinates while $\mathbf{k}$ and $\mathbf{k}_1$ are wave vectors of the incident and scattered electrons, with $\mathbf{k}' = \mathbf{k} - \mathbf{k}_1$ (Figure 8).

This operator is invariant with respect to rotation around the axis parallel to vector $\mathbf{k}'$. Therefore, if the ground state has zero orbital and spin moments, then in the frame of reference with quantization axis along $\mathbf{k}'$ only the elements having the projection of momentum on this axis will be excited.[102,141] The projection of the spin on the quantization axis may change its value $\Delta S = \pm 1$ only resulting from exchange interaction at the intercombination transitions. Hence, under impact excitation by fast electrons in the frame of reference with the OZ axis parallel to $\mathbf{k}'$, only substates with $m = 0$ for spin-conserving collisions ($\Delta S = 0$) may be excited while for intercombination transitions ($\Delta S = \pm 1$) states with $m = \pm 1$ projections wilľ be excited.

Within this $\mathbf{k}'$-oriented frame of reference the atomic ensemble may be characterized by the noncoherent superposition of states with different m values. It is therefore possible to describe this situation in terms of the cross section of excitation $\sigma(\mathbf{k}')$. Taking into consideration the angular dependence of the wave vector $\mathbf{k}' = \mathbf{k}'(\theta, \varphi)$, the cross section $\sigma(\mathbf{k}')$ is the inelastic differential cross section. The cross section of higher orders, or, in particular, the alignment cross section, can be also regarded in this case. For example, for the LS coupling approximation, the states of the electron shell may be described by a certain spin quantum number. In this approximation, the alignment cross section is

$$\sigma^{(2)}(\mathbf{k}') = \alpha(S)\sigma(\mathbf{k}')$$

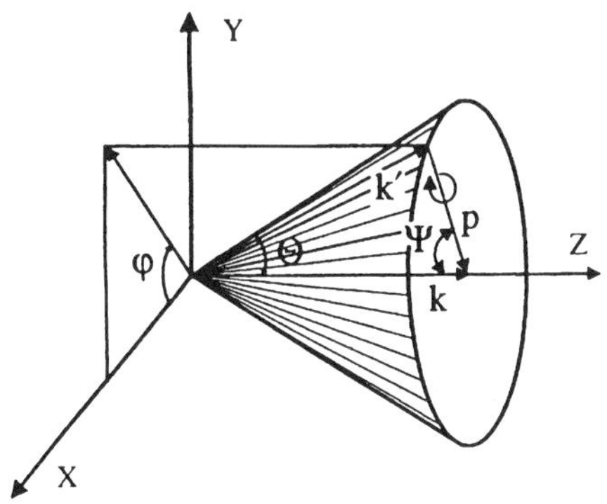

Figure 8. Calculation of the Born electron impact-alignment cross section ($\mathbf{k}$ and $\mathbf{k}'$ are the wave vectors of incident and scattered electrons).

where

$$\alpha(S) = \sqrt{5}\left[\frac{J(J+1)}{(2J+3)(2J+1)(2J-1)}\right]^{1/2}\left[\frac{3S}{J(J+1)} - 1\right]$$

Transition to the collision frame of reference with quantization axis oriented along $\mathbf{k}$ (laboratory frame of reference) is realized by a rotation matrix. After this transformation the transversal components of the polarization moments with $q \neq 0$ will also appear. These components are dependent on the asimuthal angle φ as $\exp(iq\varphi)$ and should vanish after integration over the range $[0, 2\pi]$. The total alignment cross section obtained as a result of averaging over all contributions from various directions will have only a longitudinal component

$$\sigma^{(2)}(v) = \int_0^{2\pi} d\varphi \int_0^{\pi} d\theta \sin\theta\, \sigma^{(2)}(\mathbf{k}')\left[\frac{3}{2}\cos\Psi(\theta) - \frac{1}{2}\right]$$

where $\Psi(\theta)$ is the angle between vectors $\mathbf{k}$ and $\mathbf{k}'$ (Figure 8). This angle satisfies

$$\cos^2\Psi(\theta) = \frac{1 + \delta\cos^2\theta - 2\sqrt{\delta}\cos\theta}{1 + \delta - 2\sqrt{\delta}\cos\theta}$$

Here $\delta = k_1^2/k^2 = \varepsilon_1/\varepsilon$, ε is the energy of the incident electron, $\varepsilon_1 = \varepsilon - \varepsilon_t$ is the energy of the inelasticaly scattered electron, and ε_t is the threshold energy of the studied level.

As a result, the alignment cross section under these approximations is

$$\sigma^{(2)}(v) = \pi\alpha(\Delta S)\int_0^{\pi} d\theta\, \sin\theta\sigma(\theta, v)\left[3\frac{1 + \delta\cos^2\theta - 2\sqrt{\delta}\cos\theta}{1 + \delta - 2\sqrt{\delta}\cos\theta} - 1\right] \quad (3.10.8)$$

where

$$\alpha(\Delta S) = \sqrt{5}\left[\frac{J(J+1)}{(2J+3)(2J+1)(2J-1)}\right]^{1/2}\left[\frac{3\Delta S}{J(J+1)} - 1\right] \quad (3.10.9)$$

$\delta = \varepsilon_0/\varepsilon$, $\varepsilon_0 = \varepsilon - \varepsilon_t$, and ε are the energies of the incident and scattered particles, ε_t is the threshold energy of the level under study, and ΔS is the difference in the spin numbers.

By these means for computing the alignment cross section within the Born approximation, it is possible to use the database of experimental and theoretical inelastic differential cross sections $\sigma(\theta, v)$. Unfortunately, the Born limit is only valid for high collisional energies ($E > 100$ eV for electrons and $E > 200$ keV for protons), so more elaborate methods must be employed for lower energies.

In order to estimate the alignment cross section at threshold energies, it is worthwhile to consider this question on the basis of the momentum conservation law. This approach was used in earlier works on electron impact excitation and was developed also in a number of later studies.[102,140] Let the energy of the exciting electron moving along the *OX* axis of the laboratory frame be equal to the threshold energy of an excited level under study, belonging to the energy structure of even isotopes with zero orbital and spin momenta in the ground state. Neglecting relativistic effects, the incident electron will have zero projection of orbital momentum with respect to the *OX* axis. The projection of the spin moment on the same axis will be respectively $\pm\frac{1}{2}$. According to the Percival–Seaton approach,[102,140] spin–spin and spin–orbit interactions may be neglected in a collision. In other words, no electron has its spin changed during collision. The scattered electron has zero kinetic energy and momentum. Therefore, the *OX* projection of orbital momentum is invariant under threshold collisions. Spin variation may occur only during the process of exchange of electrons. Hence, at the threshold value of the alignment cross section we may write

$$\sigma_t^2 = \alpha(S)\sigma_t$$

where σ_t is the threshold electron impact excitation cross section.

Calculations of spectroscopic collisional parameters for threshold excitation using more sophisticated models of electron–atom systems are presented in many works. For example, the basic approach, using a simple two-level scheme and nonrelativistic approximation while neglecting spin–orbit and exchange interactions, has been outlined elsewhere.[35,109,142,149,170]

In order to present some examples of the semiempirical approach to computations of polarization parameters of excited states of atomic particles, let us examine one of the most practically important cases of astrophysical and geophysical spectropolarimetric sensing techniques, which are naturally based on the hydrogen H_α line. Spectropolarimetric diagnostics of nonthermal phenomena in the solar atmosphere and, in particular, for sensing chromospheric solar flares are mostly dependent on reliable data concerning the spectroscopic characteristics of H_α line polarization.

The system of hydrogen levels and electric dipole transitions responsible for the formation of the H_α line are shown in Figure 9. Due to fine-structure splitting of the levels with $n = 2$, $n = 3$ and the Lamb shift, this line consists principally of seven components. However, for collisional ionized media, such as the solar atmosphere, a simplified energy level scheme may be used owing to energy separation between these states. Fine-structure splitting of the levels is usually neglected upon impact excitation of $3L$ ($L = S, P, D$) levels. The expression for

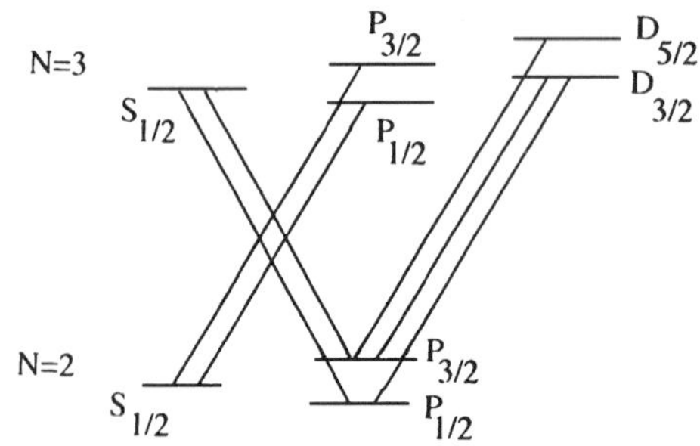

Figure 9. System of hydrogen levels and dipole transitions responsible for H_α radiation.

the degree of linear polarization of H_α radiation at right angles to the exciting particle beam may be obtained with the help of the impact excitation cross sections, as follows[147]

$$P_{H_\alpha}^{90^\circ} = 3\left(1 - \frac{\sigma_{H_\alpha}}{\sigma_{H_\alpha}^{90^\circ}}\right) \tag{3.10.10}$$

$$\sigma_{H_\alpha} = \sigma_{3S} + 0.12\sigma_{3P} + \sigma_{3D}, \qquad \sigma_{3P} = \sigma_{3P_0} + 2\sigma_{3P_1},$$
$$\sigma_{3D} = \sigma_{3D_0} + 2\sigma_{3D_1} + 3\sigma_{3D_2}$$
$$\sigma_{H_\alpha}^{90^\circ} = \sigma_{3S}^{90^\circ} + 0.12\sigma_{3P}^{90^\circ} + \sigma_{3D}^{90^\circ}, \qquad \sigma_{3S}^{90^\circ} = \sigma_{3S}, \qquad \sigma_{3P}^{90^\circ} = \frac{7\sigma_{3P_0} + 11\sigma_{3P_1}}{6} \tag{3.10.11}$$

$$\sigma_{3D}^{90^\circ} = 1.19\sigma_{3D_0} + 2.19\sigma_{3D_1} + 1.62\sigma_{3D_2} \tag{3.10.12}$$

Here $\sigma_{3L}^{90^\circ}$ ($L = S, P, D$) is the cross section for $3L$ state excitation, which is determined by $P_{3L}^{90^\circ}$, the degree of polarization of optical radiation at right angles to the exciting beam, $P_{3L}^{90^\circ}$, in such a way that

$$\sigma_{3L}^{90^\circ} = \sigma_{3L}(1 - \tfrac{1}{3}P_{3L}^{90^\circ})^{-1} \tag{3.10.13}$$

where σ_{3L} is the excitation cross section for the $3L$ state, and σ_{3Lm} ($m = 0, 1, 2$) is the cross section for impact excitation of different magnetic sublevels.

After these introductory remarks concerning the hydrogen H_α line, it is worthwhile reviewing experimental and theoretical methods to determine different atomic parameters relevant to impact polarization of hydrogen spectral emissions created by electrons or protons, as well as direct polarization parameters. A parallel analysis of experimental and theoretical results is a most useful approach to the practical needs of impact spectropolarimetric sensing. By using only computations of atomic polarization parameters in combination with reliable experimental data, as well as semiempirical experimental and theoretical results, it is possible to provide reliable parameters of an ionized medium under study via a comprehensive procedure of impact spectropolarimetric sensing.

Consider first experimental methods which can be used to determine polarization parameters of the H_α hydrogen line. We note that the characteristics of electron impact excitation of magnetic $3P_m$ and $3D_m$ substates can be inferred by measuring the angular correlation between the exciting electron and the subsequently emitted photon. Technically, these experiments are very difficult to implement, so reliable data are available only for several values of the impact energy. The electron beam modulation technique was also applied to electron impact excitation of hydroden[145,146] in order to determine the energy dependence of the excitation function and the polarization. By comparing different experimental studies, one can conclude that measurements[145] are still the most reliable experimental data.

While theoretical methods for calculating cross sections for electron impact excitation of atomic and ionic states are now being developed, there do exist only a few reliable theoretical calculations for the optical polarization parameters of the H_α line, stimulated by electron impact. Let us examine the results of theoretical calculations of H_α polarization parameters, reviewing either direct methods or computations of other atomic parameters from which the polarization data can be inferred by means of simple algebraic or normalization procedures. The simplest method for calculating H_α polarization is the first Born approximation (FBA). The scattering amplitude within this approximation is proportional to the matrix element of the impact electron–atom interaction calculated with zero-approximation wave functions, which are the product of the free-electron wave function and the atomic electron orbital (see Chapter 4).

The results of computing H_α polarization by this technique are presented elsewhere.[147,148] Comparison of these results with experimental data[145,149] (see Figure 10) shows that the theoretical results under the first Born approximation are in strong disagreement with measurements. Contrary to the first Born approximation, the distorted wave polarized orbital approximation (DWPO) is one of the more complicated methods used to calculate H_α polarization.[147] This method takes into account two additional effects: the distortion of the projectile wave function and the polarization of the atomic orbital due to the influence of the projectile. The results of the work,[147] mentioned above show that, despite the essential complication of the computational technique, there are only very small changes in the final values of the polarization, compared to the first Born approximation. Agreement with experimental data in this case was obtained only near the excitation threshold.

A close coupling approximation was carried out[150] in order to compute excitation cross sections for separated $1S - 3Lm$ ($L = S, P, D$) transitions to

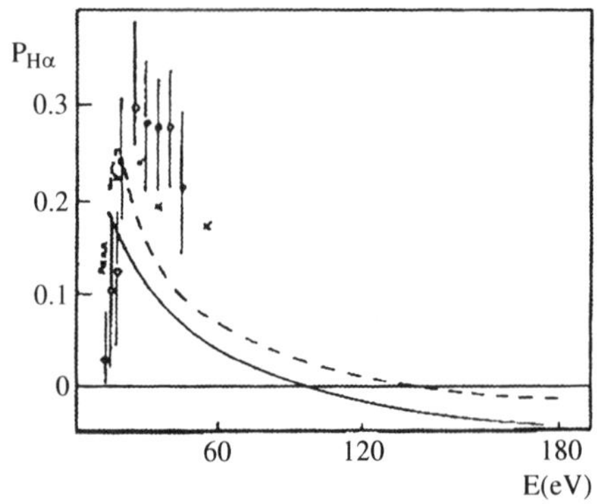

Figure 10. Measured and computed polarization fractions of the hydrogen H_α line as a function of the impact electron energy.[7]

determine H_α line polarization caused by low energy electrons. The results of these calculations are presented in Figure 10–13 for experimental and theoretical data. In contrast to the results of distorted wave polarized orbital and Born approximations, the results are in reasonable agreement with measurements of H_α line polarization.[145]

Concerning proton impact excitation, theoretical and experimental atomic data on the hydrogen quantum state polarization by protons are limited, but a large number of calculations and measurements of atomic data and excitation parameters for atomic hydrogen by protons have been performed.[144] Owing to the nature of proton–hydrogen collision, most of these studies deal with excitation

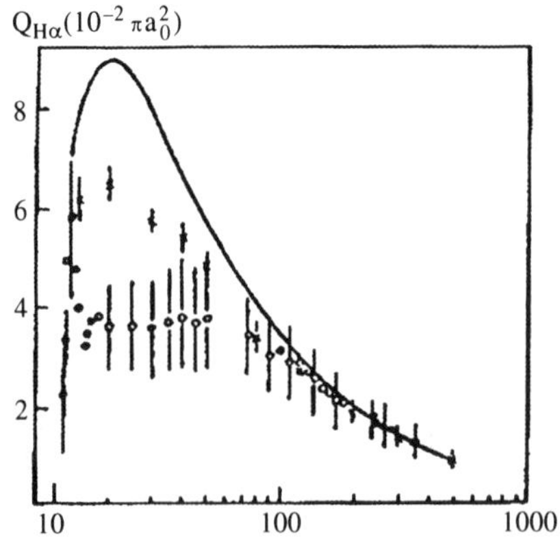

Figure 11. Total H_α excitation cross section σ_{H_α} as a function of impact electron energy,[7] including experimental data.[154]

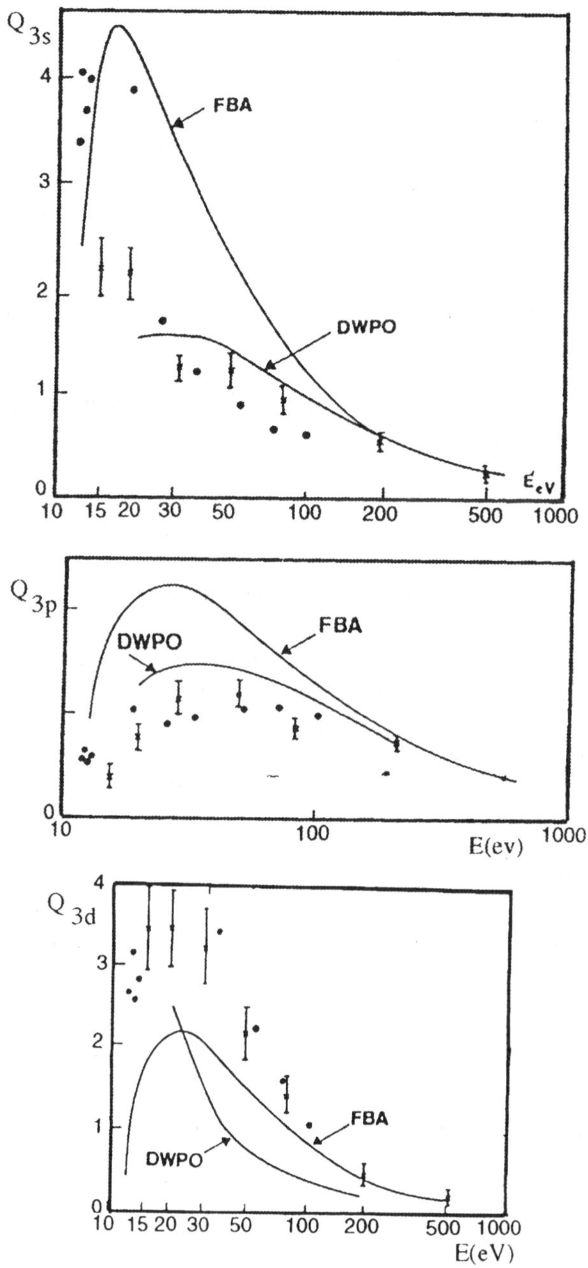

Figure 12. Experimental (dots and circles) and theoretical (FBA and DWPO) excitation cross sections for the upper 3*S* levels of the H_α line from the ground state as a function of electron energy.[7]

to the $n = 2$ state. Experimental determination of excitation cross sections for atomic hydrogen from the ground state to the $n = 2$, 3, and 4 levels has been implemented by Park *et al.*[(153)]

The measurements were based on the energy loss spectrometry method. The cross sections were obtained from an analysis of energy loss spectra of 15 to 200 keV protons by using the spectrum fitting technique. The values of H_α polarization by proton impact for the intermediate energy range have been obtained on the basis of theoretical results.[(115)] The density matrix, describing excitation of the $n = 2$ and $n = 3$ levels of atomic hydrogen, was calculated in the first Born approximation and with the close coupling scheme exmploying an expansion in terms of the target eigenstates. The transition amplitudes for excitation from the $1S$ ground state to a particular $|NLm\rangle$ state have been calculated by solving the time-dependent Schrödinger equation with the aid of the impact parameter method. At impact energies higher than 25 keV, the straight line trajectory approximation was used. The charge exchange channel was excluded in these calculations. Integral alignment of $H(2P)$ was measured for 1 to 25 keV collisions of $H - H^+$ and $He - H^+$ systems.[(151)] The experimental data are in good agreement with theoretical results.[(152)]

The degree of polarization of H_α radiation stipulated by the impact excitation of hydrogen atoms by protons has also been a subject of theoretical calculations. These computations were implemented within the Born approximation,[(50)] within the Glauber approximation,[(96)] and using the impact parameter method.[(195,196)] Calculations showed that the results of the Born approximation differ considerably from those of the Glauber approximation and the impact parameter technique. They are in agreement with the conclusions of the above-mentioned theoretical considerations.

The degree of polarization detected at right angles to the exciting beam could be calculated, by making use of relations with other atomic collisional polarization parameters. Following Blum,[(102)] the degree of polarization at right angles ($P_{3L}^{90^\circ}$) for the $n = 3$ level can be expressed in terms of the alignment parameter A_{3L}

$$P_{3P}^{90^\circ} = \frac{3}{(1 - 6A_{3P}^{-1})} \quad \text{and} \quad P_{3D}^{90^\circ} = \frac{57}{19 - 50A_{3D}^{-1}} \tag{3.10.14}$$

Excitation cross sections σ_{3L} ($L = S, P, D$) lead to the alignment parameters via the relations

$$A_{3P} = \frac{\sigma_{3P_1} - \sigma_{3P_0}}{2\sigma_{3P_1} + \sigma_{3P_0}} \quad \text{and} \quad A_{3D} = \frac{2\sigma_{3D_2} - \sigma_{3D_1} + \sigma_{3D_0}}{2\sigma_{3D_2} + 2\sigma_{3D_1} + \sigma_{3D_0}} \tag{3.10.15}$$

and have been calculated in the range of proton energies 6–25 keV. As a test, the total excitation cross section $\sigma_{(n=3)}$ was compared[(115)] with experimental data[(153)]

(Figure 13). Due to the possible influence of the charge exchange process, agreement was satisfactory only for proton energies above 200 keV.

Formulas (3.10.14) and (3.10.15) allow us to easily obtain the degree of H_α polarization. Preliminary results inferred from published data[115] are presented in Figure 14. One can see that for electron impact excitation, the H_α polarization decreases with increasing energy and changes sign at energies near 600 keV. It is evident that for relatively low energies, the charge exchange process must be included in order to provide an adequate model.

Impact excitation of hydrogen by electron beams has been studied experimentally by various techniques in the course of which many atomic data have been determined. This information is very important either for direct determination of the impact polarization parameters of the hydrogen states, or for

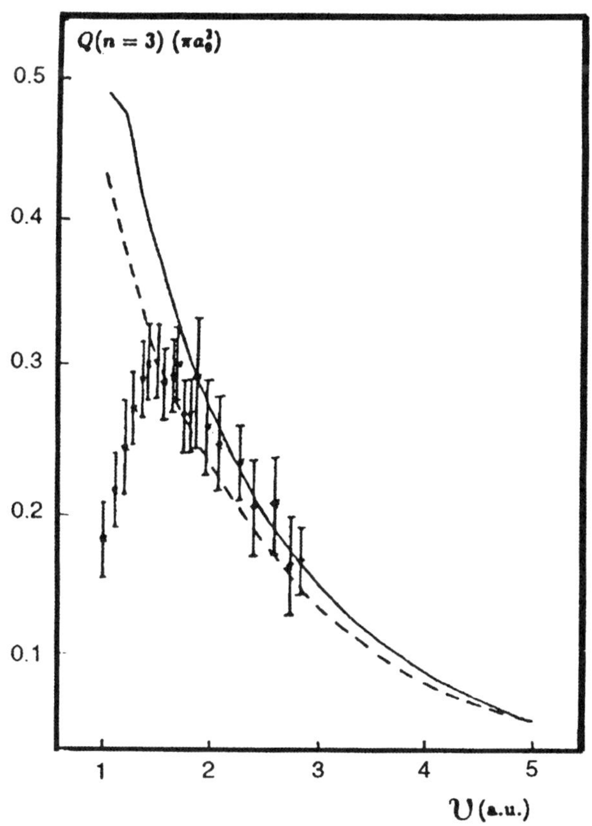

Figure 13. Energy dependence of the total cross section for excitation of an $n = 3$ array of atomic hydrogen by protons. The dotted line is the result of first Born approximation; the solid line is the close coupling approximation and points denote experimental data.[7]

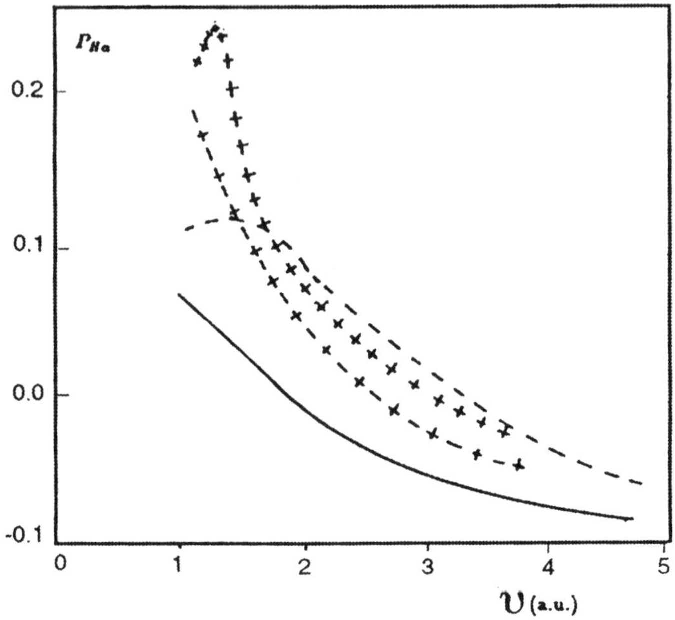

Figure 14. Polarization of H_α radiation as a function of proton or electron impact energy[7]: +-+-+-, FBA for electrons; +++, DWPO for electrons; —, FBA for protons; solid line, close coupling approximation for protons.

normalization and correction of computations. Important information about the alignment of atomic ensembles can be inferred from the cross sections of impact polarization of magnetic sublevels. Data for excitation cross sections of sublevels $|3Lm\rangle$ can be deduced from electron–photon coincidence experiments.[102] Such experiments allow one to determine a set of angular correlation atomic parameters,[155,156] related to the amplitude of impact excitation of sublevels m_l.

Simple experimental spectropolarimetric parameters relevant for a description of linearly polarized radiation emitted by excited atoms are the degree of polarization P_l and the orientation of the alignment axis, specified by angle γ. These two parameters can be expressed in terms of Stokes parameters as follows

$$P_l = (Q^2 + U^2)^{1/2} \tag{3.10.16}$$

$$\gamma = \frac{1}{2}\arctan\frac{U}{Q} \tag{3.10.17}$$

The total impact alignment cross section, $\sigma^{(2)}(v)$, may be easily deduced from these parameters

$$\sigma^{(2)}(v) = 2\pi\int_0^\pi d\theta \sin\theta\, \sigma^{(2)}(\theta, v)$$

where $\sigma^{(2)}(\theta, v)$ is the differential alignment cross section, expressed in terms of the differential excitation cross section $\sigma(\theta, v)$

$$\sigma^{(2)}(\theta, v) = \frac{1}{2\sqrt{6}}\sigma(\theta, v)[1 + 3P_l(v)\cos\gamma] \tag{3.10.18}$$

In the simplest case of $1S$-$3P$ excitation, the following parameter is very useful for applications

$$\lambda = \frac{\sigma_{3p_0}(\theta, v)}{\sigma_{3p}(\theta, v)} \tag{3.10.19}$$

where $\sigma_{3p_0}(\theta, v)$ is the differential cross section corresponding to excitation of the $m_l = 0$ sublevel, and $\sigma_{3p}(\theta, v)$ is the total differential cross section for excitation of the $3P$ level.

A relation between parameters λ, P_l, and γ takes the form

$$\lambda = \frac{P_l(v)\cos 2\gamma + 1}{2} \tag{3.10.20}$$

The differential cross section is

$$\sigma^{(2)}(\theta, v) = -\frac{1}{\sqrt{6}}\sigma(\theta, v)[1 - 3\lambda(\theta, v)] \tag{3.10.21}$$

Information on $3P_J$ excitation was obtained in a straightforward manner from the detection of L_β photons coinciding with electrons scattered for a corresponding loss in energy. Measurements of parameter λ were carried out for exciting electron energies of 53.4 eV and 100 eV and for electron scattering angles of 20°and 25° (Table 1). The theoretical values of λ calculated in the first Born approximation are also included in Table 1.

Due to general experimental difficulties arising from the small excitation cross-section of this state[146] and the low level of electron and optical signals, only a few angular correlation experiments have been implemented to measure the parameters of electron impact excitation of the $3D_J$ states of atomic hydrogen. An experimental study of the $3D_J$ state has been conducted.[155,157] On the other hand, experimental limitations on the resolution power of the energy-loss technique does not allow one to separate scattered electrons in three different channels (after $3S$, $3P$, and $3D$ excitations). From the experimental point of view, ordinary spectroscopy turns out to be a rough instrument for the resolution of different fine-structure components forming the H_α radiation. The only way to reliably measure collision parameters for the $3D_J$ state is to use the specific decay scheme of this state. The idea is that information about the $3D_J$ state is carried by

TABLE 1
Experimental Values of Parameter λ for the $3P_J$ State at Incident Energies E of 54.4 and 100 eV and Electron Scattering Angles $\theta = 20°$ and 25° Obtained[155] Together with Theoretical Data λ_{FBA} in the First Born Approximation (FBA)

E (eV)	$\theta°$	λ	λ_{FBA}
54.4	20	0.644 ± 0.042	0.244
54.4	25	0.444 ± 0.047	0.225
100	20	0.495 ± 0.089	0.121
100	25	0.132 ± 0.045	0.126

L_α photons emitted by the decay of the $2P_J$ state, which is populated by cascades from $3S$ and $3D_J$ parent levels. Two experimental versions of this idea have been realized.[155,157] The results show significant deviations from theoretical estimates using the first Born approximation (Table 2).

Another more elaborate method, known as the electron beam modulation technique, was applied to H_α electron impact excitation studies. This method was used[145] to measure the energy dependence of the excitation function and polarization. These measurements remain the only very pure experimental determination of the H_α degree of impact polarization.

The same technique was used to measure the hydrogen $3S$, $3P$, and $3D$ impact excitation cross section.[146] A complete analysis of the output signal as a function of the modulation frequency allowed separation and extraction of three different components with an error of less than 10%. As a consequence, the energy dependence of the cross sections σ_{3L} ($L = S, P, D$) has been obtained for the energy range 5–13.8 eV. Correction for cascading was also introduced.

TABLE 2
Experimental Values of Parameters P_l and γ for the $3D_J$ State at Incident Energies of 53.4 and 100 eV and Electron Scattering Angles $\theta = 20°$ and 25°; γ_{FBA} is a Theoretical Value Calculated in the First Born Approximation (FBA)[7,155]

E (eV)	$\theta°$	P_l	$\gamma°$	$\gamma°_{FBA}$
54.4	20	0.0	-45.0 ± 2.2	-60.4
54.4	25	0.057	-52.6 ± 3.0	-61.8
100	20	0.063	-54.0 ± 1.3	69.6
100	25	0.090	-73.2 ± 3.9	-69.2

The total H_α excitation cross section of atomic hydrogen, calculated as a function of the electron energy, is presented in Figure 10 together with experimental results.[145] An evident difference in the energy dependence of the total cross section, σ_{H_α}, has been explained,[146] by suggesting that this difference could be due to the electric-field mixing of the $3L$ states in the experiment.[145] The result of Mahan *et al.*[146] has been used elsewhere[147] to calculate the H_α line polarization (Table 3).

The excitation cross sections of magnetic sublevels, σ_{3Lm}, which are indispensable in such a calculation, were computed theoretically in the Born approximation. The computed[147] and measured[145] energy dependence of the H_α line polarrization are presented in Figure 10 and show quite a pronounced difference.

A different way to obtain the cross sections of separate $3L$ components has been suggested.[158] It was based on the expression for the H_α line polarization fraction perpendicular to the propagation direction of an electron beam:

$$P_{H_\alpha}^{90^\circ}(E) = \frac{I_{3P}(E)P_{3P}(E) + I_{3D}(E)P_{3D}(E)}{I_{H_\alpha}(E)} \tag{3.10.22}$$

The energy dependence of P_{3P} and P_{3D} can be estimated using earlier results.[102,133,140] Since the intensities I_{3P} and I_{3D} are proportional to the cross sections σ_{3P} and σ_{3S}, it was proposed[158] that these quantities be derived by means of a two-parameter fit of the measured energy dependence of the H_α polarization fraction. Unfortunately, due to the unavailability at that time of high-density sources of atomic hydrogen, the accuracy of H_α polarization measurements did not justify the procedure. Therefore, by making use of the Eq. (3.10.20), this method was used[158] to check only agreement between measurements[146] of parameters $I_{3P}(E)/I_{H_\alpha}(E)$ and $I_{3D}(E)/I_{H_\alpha}(E)$ on the one hand and the measurement[145] of the H_α polarization fraction on the other. A rather good degree of consistency was found.

Practically important relations between different sets of atomic collisional parameters relating to electron–photon angular correlation experiments for *S–3P* and *S–3D* excitation modes for computing the differential impact alignment cross

TABLE 3
Theoretical Values of Hydrogen H_α Polarization Calculated in the First Born Approximation (FBA), Distorted Wave Polarized Orbital (DWPO),[147] and DWPA[156]

E (eV)	FBA	DWPO	DWPA
54.4	5.19×10^{-2}	7.19×10^{-2}	5×10^{-2}
100	2.59×10^{-2}	2.62×10^{-2}	9×10^{-2}

section $\sigma^{(2)}(\theta, v)$ (θ is the exciting electron scattering angle) are given in Section 3.12. The results of many original and review papers on this subject(7,9,142,143) are presented there.

3.11. Anisotropic Collisional Relaxation of Excited Hydrogen Atoms

Collisional relaxation within a system of energy levels of atomic hydrogen is a more complicated problem than in the ordinary case. A description of the collisional relaxation of an excited hydrogen atom requires one to include collisional mixing not only of states with different magnetic quantum numbers m, but of states with different orbital quantum numbers l, which have equal energies for a level with given principal quantum number n as well. This means that a random degeneracy of energy levels of a one-electron atom caused by four-dimensional Fock symmetry(124,125,162,163) is a complicated picture of collisional relaxation of atomic states, when compared to a commonly considered case where energy degeneracy in the orbital quantum number l is absent.

We shall investigate the characteristic features of the relaxation of atomic polarization moments and of accompanying polarization phenomena in the emission spectra, which are associated with collisional mixing of different l states within a particular Rydberg energy array, specified by the principal quantum number n. Consider the simplest case of relaxation of an excited group of states of a hydrogen atom with $n = 2$ under collisions with a monokinetic beam of charged particles (protons). A group of degenerate states of the hydrogen atom for $n = 2$ contains four states, described by wave functions $\psi_{00}, \psi_{1-1}, \psi_{01}$, and ψ_{11} in a factorized form with separate radial and angular variables, as follows

$$\psi_{lm} = R_{2l}(r)Y_{lm}(\Theta, \phi) \tag{3.11.1}$$

Here, $R_{2l}(r)$ are radial wave functions for the state $n = 2$ and $Y_{lm}(\Theta, \phi)$ are spherical functions; l and m are, respectively, the orbital and magnetic quantum numbers.

Functions ψ_{lm} correspond to quantization of the projection of angular momentum with respect to the laboratory OZ axis, or the direction of the incident proton beam. The fine structure and Lamb shift of energy levels will be neglected. The ψ_{00}, ψ_{1-1}, ψ_{01}, and ψ_{11} states of a free hydrogen atom have identical energies

$$E_2 = -\frac{m_e e^4}{8\hbar^2}$$

where m_e is the electron mass.

In general, an excited atomic state is a superposition of the four given l, m states, described by means of the following linear combination

$$\Psi(\mathbf{r}, t) = \sum_{l,m} c_{l,m}(t)\psi_{lm}(r)\exp\left(-\frac{iE_0 t}{\hbar}\right) \tag{3.11.2}$$

An ensemble of excited hydrogen atoms within the l, m presentation is described in the most general manner by the density matrix

$$\rho_{mm_1}(ll_1) = \overline{c_{lm}c^*_{l_1 m_1}} \tag{3.11.3}$$

where c_{lm} are coefficients from expression (3.11.2) for the wave functions of individual atoms and the overline denotes averaging over the ensemble.

Following the adopted procedure, let us introduce the polarization moments of the density matrix of an ensemble of excited hydrogen atoms

$$\rho^k_q(ll_1) = \sum_{mm_1}(-1)^{l_1-m_1}\begin{bmatrix} l & l_1 & \kappa \\ m & -m_1 & q \end{bmatrix}\rho_{mm_1}(ll_1) \tag{3.11.4}$$

The rank κ of the polarization moment $\rho^k_q(ll_1)$, taking integral values between $|l - l_1|$ and $(l + l_1)$, determines an irreducible representation of the group of three-dimensional rotations. Its projection q takes integral values from $-\kappa$ to $+\kappa$.

As the orbital quantum numbers l and l_1 for the hydrogen atoms in the state with $n = 2$ take values 0 and 1, two polarization moments $\rho^0_0(0, 0)$ and $\rho^0_0(1, 1)$ of zero rank are scalars proportional to populations of states with $l = 0$ and $l = 1$

$$n_l = \sqrt{2l + 1}\rho^0_0(l, l) \tag{3.11.5}$$

There are also three nonzero polarization moments of first rank, including the axial vector of orientation of the "pure" state with $l = 1$, or $\rho^1_q(1, 1)$; two polar vectors of mixed orientation of states with $l = 0$, $l = 1$, or $\rho^1_q(1, 0)$, $\rho^1_q(0, 1)$; and one polarization moment of second rank, or the alignment tensor $\rho^2_q(1, 1)$.

Owing to the preferential direction of particle collision (OZ axis), the collisional relaxation of polarization moments $\rho^k_q(ll_1)$ will be determined in terms of irreducible representations of the $C_{\infty v}$ symmetry group.[163] This group has two one-dimensional irreducible representations (A_1 and A_2) and two two-dimensional irreducible representations (E_1 and E_2). The one-dimensional irreducible representation A_1 of this group includes scalars $\rho^0_0(0, 0)$, $\rho^0_0(1, 1)$, the longitudinal alignment component $\rho^2_0(1, 1)$ of the $l = 1$ state, the longitudinal components of the polar vectors of "mixed" orientation of states $l = 1$ and $l = 0$, or $\rho^1_0(1, 0)$ and $\rho^1_0(0, 1)$. The one-dimensional representation A_2 includes the longitudinal component of the axial orientation vector $\rho_0^1(1, 1)$ of the pure $l = 1$ state. The two-dimensional irreducible E_1 representation includes components of orientation $\rho_{\pm1}^1(1, 1)$ and ordering $\rho_{\pm1}^1(1, 1)$ of the pure $l = 1$ state and also of the mixed orientations $\rho_{\pm1}^1(1, 0)$ and $\rho_{\pm1}^1(0, 1)$ with respect to the axis, inclined to the OZ axis. The two-dimensional representation E_2 includes the transverse alignment components $\rho_{\pm2}^2(1, 1)$ of the pure $l = 1$ state.

Thus the components of the polarization moments of an ensemble of hydrogen atoms in the $n = 2$ state for different irreducible representations of the symmetry group, specific for the collisional problem, can be classified by the symmetry group $C_{\infty v}$ in the following way

$$\rho_0^0(0,0), \quad \rho_0^0(1,1), \quad \rho_0^2(1,1), \quad \rho_0^1(1,0), \quad \rho_0^1(0,1), \quad \rho_0^1(1,1)$$

$$\rho_{\pm1}^1(1,1), \quad \rho_{\pm1}^2(1,1), \quad \rho_{\pm1}^1(1,0), \quad \rho_{\pm1}^1(0,1), \quad \rho_{\pm2}^2(1,1)$$

Components of polarization moments positioned in different lines undergo anisotropic collisional relaxation independently of each other. At the same time, their components within each line relax jointly. This is known as the collisional interconversion of polarization moments, characteristic for anisotropic collisions specified by a preferential direction of interaction. Due to the properties of irreducible representations A_2 and E_2, the longitudinal components of orientation, $\rho_0^1(1, 1)$, and alignment, $\rho_0^2(1, 0)$, of the pure $l = 1$ state undergo collisional relaxation independently of all other polarization moments. Therefore, collisional relaxation of these moments is specified by a one-exponent decay law. Relaxation of the remaining components of polarization moments is accompanied by their interconversion or mixing, but strictly within the limits determined by their classification in terms of irreducible representations of the $C_{\infty v}$ group. Therefore, their relaxation is described by multi-exponent decay laws (a five-exponent law in the case of polarization moments of A_1 symmetry and a four-exponent law in the case of E_1 symmetry).

Let us express the collisional variation of polarization moments in terms of the cross sections of this process. The cross section $\sigma_q^{kk'}(ll_1, l'l_1'; v)$ describing the contribution made by collisions with protons of velocity v to the time variation of polarization moments of excited hydrogen atoms can be introduced with the help of the following equation

$$\left[\frac{\partial \rho_q^k(ll')}{\partial t}\right]_{coll} = n_0 v \sum_{k'l'l_1'} \sigma_q^{kk'}(ll_1, l'l_1'; v)\rho_q^{k'}(l_1 l_1') \tag{3.11.6}$$

The cross section is diagonal with respect to q and splits into blocks according to the classification of components of polarization moments in terms of irreducible representations of the $C_{\infty v}$ group, presented above. These blocks have dimensions 5, 4, 1, and 1. The total number of elements of the relaxation matrix in the blocks is equal to $5 \times 5 + 4 \times 4 + 1 \times 1 + 1 \times 1 = 43$.

Consider now the most common case for astrophysical applications, when the relaxation of the excited atomic hydrogen is a result of collisions with protons. Here, one can model the motion of a proton with respect to a hydrogen atom by a trajectory, which consists of two semilinear parts (Figure 15).

Collisions can be described by introducing two coordinate systems: the collision system $(\tilde{X}, \tilde{Y}, \tilde{Z})$ with axis $O\tilde{Y}$ perpendicular to the collision plane and $O\tilde{Z}$ parallel to the trajectory of the proton, and the system $(\tilde{\tilde{X}}, \tilde{\tilde{Y}}, \tilde{\tilde{Z}})$, where axis $O\tilde{\tilde{Y}}$ coincides with axis $O\tilde{Y}$ and axis $O\tilde{\tilde{Z}}$ axis is directed toward the projectile. The angle between $O\tilde{Z}$ and $O\tilde{\tilde{Z}}$ will be designated as α.

The state of a hydrogen atom interacting with an incident proton is described by the Schrödinger equation

$$i\hbar \frac{\partial \Psi}{\partial t} = (H_0 + W)\Psi \tag{3.11.7}$$

where H_0 is the Hamiltonian of the free hydrogen atom and the term

$$W = -\frac{e^2}{r'} = -\frac{e^2}{|\mathbf{R} - \mathbf{r}|} \tag{3.11.8}$$

denotes the interaction of the hydrogen atom with the precipitating proton.

An approximate wave function of the atomic hydrogen at state $n = 2$ is sought as an expansion in terms of eigenfunctions of the $\hat{l}^2$ and $\hat{l}_z$ operator:

$$\Psi(\mathbf{r}, t) = \sum_{l,m} \tilde{\tilde{c}}_{l,m}(t) \tilde{\tilde{\psi}}_{l,m}(\mathbf{r}, t) \exp\left(-\frac{E_2 t}{\hbar}\right) \tag{3.11.9}$$

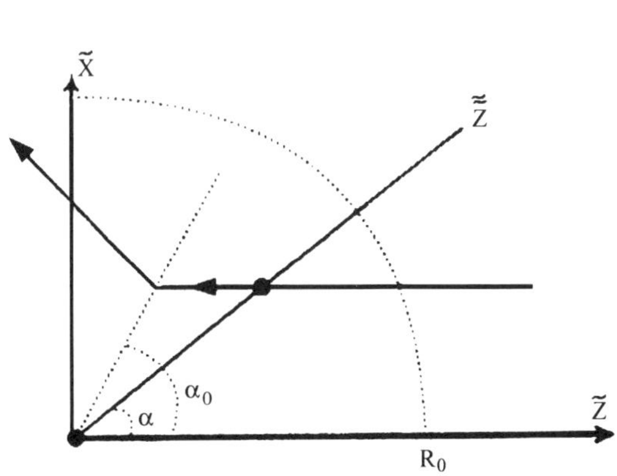

Figure 15. Collision of a hydrogen atom with an incident proton.

where E_2 is the energy of the $n = 2$ state of atomic hydrogen and the sign $\approx$ means that the corresponding value relates to the following reference system. This relation differs from Eq. (3.11.2), because of the difference in the reference systems for which these relations are valid: Eq. (3.11.2) is connected to the laboratory frame of reference while Eq. (3.11.9) is written in the following one.

On substituting Eq. (3.11.9) into (3.11.7) and taking into consideration the rotation of the following reference system, the parameters

$$\varepsilon_2 = \frac{1}{2a_0} \qquad \text{and} \qquad \eta = \frac{e^2}{\hbar v \rho \varepsilon_2}$$

enable a system of differential equations to be obtained for the coefficients of expansion (3.11.9)

$$\frac{d}{d\alpha}\tilde{\tilde{c}}_{lm} = -i\eta \sum_{l'} \langle lm|W|l'm\rangle \tilde{\tilde{c}}_{l'm} + \tfrac{1}{2}\left[\sqrt{(l-m)(l+m+1)}\,\tilde{\tilde{c}}_{lm+1} - \sqrt{(l+m)(l-m+1)}\,\tilde{\tilde{c}}_{lm-1}\right] \tag{3.11.10}$$

Expansion in terms of Legendre polynomials allows the matrix elements of interaction operator (3.11.9) to be expressed in the form

$$\langle lm|W|l'm\rangle = \varepsilon_2 \sqrt{\frac{2l'+1}{2l+1}} \sum_k \begin{bmatrix} l' & k & l \\ 0 & 0 & 0 \end{bmatrix} \begin{bmatrix} l' & k & l \\ m & 0 & m \end{bmatrix} f^{(k)}_{nlnl'} \tag{3.11.11}$$

where, in terms of the dimensionless quantity $x = \varepsilon_2 R$, the functions $f^{(k)}_{nlnl'}$ take the integral form

$$f^{(k)}_{nlnl'} = \frac{1}{\varepsilon_2^3}\left[\frac{1}{x^{k-1}} \int_0^x t^{k+1} R_{nl}\left(\frac{t}{\varepsilon_2}\right) R_{nl'}\left(\frac{t}{\varepsilon_2}\right) dt + x^k \int_x^\infty t^{1-k} R_{nl}\left(\frac{t}{\varepsilon_2}\right) R_{nl'}\left(\frac{t}{\varepsilon_2}\right) dt\right] \tag{3.11.12}$$

Symmetry properties of the Clebsh–Gordan coefficients on the right hand side of expression (3.11.11) restrict the number of terms of the sum to those which satisfy the condition

$$|l - l'| < k < l + l'$$

In the case $n = 2$ only four matrix elements from relationship (3.11.12) meet this condition. In terms of the dimensionless parameter x they are expressed as follows

$$
\begin{aligned}
f_{2020}^{(0)} &= \frac{(x^3 + x^2 + 1.5x + 1)e^{-2x} - 1}{x} \\
f_{2120}^{(1)} &= \frac{\sqrt{3}[3 - (2x^4 + 4x^3 + 6x^2 + 6x + 3)e^{-2x}]}{2x^2} \\
f_{2121}^{(0)} &= \frac{(2x^3 + 6x^2 + 9x + 6)e^{-2x} - 6}{6x} \\
f_{2121}^{(2)} &= \frac{5[(2x^5 + 6x^4 + 12x^3 + 18x^2 + 18x + 9)e^{-2x} - 9]}{6x^3}
\end{aligned}
\tag{3.11.13}
$$

The x-dependence of matrix elements $\langle 200|W|200\rangle$, $\langle 200|W|210\rangle$, and $\langle 21m|W|21m\rangle$ $(m = 0, \pm 1)$ is presented later in Figure 16. The diagonal matrix elements for large x decrease as x^{-1}, while matrix element $\langle 200|W|210\rangle$ decreases as x^{-2}.

For small internuclear distances the behavior of matrix elements differs substantially. The diagonal matrix elements at $R \to 0$ tend to the quantity $\langle 1/r\rangle$ for the corresponding state of the unified atom and the matrix element $\langle 200|W|210\rangle$ is equal to zero due to the orthogonality of functions ψ_{00} and ψ_{10}.

System (3.11.10) for the case $n = 2$ can be written in the form

$$
\begin{aligned}
\frac{d\tilde{\tilde{c}}_{00}}{d\alpha} &= -i\eta\left(f_{2020}^{(0)}\tilde{\tilde{c}}_{00} + f_{2021}^{(1)}\frac{\tilde{\tilde{c}}_{10}}{\sqrt{3}}\right) \\
\frac{d\tilde{\tilde{c}}_{1-1}}{d\alpha} &= -i\eta[(f_{2121}^{(0)} - \tfrac{1}{5}f_{2121}^{(2)})\tilde{\tilde{c}}_{1-1}] + \frac{\tilde{\tilde{c}}_{10}}{\sqrt{2}} \\
\frac{d\tilde{\tilde{c}}_{10}}{d\alpha} &= -i\eta\left[(f_{2121}^{(0)} - \tfrac{2}{5}f_{2121}^{(2)})\tilde{\tilde{c}}_{10} + f_{2021}^{(1)}\frac{\tilde{\tilde{c}}_{00}}{\sqrt{3}}\right] + \frac{\tilde{\tilde{c}}_{11} - \tilde{\tilde{c}}_{1-1}}{\sqrt{2}} \\
\frac{d\tilde{\tilde{c}}_{11}}{d\alpha} &= -i\eta[(f_{2121}^{(0)} - \tfrac{1}{5}f_{2121}^{(2)})\tilde{\tilde{c}}_{11}] - \frac{\tilde{\tilde{c}}_{10}}{\sqrt{2}}
\end{aligned}
\tag{3.11.14}
$$

The Coulomb terms on the right hand side of system (3.11.14) can be excluded because they do not depend on the magnetic quantum number and, as a result, the principal contribution to the solution of this system for large x will be provided by terms which decrease as x^{-2} for large x. Substitution of function $f_{2021}^{(1)}$ by its long-range asymptotics and neglect of the other coefficients $f_{nln'l'}^{(k)}(k \neq 1)$ leads to the dipole approximation in the description of the interaction. This approximation was used to integrate system (3.11.14) elsewhere.[84,94,123]

This approximation is convenient in that system (3.11.14) is reduced to a system with constant coefficients, the fundamental solution of which can be easily represented analytically. On the other hand, it can be tested by direct calculation that the logarithmically diverging integrals arise under averaging over the impact parameter within the framework of this approximation. The latter difficulty can be removed only avoiding the integration process with the aid of some model considerations. This approach was employed[94] when the variation in the integral due to the addition of new points of division became small.

It is reasonable to choose the upper integration limit based on the following considerations. It is known[33,36] that the $2^2P_{1/2,3/2}$ state of an isolated hydrogen atom is split with a splitting energy of about $\Delta = 0.33\ \text{cm}^{-1}$. A free hydrogen atom in state $2^2P_j, n = 2$, can be described by the six wave functions $\psi_{1/2\pm1/2}$, $\psi_{3/2\pm1/2}$, $\psi_{3/2\pm3/2}$.

We also assume that for small internuclear distances ($R < R_0$), when Stark spliting at the $n = 2$ level is more than Δ, the L S coupling is breaking down and variation in the magnetic sublevel populations due to collisions is described by system (3.11.14). At $R = R_0$, when Stark splitting is equal to fine-structure splitting, the L–S coupling is reestablished. In this situation, collisional variation in hydrogen states with $n = 2$ can be described by a wave function in the form of an expansion in terms of eigenfunctions of the square of the total momentum and its projection

$$\Psi(\mathbf{r}, t) = \sum_{JM} \tilde{\tilde{a}}_{JM}(t)\tilde{\tilde{\psi}}_{JM}(\mathbf{r}, t) \exp\left(-i\frac{E_J t}{\hbar}\right) \tag{3.11.15}$$

where summation is implemented only over J numbers, which are obtained by vector addition of the orbital moment $l = 1$ with the spin moment.

The only possible kind of collisional ordering for a state with $l = 0$ (the total angular momentum can only be $J = \frac{1}{2}$ in this case) is the population and the orientation. By substituting expansion (3.11.15) into Eq. (3.11.7), one obtains a system of differential equations for the expansion coefficients $\tilde{\tilde{a}}_{Jm}$, similar to Eq. (3.11.10), in the following form

$$\frac{d\tilde{\tilde{a}}_{JM}}{d\alpha} = -i\eta \sum_{J'l'} \langle JM(l)|W|J'M(l')\rangle \tilde{\tilde{a}}_{J'M} + \tfrac{1}{2}\left[\sqrt{(J-M)(J+M+1)}\,\tilde{\tilde{a}}_{JM+1} - \sqrt{(J+M)(J-M+1)}\,\tilde{\tilde{a}}_{JM-1}\right] \tag{3.11.16}$$

The interaction matrix elements from Eq. (3.11.16) with the quantum mechanical approach of angular momenta coupling can be expressed as

$$\langle JM(l)|W|J'M(l')\rangle = (-1)^{l+J+S}\sqrt{2J'+1}\sum_p \sqrt{2p+1}\begin{Bmatrix} l & l' & p \\ J' & J & \frac{1}{2} \end{Bmatrix} \cdot \begin{bmatrix} p & J' & J \\ 0 & M & M \end{bmatrix} W^{(p)}(ll') \tag{3.11.17}$$

Here, quantity $W^{(p)}(ll')$ is related to the matrix elements by the following relation:

$$W^{(p)}(ll') = \sum_m (-1)^{l'-m}\begin{bmatrix} l & l' & p \\ m & -m & 0 \end{bmatrix}\langle lm|W|l'm\rangle \tag{3.11.18}$$

When $l = 1$, the last two expressions and expressions (3.11.13) for quantities $f^{(k)}_{lml'm)}$ allow system (3.11.17) to be characterized by the following nonzero matrix elements

$$\begin{aligned} \langle\tfrac{3}{2}\mp\tfrac{1}{2}(1)|W|\tfrac{1}{2}\mp\tfrac{1}{2}(1)\rangle &= \sqrt{\tfrac{2}{5}}f^{(2)}_{2121} \\ \langle\tfrac{3}{2}\pm\tfrac{1}{2}(1)|W|\tfrac{3}{2}\pm\tfrac{1}{2}(1)\rangle &= f^{(0)}_{2121} + \frac{f^{(2)}_{2121}}{5} \\ \langle\tfrac{3}{2}\pm\tfrac{3}{2}(1)|W|\tfrac{3}{2}\pm\tfrac{3}{2}(1)\rangle &= f^{(0)}_{2121} - \frac{f^{(2)}_{2121}}{\sqrt{5}} \end{aligned} \tag{3.11.19}$$

It is clear from these expressions that if one excludes terms describing Coulomb interaction in system (3.11.16), the main contribution to the solution of this system (3.11.16) is provided by terms of order $1/x^3$ and the integral over the impact parameter converges. The matrix elements as functions of a dimensionless internuclear parameter are given in Figure 16.

Divergence of integrals is a particular problem arising in different fields of physics where Coulomb interaction between particles plays a leading role. Logarithmic divergence of integrals is an important problem in the theory of a Coulomb plasma. The usual way to remove this difficulty is to cut out the integration process for impact parameters exceeding the Debye radius r_D[183]

$$r_D = \sqrt{\frac{kT_pT}{4\pi e^2 Z(n_p^{(0)}T_p + n^{(0)}T)}}$$

where $n_p^{(0)}$, T_p and $n^{(0)}$, T are, respectively, the density and temperature of the incident and "field" particles; Z is the ion charge and k is the Boltzmann constant.

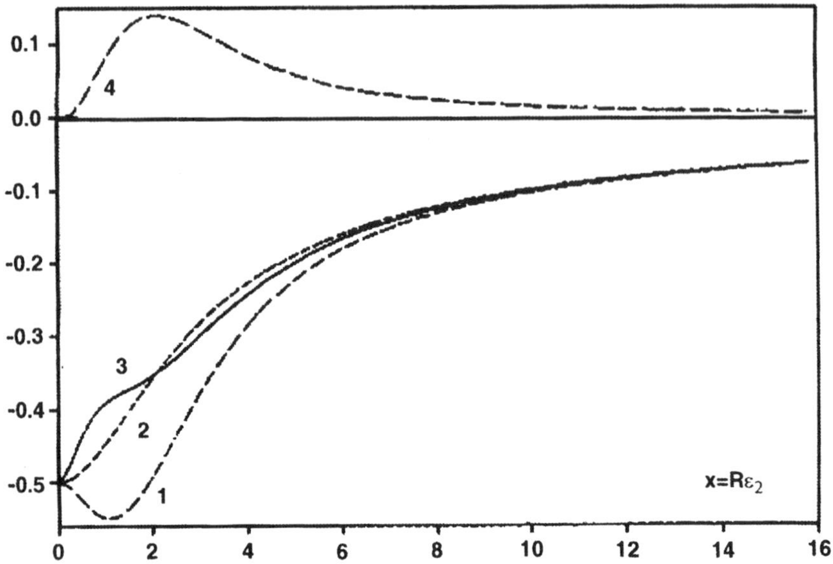

Figure 16. Dependencies of matrix elements $\langle 210|W|210\rangle$ (1), $\langle 211|W|211\rangle$ (2), $\langle 200|W|200\rangle$ (3), $\langle 200|W|210\rangle$ (4) on dimensionless internuclear distance $x = \varepsilon_2 R$. The energy is plotted in units of e^2/a_0.

Let us substitute densities and temperatures typical of the optically active region of the solar chromosphere as one example of the application of this theory: $n^{(0)} \sim n_p^{(0)} \sim 10^{12}\,\text{cm}^{-3}$, $T_p \sim T \sim 10^4$ K. This yields $r_D \sim 4.8 \times 10^{-4}$ cm, which exceeds by more than an order of magnitude the distance R_0 for which Stark splitting is equal to the fine-structure splitting.

In order to determine the scattering matrix in the field $R > R_0$, the solution of system (3.11.16) was joined at the point $R = R_0$ with the solution of system (3.11.14). It was assumed that the incoming atom is moving along the linear trajectory (Figure 15) which, for $\alpha = \alpha_0$, bends at an angle $2\alpha_0$. This angle is determined by the known Rutherford formula[169]

$$\alpha_0 = \arcsin\left[1 + \left(\frac{e^2}{mv^2 b}\right)^2\right]^{1/2}$$

As a result, the diagonal with respect to the first and second pairs of cross sections describes the collisional transfer, or destruction of ordering, of angular momenta of a certain type. These cross sections, calculated in the framework of this model, are presented in Table 4 for relative velocities of 200, 600, and 1000 m/s in units of a_0^2. The values of the cross sections do not change much within the given interval of the velocity variation. The cross section σ_0^{11}, describing the collisional destruction of the longitudinal component of the orientation vector, is much less than the cross section σ_1^{11} responsible for the destruction of the transverse component of the orientation vector. Most effective is the collisional destruction of the inclined component ρ_1^2 of the alignment tensor and slowest is the collisional destruction of the transverse component ρ_2^2. Table 4 shows, that an extension of the dipole approximation leads to a nonzero value of the cross section of the mutual conversion of orientation and alignment, although this value is much smaller than the other cross sections. The value of this cross section is strongly dependent on the relative velocity, changing sign in the interval of 600–1000 m/s.

The effective cross sections $\sigma_q^{kk_1}(ll_1, l'l'_1; v)$ allow one to calculate the effectiveness of the processes of collisional relaxation of polarization moments of hydrogen atoms in state with $n = 2$ and to determine the corresponding polarization of light under various physical conditions. The intensities of linearly polarized components of optical emission for the $2p$–$1s$ transition are given by the formulas

$$\begin{aligned} [I_x - I_y]_z &= 2C \operatorname{Re} \rho_2^2(1,1) \\ [I_z - I_x]_y &= C\sqrt{6}[\operatorname{Re} \rho_2^2(1,1) - \sqrt{(\tfrac{3}{2})}\rho_0^2(1,1)] \\ [I_{xz^+} - I_{xz^-}] &= C\sqrt{2} \operatorname{Re} \rho_1^2(1,1) \end{aligned} \tag{3.11.20}$$

where subscripts attached to square brackets indicate the direction of observation while, xz^+ and xz^- denote the directions coinciding with the diagonals of the first

TABLE 4
Effective Cross Sections $\sigma_q^{kk_1}(ll_1, l'l'_1; v)$ (in 10^{-9} cm^2) for Collisional Relaxation of Polarization Moments $\rho_q^k(ll')$ of the Excited State of Hydrogen Atoms with $n = 2$ under Collisions with a Proton Beam

k	k'	q	l	l'	l_1	l'_1	$v = 200$ m/s	$v = 600$ m/s	$v = 1000$ m/s
0	0	0	0	0	0	0	−3.57E + 06	−3.78E + 06	−3.55E + 06
0	0	0	0	0	1	1	2.06E + 06	2.16E + 06	2.05E + 06
0	0	0	1	1	1	1	−1.19E + 06	−1.26E + 06	−1.18E + 06
0	2	0	0	0	1	1	−2.88E + 06	−2.87E + 06	−2.48E + 06
0	2	0	1	1	1	1	1.66E + 06	1.67E + 06	1.43E + 06
1	1	0	1	1	1	1	−1.94E + 04	−1.78E + 05	−3.38E + 05
2	2	0	1	1	1	1	−2.39E + 06	−2.73E + 06	−2.64E + 06
1	1	1	1	1	1	1	−7.26E + 06	−7.71E + 06	−6.60E + 06
1	2	1	1	1	1	1	−8.73E + 03	−1.02E + 04	2.45E + 04
2	2	1	1	1	1	1	−7.24E + 06	−7.52E + 06	−6.30E + 06
2	2	2	1	1	1	1	−1.92E + 04	−1.75E + 05	−3.26E + 05

and second quarters of the xz plane. The differences in the intensities of optical components with left and right hand polarizations are given by

$$\begin{aligned}
[I_+ - I_-]_z &= C\rho_0^1(1,1) \\
[I_+ - I_-]_x &= C \operatorname{Re} \rho_1^1(1,1) \\
[I_+ - I_-]_z &= C \operatorname{Im} \rho_1^1(1,1)
\end{aligned} \tag{3.11.21}$$

The total intensity of emitted light is equal to

$$I = I_x + I_y + I_z = C\sqrt{18}\rho_0^0(1,1) \tag{3.11.22}$$

Both formulas (3.11.21) and (3.11.22) comprise the same proportionality factor C, which incorporates universal constants and reduced matrix elements.

We now employ an analysis of relaxation cross sections under collisions to examine one important spectropolarimetric effect resulting from the collisional transfer of populations between $2s$ and $2p$ hydrogen states in a plasma. The lifetimes of $2p$ and $2s$ states of a hydrogen atom differ by eight orders of magnitude ($\tau_{2p} = 1.6 \cdot 10^{-9}$ s, $\tau_{2s} = 0.14$ s.[164] Therefore, under certain special conditions, specifically in an interstellar gas,[34] the excited hydrogen atoms can be accumulated in the metastable $2s$ state and the L_α radiation emitted by an ensemble of hydrogen atoms could arise exclusively as a result of the collisional transfer of a population between $2s$ and $2p$ states. In this case, collisions not only populate the $2p$ state, but they also induce its alignment along the preferential direction of particle collisions. This effect is observed in the polarization of light emitted by hydrogen atoms in the $2p$ state under conditions of collisional depopulation of the metastable $2s$ state.

A complete set of processes, valid under collisional interaction of metastable hydrogen atoms with a proton beam, includes collisional population transfer, as well as the generation and destruction of polarization moments. It is described by a system of five relaxation equations of symmetry A_1 for the quantities $\rho_0^0(0,0)$, $\rho_0^0(1,1)$, $\rho_0^2(1,1)$, $\rho_0^1(1,0)$, and $\rho_0^1(0,1)$.

In view of Eq. (3.11.6), the rate constants of the processes of radiative decay and population of the metastable $2s$ state, this system of equations takes the form

$$\frac{\partial\rho_0^k(ll_1)}{\partial t} = -\sum_{k'l'l'_1} n_0 v \sigma_0^{kk'}(ll_1, l'l'_1; v)\rho_0^{k'}(ll'_1) - \frac{\Gamma_l + \Gamma_{l_1}}{2}\rho_0^k(ll') + \delta_{k0}\delta_{l0}\delta_{l'0}\delta(t) \tag{3.11.23}$$

where the set of indices (κ, l, l') take values $(0,0,0)$, $(0,1,1)$, $(2,1,1)$, $(1,1,0)$, $(1,0,1)$ and $\Gamma_l = 1/\tau_l$ denotes the reciprocal of the lifetime of the level with quantum number l ($\tau_l = \tau_{2s}$ or τ_{2p}).

System of equations (3.11.23) corresponds to the case for which an initial population $\rho_0^0(0, 0)$ of the metastable 2*s* state is not zero, while all the remaining polarization moments have zero initial values. As a result, the collisional relaxation creates both the population and longitudinal alignment of the 2*p* state. On the spectroscopic side, the linear polarization of light reflects the degree of alignment of the upper state of transition

$$\Delta I(t) = [I(t)_z - I(t)_x]_y = -3C\rho_0^2(1, 1) \tag{3.11.24}$$

The corresponding degree of linear polarization takes the form

$$P(t) = \frac{[I(t)_z - I(t)_x]_y}{[I(t)_x - I(t)_z]_y} = -\frac{3\rho_0^2(1, 1)}{2\sqrt{2}\rho_0^0(1, 1) - \rho_0^2(1, 1)} \tag{3.11.25}$$

Figures 17–20 present time dependences of the alignment signal (3.11.23) and degree of polarization (3.11.25) calculated by the solution of Eqs. (3.11.14) for different velocities v and proton-beam densities n.

At the initial instant $t = 0$, the radiative 2*p* state is not populated and therefore the alignment signal is zero. At $t > 0$, the collisional depopulation of the metastable 2*s* state begins. This process is accompanied by collisional population of the 2*p* state and induces its alignment. This situation is presented in Figures 17 and 18 by the initial parts of the curves corresponding to the linear growth of the alignment signals. Because of a low population of the 2*p* state, the

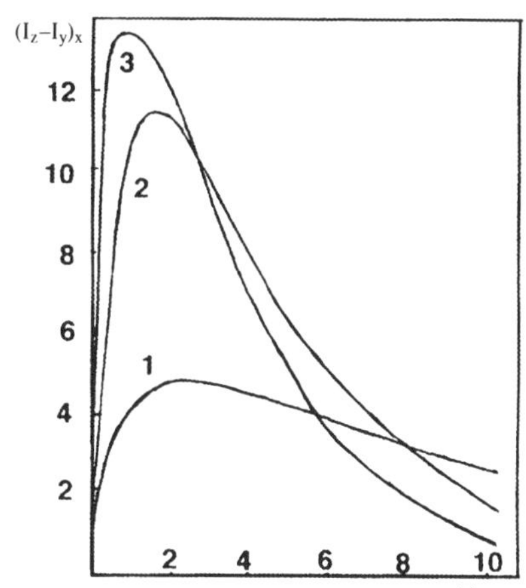

Figure 17. Alignment signal $\Delta I(t) = [I_z(t) - I_x(t)]_y$ of the 2*p* hydrogen state as a function of dimensionless time t/τ_{2p}, for a proton density in a beam of $n = 3 \times 10^{13}\ \text{cm}^{-1}$ and different proton velocities $v = 200$ m/s (1), 600 m/s (2), 1000 m/s (3).

alignment signal $\Delta I(t)$ at this stage coincides with its value obtained from the second formula of Eqs. (3.11.20) with

$$\rho_2^2(1,1) = 0 \quad \text{and} \quad \rho_0^2(1,1) = \sigma_2^{20}(1,1;0,0)n_0vt \tag{3.11.26}$$

It is a linear function of t.

As t increases, the collisional depolarization begins and destroys the alignment. As a result of competition between the generations, destruction of the alignment, and radiative depopulation of the $2p$ state, the alignment signal attains

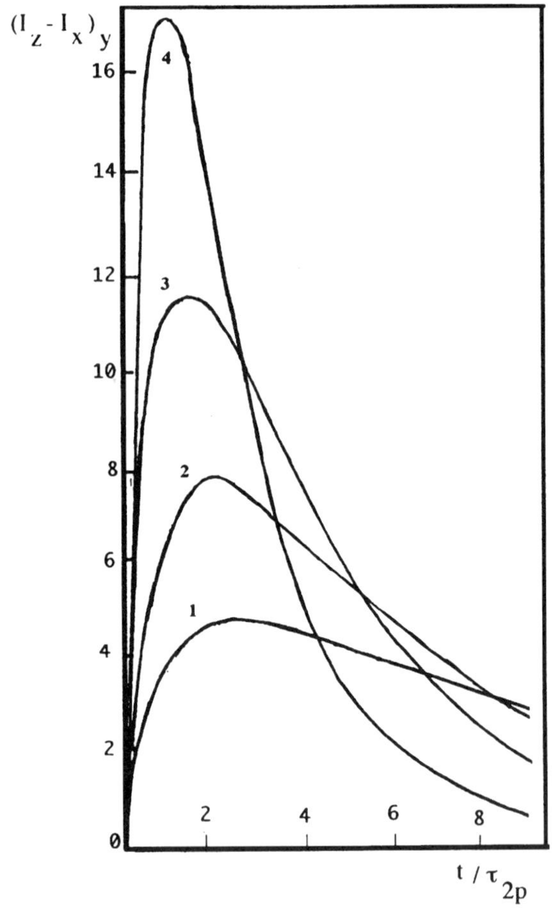

Figure 18. Alignment signal $\Delta I(t) = [I_z(t) - I_x(t)]_y$ of the $2p$ hydrogen state as a function of dimensionless time t/τ_{2p}, for a proton beam velocity of $v = 200\,\text{cm/s}$ and different proton densties $n = 3 \times 10^{13}\,\text{cm}^{-3}$ (1), $6 \times 10^{13}\,\text{cm}^{-3}$ (2), $12 \times 10^{13}\,\text{cm}^{-3}$ (3), $30 \times 10^{13}\,\text{cm}^{-3}$ (4).

maxima. At large t, radiative depopulation of the $2p$ state dominates and the alignment signals approach zero.

Generally, the higher the efficiency of anisotropic collisions in comparison with radiative decay, the narrower the alignment signals and the higher their amplitudes. Following formula (3.11.6), this efficiency increases with increasing density of the incident particles, so decreasing their velocity.

Figures 19 and 20 present time dependences of the degree of linear polarization of light, $P(t)$, emitted by hydrogen atoms at $2p$–$1s$ transitions. They are calculated using formula (3.11.25). At small t, the alignment signal (3.11.26) is proportional to $n_0 vt$. Therefore, the initial value of the degree of polarization at $t = 0$ is given by

$$P(t) = \frac{\mathrm{Re}\,\sigma_2^{20}(1,1;0,0) - \sqrt{(\tfrac{3}{2})}\sigma_0^{20}(1,1;0,0)}{\sqrt{3}\sigma_0^{00}(1,1;0,0)} \tag{3.11.27}$$

The degree of polarization depends on the proton velocity being independent of the beam density n_0. The dependence on proton velocity is associated with the velocity dependency of the effective cross sections σ_2^{20}, σ_0^{20}, and σ_0^{00}.

In accordance with the formula, all curves in Figures 19 and 20 begin from the same point. As time t increases, the degree of polarization $P(t)$ decreases due to the collisional destruction of the alignment. At large t, it tends to some limit value characteristic for the given v and N. Of particular interest for experimental observation is the value of the degree of polarization $P(t)$ at an instant t corresponding to the maximum alignment signal $\Delta I(t)$.

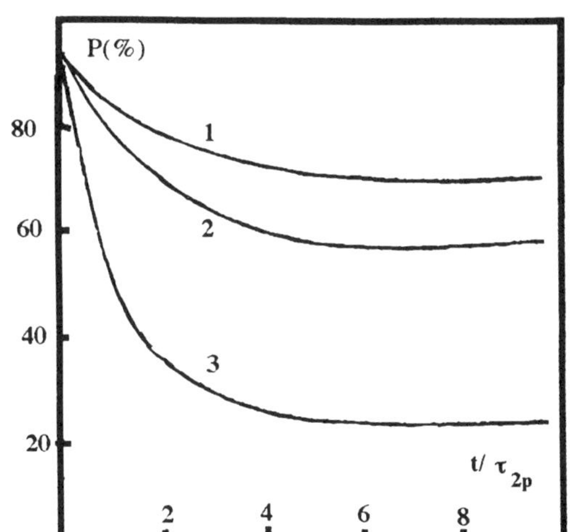

Figure 19. Degree of polarization of the L_α hydrogen line as a function of dimensionless time t/τ_{2p}, for a proton density in a beam of $n = 3 \times 10^{13}\ \mathrm{cm}^{-1}$ and different proton velocities $v = 200\ \mathrm{m/s}$ (1), 600 m/sec (2), 1000 m/s (3).

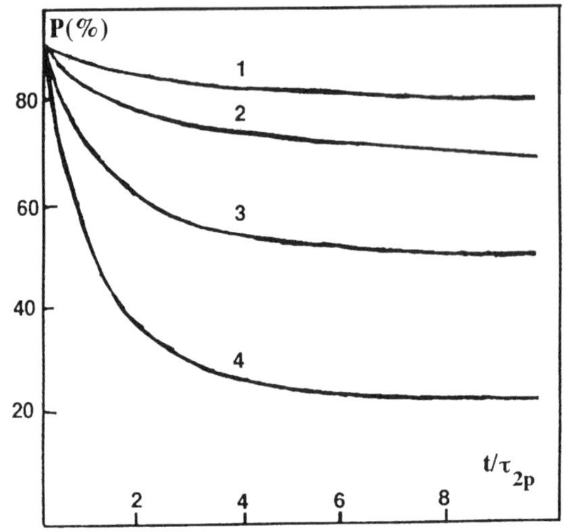

Figure 20. Degree of polarization of the L_α hydrogen line as a function of dimensionless time t/τ_{2p}, for a proton beam velocity of $v = 200\,\text{m/s}$ and different proton densities $n = 3 \times 10^{13}\,\text{cm}^{-3}$ (1), $6 \times 10^{13}\,\text{cm}^{-3}$ (2), $12 \times 10^{13}\,\text{cm}^{-3}$ (3) $30 \times 10^{13}\,\text{cm}^{-3}$ (4).

Comparison of Figures 17–20 shows that in all cases under consideration, the degree of linear polarization at the maximum alignment signal is close to 80%. This means that measurement of the degree of polarization of the Lyman α line allows one to detect collisions of metastable hydrogen atoms with fluxes of charged particles.

Analysis of collisional relaxation shows that the processes of mixing of states with different orbital quantum numbers l, possessing equal energies in the hydrogen-like atoms, complicates the picture of anisotropic collisional relaxation of polarization moments. For the group of degenerate hydrogen states with principal quantum number $n = 2$, studied above, most characteristic are new processes of anisotropic relaxation of the population and alignment of the radiative $2p$ state with account for depopulation of the metastable $2s$ state. Another result worth of mention, for the group of hydrogen states with $n = 2$, is the disappearance of the anisotropic collisional interconversion of alignment and orientation as a result of the general symmetry rules.

3.12. *Spectropolarimetric Effects for Angular Correlation Experiments*

There exist a large number of experiments in atomic physics in which selective recording of radiation is carried out from a part of the ensemble of atomic particles excited by projectiles scattered at some given angle. These angular correlation experiments provide much information about atomic parameters and can be used for the direct implementation of impact spectropolarimetric sensing as well as for experimental verification of computations of collisional alignment cross sections for studied atomic transitions.

The scheme of such an experiment is shown in Figure 21. A structure-free particle with initial velocity vector oriented along the OZ axis of the collision frame of reference encounters an atomic particle located at the origin of the reference system. It is then scattered at angle α, specified by a detector located in the plane of the collision. The receiver of an optical radiation detects simultaneously the emission of the excited atomic particles within the solid angle $d\Omega = \sin\theta\, d\theta\, d\varphi$.

Despite experimental difficulties, there exist a number of angular correlation experiments between the optical emission and the scattered particle.[86,104–109,158–160] It will be shown below that these experiments enable one to provide comprehensive information on the nature of impact ordering of the angular momenta of an excited atomic ensemble. A general expression for the polarization moments created in the ensemble under the impact excitation by structure-free particles scattered after that at a given angle was obtained in Section 3.8.

In order to illustrate the geometry of angular correlation experiments, we select mutually orthogonal orths $\mathbf{e}_1$ and $\mathbf{e}_2$ lying in the plane orthogonal to the OZ' axis, as shown in Figure 22. The orth $\mathbf{e}_1$ lies in the ZOZ' plane, making an angle $\pi/2 - \theta$ with the OZ axis and $\pi + \varphi$ with the OX axis, while the orth $\mathbf{e}_2$ is orthogonal to the OZ' axis and makes an angle $\pi/2 + \varphi$ with the OX axis. Besides

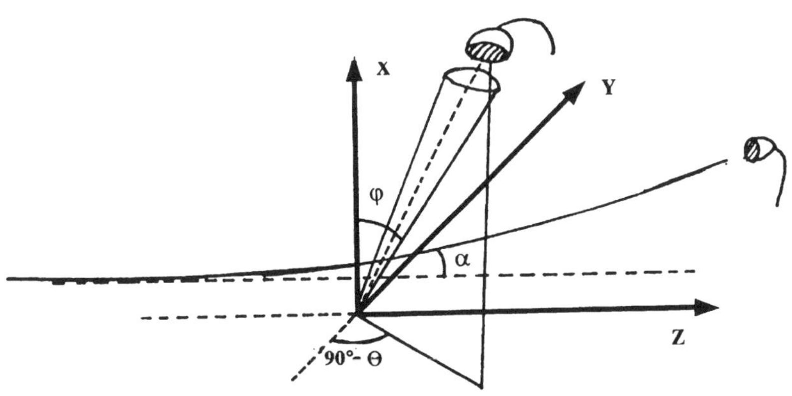

Figure 21. Principal scheme of an angular correlation experiment.

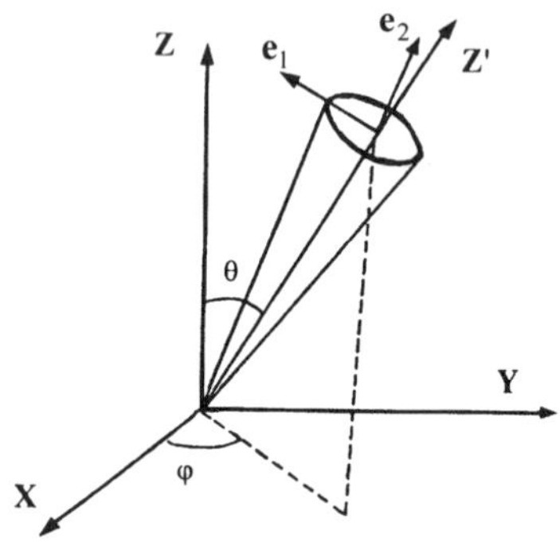

Figure 22. Orientation of orths in a laboratory frame of reference for the description of angular correlation experiments.

orths $\mathbf{e}_{1,2}$ let us introduce a pair of mutually orthogonal orths $\mathbf{e}'_{1,2}$, oriented at an angle $\pi/4$ with respect to the pair $\mathbf{e}_{1,2}$

$$\mathbf{e}'_1 = \frac{\mathbf{e}_1 + \mathbf{e}_2}{\sqrt{2}} \qquad \text{and} \qquad \mathbf{e}'_2 = \frac{-\mathbf{e}_1 + \mathbf{e}_2}{\sqrt{2}} \tag{3.12.1}$$

From a geometrical consideration of Figure 22 and formulas (3.12.1) one sees that the tensor $\Phi_{kq}(\mathbf{e}_\lambda)$, determined by formula (3.6.4), specifies the geometrical factor in the expression for the polarized light intensity. For each unit vector we can write

$$\mathbf{e}_1 \quad \Phi_{20} = \frac{1 - 3\sin^2\theta}{\sqrt{6}}, \qquad \Phi_{2-1} = -\frac{\sin 2\theta e^{i\varphi}}{2}, \qquad \Phi_{2-2} = -\frac{\cos^2\theta e^{2i\varphi}}{2}$$

$$\mathbf{e}_2 \quad \Phi_{20} = \frac{1}{\sqrt{6}}, \qquad \Phi_{2-1} = 0, \qquad \Phi_{2-2} = \frac{e^{2i\varphi}}{2} \tag{3.12.2}$$

$$\mathbf{e}'_1 \quad \Phi_{20} = \frac{2(3\cos^2\theta - 1)}{\sqrt{6}}, \qquad \Phi_{2-1} = \frac{(i\sin\theta - \frac{1}{2}\sin 2\theta)e^{i\varphi}}{2},$$

$$\Phi_{2-2} = \frac{(\sin^2\theta + 2i\cos\theta)e^{2i\varphi}}{4}$$

$$\mathbf{e}'_2 \quad \Phi_{20} = \frac{3\cos^2\theta - 1}{2\sqrt{6}}, \qquad \Phi_{2-1} = \frac{-(i\sin\theta + \frac{1}{2}\sin 2\theta)e^{i\varphi}}{2},$$

$$\Phi_{2-2} = \frac{(\sin^2\theta - 2i\cos\theta)e^{2i\varphi}}{4} \tag{3.12.3}$$

These expressions, as well as formulae (3.6.9) representing the intensity of a circularly polarized light beam propagating in a direction determined by polar angles θ and φ, enable one to write expressions for the total intensity of radiation within a solid angle $d\Omega$ and for the three Stokes parameters:

$$
\begin{aligned}
I = c_{jj_0}\Bigg\{&\frac{2}{3}\frac{(-1)^{j+j_0+1}}{\sqrt{2j+1}}\rho_0^0(j) + \begin{Bmatrix} 1 & 1 & 2 \\ j & j & j_0 \end{Bmatrix} \\
&\cdot\left[\frac{3\cos^2\theta - 1}{\sqrt{6}}\rho_0^2(j) - \operatorname{Re}\left[e^{i\varphi}\sin 2\theta\rho_1^2(j)\right]\right. \\
&\left.+ \operatorname{Re}\left[\sin^2\theta e^{2i\varphi}\rho_1^2(j)\right]\right]\Bigg\}
\end{aligned} \tag{3.12.3a}
$$

$$
\begin{aligned}
IQ = I_{\mathbf{e}_1} - I_{\mathbf{e}_2} = -c_{jj_0}&\begin{Bmatrix} 1 & 1 & 2 \\ j & j & j_0 \end{Bmatrix} \\
&\cdot\{3\sin^2\theta\rho_0^2(j) + \operatorname{Re}\left[\sin 2\theta e^{i\varphi}\rho_1^2(j)\right] \\
&+ \operatorname{Re}\left[(\cos^2\theta + 1)e^{2i\varphi}\rho_1^2(j)\right]\}
\end{aligned} \tag{3.12.3b}
$$

$$
\begin{aligned}
IU = I_{\mathbf{e}'_1} - I_{\mathbf{e}'_2} = -2c_{jj_0}&\begin{Bmatrix} 1 & 1 & 2 \\ j & j & j_0 \end{Bmatrix} \\
&\cdot\{\operatorname{Re}\left[i\sin 2\theta e^{i\varphi}\rho_1^2(j)\right] + \operatorname{Re}\left[i\cos\theta e^{2i\varphi}\rho_2^2(j)\right]\}
\end{aligned} \tag{3.12.3c}
$$

$$
IV = (I_+ - I_-)_{\theta\varphi} = 2c_{jj_0}\begin{Bmatrix} 1 & 1 & 1 \\ j & j & j_0 \end{Bmatrix}\left\{\frac{\cos\theta}{\sqrt{2}}\rho_0^1(j) + \operatorname{Re}\left[\sin\theta e^{i\varphi}\rho_1^1(j)\right]\right\} \tag{3.12.3d}
$$

where c_{jj_0} is a constant incorporating the fundamental constants, frequency, and reduced matrix elements of the transition.

The polarization characteristics of an excited atomic ensemble, or the set of spectropolarimetric parameters I, IQ, IU, IV for angular correlation experiments are determined completely by the set of six polarization moments $\rho_q^k(j)(|q| \leq k = 0, 1, 2)$. For this particular way of recording data, the problem is characterized by a specific symmetry stipulated by the invariance of the collisional process with respect to reflections in the collision plane (*ZX* plane, Figure 22). The requirement for polarization moment invariance with respect to reflections in the collision plane imposes specific restrictions on the ranks of the polarization moments in the right hand side of Eqs. (3.12.3).

Indeed, if a reflection in the collision plane transforms the elements of the density matrix in the representation of the magnetic quantum numbers as follows

$$
\rho_{mm'} \Rightarrow (-1)^{m+m'}\rho_{-m-m'}
$$

then the transformation of the polarization moment resulting from this operation can be written in the form

$$\hat{\sigma}_{xz}\rho_q^\kappa(j) = \sum_{mm'}(-1)^{j-m}\begin{bmatrix} j & j & \kappa \\ m & -m' & q \end{bmatrix}\rho_{-m-m'}$$

$$= \sum_{mm'}(-1)^{j-m'+q+\kappa}\begin{bmatrix} j & j & \kappa \\ m & -m' & -q \end{bmatrix}\rho_{mm'}$$

$$= (-1)^{\kappa+q}\rho_{-q}^\kappa(j) \tag{3.12.4}$$

Therefore, an additional condition imposed on the polarization moments in the right hand side of Eq. (3.12.3) takes the form

$$\rho_q^\kappa(j) = (-1)^{\kappa+q}\rho_{-q}^\kappa(j) \tag{3.12.5}$$

For polarization moments with $k \leq 2$ we have:

$$\kappa = \mathbf{0} \quad \rho_0^0(j) = \rho_0^0(j)$$

$$\kappa = \mathbf{1} \quad \rho_0^1(j) = -\rho_0^1(j), \qquad \rho_1^1(j) = \rho_{-1}^1(j)$$

$$\kappa = \mathbf{2} \quad \rho_0^2(j) = \rho_0^2(j), \qquad \rho_1^2(j) = -\rho_{-1}^2(j), \qquad \rho_2^2(j) = \rho_{-2}^2(j) \tag{3.12.6}$$

If the condition

$$\rho_q^{k^*}(j) = (-1)^q\rho_{-q}^k(j)$$

is added to relations (3.12.6), expressing the fact that the density matrix is Hermitian and satisfies (3.4.12), we have

$$\rho_0^1(j) = 0$$

the polarization moments ρ_0^0, ρ_0^2, ρ_1^2, ρ_2^2 are real, and the polarization moment ρ_1^1 is purely imaginary.

The last result is valid for angular correlation experiments, so when the collisional problem is specified by a plane of symmetry, conditions (3.12.3) can be written as

$$I = c_{jj_0}\left\{\frac{2}{\sqrt{3}}\frac{(-1)^{j+j_0+1}}{\sqrt{2j+1}}\rho_0^0(j) + \begin{Bmatrix} 1 & 1 & 2 \\ j & j & j_0 \end{Bmatrix}\left[\frac{3\cos^2\theta - 1}{\sqrt{6}}\rho_0^2(j)\right.\right.$$
$$\left.\left. - \rho_1^2(j)\sin 2\theta\cos\varphi + \rho_2^2(j)\sin^2\theta\cos 2\varphi\right]\right\} \tag{3.12.7a}$$

$$IQ = -c_{jj_0}\begin{Bmatrix} 1 & 1 & 2 \\ j & j & j_0 \end{Bmatrix}[3\sin^2\theta\rho_0^2(j) + \rho_1^2 \sin 2\theta\cos\varphi \\ + \rho_2^2(j)(\cos^2\theta + 1)\cos 2\varphi] \tag{3.12.7b}$$

$$IU = 2c_{jj_0}\begin{Bmatrix} 1 & 1 & 2 \\ j & j & j_0 \end{Bmatrix}[\sin 2\theta \sin\varphi\rho_1^2(j) + \rho_2^2(j)\cos\theta\sin 2\varphi] \tag{3.12.7c}$$

$$IV = -2c_{jj_0}\begin{Bmatrix} 1 & 1 & 1 \\ j & j & j_0 \end{Bmatrix}\sin\theta\sin\varphi\,\mathrm{Im}[\rho_1^1(j)] \tag{3.12.7d}$$

Hence the set of five polarization moments ρ_0^0, ρ_1^1, ρ_0^2, ρ_1^2, ρ_2^2 totally specify the polarization properties of an ensemble of excited atomic particles resulting from collisions, characterized by a plane of symmetry.

Apart from the problem of determining the spectroscopic polarization characteristics of an atomic ensemble, one can pose the inverse problem of determining the polarization moments, or quantities specifying the types of ordering of momenta of an excited atomic ensemble, making use of experimentally determined Stokes parameters detected at some angle to the collisional excitation direction. It can be done in different ways. For example, to determine the second-order polarization moments ($k = 2$), it is enough to measure parameter IQ for three angles; the moment with $k = 1$ can be determined by measuring parameter IV at angles θ and $\varphi \neq 0$. As proposed elsewhere,[(86)] it is also possible to measure a set of three Stokes parameters along the OY axis and one Stokes parameter in the direction of the OX axis. The first group of measurements along OY will give

$$IQ = (I_z - I_x)_y = c_{jj_0}\begin{Bmatrix} 1 & 1 & 2 \\ j & j & j_0 \end{Bmatrix}\left(-\frac{3}{\sqrt{6}}\rho_0^2 + \rho_2^2\right) \\ = K_{jj_0}^{(2)}\sqrt{30}(\langle j_\pm^2\rangle - \langle 3j_z^2 - j^2\rangle) \tag{3.12.8a}$$

$$IU = (I_{zx+} - I_{zx-})_y = 2c_{jj_0}\begin{Bmatrix} 1 & 1 & 2 \\ j & j & j_0 \end{Bmatrix}\rho_1^2 = 4K_{jj_0}^{(2)}\sqrt{15}\langle j_z j_- + j_- j_z\rangle \tag{3.12.8b}$$

$$IV = (I_+ - I_-)_y = 2c_{jj_0}\begin{Bmatrix} 1 & 1 & 1 \\ j & j & j_0 \end{Bmatrix}\mathrm{Re}\,(i\rho_1^1) = -2\sqrt{6}K_{jj_0}^{(1)}\langle j_y\rangle \tag{3.12.8c}$$

The result of the single measurement along OX axis is given by

$$IQ' = (I_z - I_y)_x = c_{jj_0}\begin{Bmatrix} 1 & 1 & 2 \\ j & j & j_0 \end{Bmatrix}\left(-\frac{3}{\sqrt{6}}\rho_0^2 - \rho_2^2\right) \\ = -K_{jj_0}^{(2)}\sqrt{30}(\langle j_\pm^2\rangle + \langle 3j_z^2 - j^2\rangle) \tag{3.12.8d}$$

where the following notation was used:

$$K^{(1)}_{jj_0} = c_{jj_0}\begin{Bmatrix} 1 & 1 & 1 \\ j & j & j_0 \end{Bmatrix}[2j(2j+1)(2j+2)]^{-1/2}$$
$$K^{(2)}_{jj_0} = c_{jj_0}\begin{Bmatrix} 1 & 1 & 2 \\ j & j & j_0 \end{Bmatrix}[(2j-1)2j(2j+1)(2j+2)(2j+3)]^{-1/2} \qquad (3.12.9)$$

In the case of certain atomic transitions the number of independent parameters may be reduced. For example, for the impact excitation $j_0 = 0 \rightarrow j = 1$ this number is equal to three. Indeed, for the state with moment $j = 1$ there exist three eigenfunctions of the square of the momentum and its projection: $\psi_{1-1}(r)$, $\psi_{10}(r)$, $\psi_{11}(r)$.

Three combinations, possessing a particular parity with respect to reflections in the collision plane, can be composed from these functions (Figure 23)

$$\varphi_z = \psi_{10}, \qquad \varphi_x = \frac{\psi_{11} - \psi_{1-1}}{\sqrt{2}}, \qquad \varphi_y = \frac{\psi_{11} + \psi_{1-1}}{\sqrt{2}} \qquad (3.12.10)$$

The ground state wave function $\psi_{00}(r)$ will be mixed only with functions φ_z and φ_x, while function φ_y will be invariant in the course of collisions. Therefore, if c_{jm} is the coefficient of expansion of the wave function in terms of the eigenfunctions of the square of the momentum and its projection, due to the conservation of φ_y, it is true that in the process of collisions

$$c_{11} = -c_{1-1} \qquad (3.12.11)$$

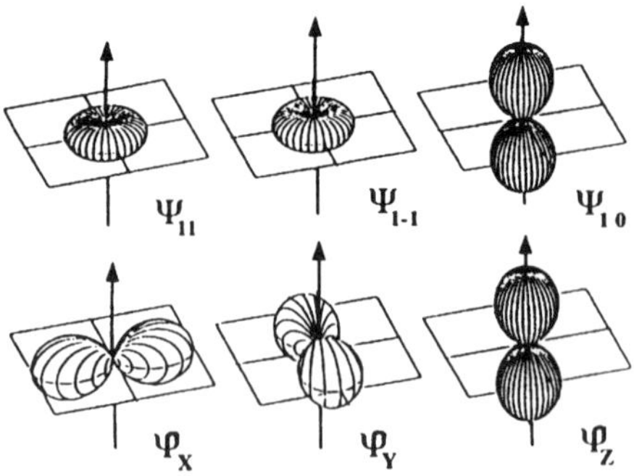

Figure 23. Geometrical representation of basic wave functions $\psi_{1-1}(r)$, $\psi_{10}(r)$, $\psi_{11}(r)$ and their linear combinations φ_x, φ_y, φ_z with special parity features under reflections in the collision plane.

If the angle brackets denote ensemble averaging, the following conditions for the elements of the density matrix

$$\rho = \begin{pmatrix} \langle c_{-1}c_{-1}^* \rangle & \langle c_{-1}c_0^* \rangle & \langle c_{-1}c_1^* \rangle \\ \langle c_0c_{-1}^* \rangle & \langle c_0c_0^* \rangle & \langle c_0c_1^* \rangle \\ \langle c_1c_{-1}^* \rangle & \langle c_1c_0^* \rangle & \langle c_1c_1^* \rangle \end{pmatrix} \tag{3.12.12}$$

are valid

$$\langle c_1c_1^* \rangle = \langle c_{-1}c_{-1}^* \rangle = -\langle c_1c_{-1}^* \rangle = -\langle c_{-1}c_1^* \rangle \tag{3.12.13}$$

and

$$\langle c_{-1}c_0^* \rangle = -\langle c_1c_0^* \rangle = \langle c_0c_{-1}^* \rangle^* = -\langle c_0c_1^* \rangle^* \tag{3.12.14}$$

Hence the density matrix for state $j = 1$ consists of three independent elements

$$\rho_{11} = \langle c_1c_1^* \rangle, \qquad \rho_{00} = \langle c_0c_0^* \rangle, \qquad \rho_{10} = \langle c_1c_0^* \rangle \tag{3.12.15}$$

As quantities ρ_{11} and ρ_{00} are real and ρ_{10} is complex, conditions (3.12.15) comprise only four independent parameters.

By expressing the polarization moments $\rho_0^0(1), \rho_1^1(1), \rho_0^2(1), \rho_1^2(1), \rho_2^2(1)$ in terms of density matrix elements, we have

$$\rho_0^0(1) = \frac{\rho_{00} + 2\rho_{11}}{3}, \qquad \rho_0^2(1) = \frac{\sqrt{2}(2\rho_{11} - \rho_{00})}{\sqrt{3}}$$

$$\rho_1^2(1) = -\sqrt{2}\,\mathrm{Re}\,\rho_{10}, \qquad \rho_2^2(1) = \rho_{-1-1}, \qquad \rho_1^1(1) = \sqrt{2}i\,\mathrm{Im}\,\rho_{10} \tag{3.12.16}$$

It is evident that only four polarization moments will be independent, such as $\rho_0^0(1), \rho_0^2(1), \rho_1^2(1), \rho_1^1(1)$.

According to Eq. (2.11.7), four independent measurements should be undertaken to determine these parameters, for example, the intensity in the *OY* direction and the three Stokes parameters *IQ*, *IU*, *IV*, as suggested elsewhere.[86] In case of a pure state with the density matrix normalized to unity, namely

$$\rho_0^0(1) = \frac{Sp\rho}{\sqrt{3}} = \frac{1}{\sqrt{3}}$$

the number of independent parameters is reduced to three and only three measurements of Stokes parameters IQ, IU, IV in the direction of the OY axis are needed for the spectroscopic determination.

Angular correlation experiments can provide information on the impact alignment cross section. Therefore, it is worth discussing the results of these experiments and the parametrization of these data for application to practical problems of spectropolarimetric sensing. Indeed, for a wide group of states the cross section of the electron impact alignment may be obtained by employing the bank of experimental and theoretical data on angular correlation parameters.[86,104–109,159,160] In such experiments the coincidence of inelastically scattered electrons and polarization or angular characteristics of emitted photons is studied. For experiments of this type, the axial symmetry of the system with respect to the beam direction is violated. The polarization of an atomic ensemble is not described only in terms of the longitudinal components of polarization moments of the atomic density matrix. Polarization moments of odd ranks with q not equal to zero will also be present.

Consider the commonly used scheme for experimental measurement of electron photon angular correlation spectropolarimetric effects observed at right angles to the scattering plane. Besides Stokes parameters P, Q, and C, the following set of parameters is exploited

$$P_1 = (P^2 + Q^2)^{1/2}$$

or the maximal degree of polarization attained at an angle of the analyzer given by

$$\gamma_0 = (\tfrac{1}{2}) \arctan(Q/P)$$

with the mean z-projection of the total angular momentum of the electron shell $\langle J_\perp \rangle$ proportional to Stokes parameter C. Fano parameters are also used to describe experiments on electron photon angular correlations[102,141]

$$O_{1-} = \frac{\langle J_z \rangle}{J(J+1)}, \qquad A_{1+} = \frac{\langle J_y J_x + J_x J_y \rangle}{J(J+1)}$$
$$A_0 = \frac{\langle 3J_x^2 - J^2 \rangle}{J(J+1)}, \qquad A_{2+} = \frac{\langle J_y^2 - J_z^2 \rangle}{J(J+1)} \tag{3.12.17}$$

Polarization of states with total angular moment $J = 1$ is described completely by the following parameters[102,109]

$$\lambda = \frac{|c_0|^2}{\sum_m |c_m|^2} = \frac{|c_0|^2}{|c_0|^2 + 2|c_1|^2} \qquad \text{and} \qquad \chi = \arg\left(\frac{c_1}{a_0}\right)$$

Table 5 gives formulas connecting different parameters used to describe the spectropolarimetric effects in electron–photon angular correlation experiments. Using these relationships and the definition of $\sigma^{(2)}$ in Eq. (3.10.8), one can obtain expressions for the electron impact alignment cross section in terms of the angular correlations parameters.

For $J = 1$ we have

$$\begin{aligned}
\sigma^{(2)}(v) &= \frac{2\pi}{\sqrt{6}}\int_0^\pi d\theta \sin\theta\, \sigma(\theta, v)[1 - 3\lambda(\theta, v)] \\
\sigma^{(2)}(v) &= \frac{\pi}{\sqrt{6}}\int_0^\pi d\theta \sin\theta\, \sigma(\theta, v)[1 + 3P(\theta, v)] \\
\sigma^{(2)}(v) &= \frac{\pi}{\sqrt{6}}\int_0^\pi d\theta \sin\theta\, \sigma(\theta, v)[1 + 3P_l(\theta, v)\cos 2\,\gamma_0(\theta, v)]
\end{aligned} \tag{3.12.18}$$

TABLE 5
Relationship between Spectropolarimetric Parameters used in Electron–Photon Angular Correlation Experiments for Different Reference Frames[a]

	PQC	$P_e\gamma\langle I_\perp\rangle$	$\lambda\chi$	$A_0A_{1+}A_{2+}O_{1-}$
P		$P_e\cos 2\gamma_0$	$2\lambda - 1$	$-(4A_0 + 1)/3$
Q		$P_e\sin 2\gamma_0$	$-2\sqrt{\lambda(1-\lambda)}\cos\chi$	$-2A_{1+}$
C		$-\langle J_\perp\rangle$	$2\sqrt{\lambda(1-\lambda)}\sin\chi$	$-2O_{1-}$
P_e^2	$P^2 + Q^2$		$(2\lambda - 1)^2 + 4\lambda(1-\lambda)\cos^2\chi$	$\frac{1}{9}(4A_0 + 1)^2 + 4A_{1+}$
$\tan 2\gamma_0$	Q/P		$-2\sqrt{\lambda(1-\lambda)}/(2\lambda - 1)$	$6A_{1+}/(4A_0 + 1)$
$\langle J_\perp\rangle$	$-C$		$-2\sqrt{\lambda(1-\lambda)}\sin\chi$	$2O_{1-}$
λ	$(P + 1)/2$	$\frac{1}{2}(P_e\cos 2\gamma_0 + 1)$		$-(2A_0 - 1)/3$
$\tan\chi$	C/Q	$-\langle J_\perp\rangle/P_e\sin 2\gamma_0$		O_{1-}/A_{1+}
A_0	$-(1 + 3P)/4$	$-\frac{1}{4}(1 + P_e\cos 2\gamma_0)$	$(1 - 3\lambda)/2$	
A_{1+}	$-Q/2$	$-\frac{1}{2}\sin 2\gamma_0$	$\sqrt{\lambda(1-\lambda)}\cos\chi$	
A_{2+}	$(P - 1)/4$	$(P_e\cos\gamma_0 - 1)/4$	$(\lambda - 1)/2$	
O_{1-}	$-C/2$	$\langle J_\perp\rangle/J(J+1)$	$-\sqrt{\lambda(1-\lambda)}\sin\chi$	

[a] (P, Q, C), $(P_e, \gamma_0, \langle J_\perp\rangle)$—detector frame of reference with OZ axis orthogonal to the scattering plane; (λ, χ), (A_q, O_q)—collision frame of reference for $J_0 = 0 - J = 1$ spectral transition.

For $J \neq 1$ one may write

$$\sigma^{(2)}(v) = 2\pi\sqrt{5}\left[\frac{J(J+1)}{(2J+3)(2J+1)(2J-1)}\right]^{1/2} \int_0^{\pi} d\theta \sin\theta\, \sigma(\theta, v) A_0(\theta, v) \tag{3.12.19}$$

where $\sigma(\theta, v)$ is the differential scattering cross section.

3.13. Concluding Remarks

This chapter demonstrates that the most effective detailed description of the polarization of an atomic ensemble is implemented using the density matrix in the representation of polarization moments. It is shown that the symmetry of an atomic system is directly reflected by spectropolarimetric observables. Very important relations between the set of Stokes parameters and the polarization moments are of the utmost significance in the practical sensing problems, and they are obtained and analyzed.

Different basic physical processes in atomic ensembles (relaxation of excited states, resonance interaction with the electromagnetic field, collisions) take place within an ensemble of free particles under typical conditions of natural (solar atmosphere, ionosphere of the Earth) and artificial (gas discharges) ionized media. The processes are described and form a framework within which remote sensing techniques can be implemented. Kinetics of the polarization moments of atomic ensembles and subsequent variation in the spectroscopic observables during different processes have been studied in this chapter. They provide the principal theoretical basis on which to construct practical methods to realize spectropolarimetric sensing of ionized entities.

4

Fast Particle Collision with a Heavy Atomic Target

4.1. Introduction

This chapter is devoted to general theoretical considerations and the basic equations describing the variation in polarization moments of atomic ensembles due to collisions with foreign atoms. The problem of the collisional variation of polarization moments is treated in terms of relaxation matrix (3.8.10) for low relative velocities. For high relative velocities, this problem is reduced to the determination of the matrix of cross sections of the collisional creation of momenta ordering.

In case of collisions of heavy atomic particles, these problems are traditionally solved within the framework of the impact parameter method. The relative motion of particles in this approach is described classically and the interaction of electron shells is treated using quantum mechanical methods. The wave function of the system of colliding particles is constructed as a resolution in terms of the set of basic functions. Three principal conditions must be satisfied in choosing the set of these functions, which must be most comprehensive. It should comprise functions corresponding to the initial and final states of the system and, finally, the calculation of matrix elements employing this basis should not be too complicated. There exists different ways to set up this basis.[101] For instance, atomic orbitals with effective parameters,[97] shifted atomic orbitals,[98] a multielectron wave function,[103] or Hartry–Fock functions[110] could be used as basic functions. The impact parameter method was effectively utilized for a problem of astrophysical significance related to the description of spectro-polarimetric effects in H_α, H_β, and L_α hydrogen emissions during direct excitation of hydrogen atoms by protons.[94–96]

Apart from the impact parameter method, this chapter describes other, less precise methods of calculations such as the Born method or its modifications [111–115]. The Born method may be regarded as an approximate solution of the equation governing the impact parameter method to first order subject to resolution in terms of inverse velocity powers.

Another possibility for the theoretical description of impact polarization effects is the Glauber method[44–46, 116–118] related to the eikonal technique.[44] The advantages of the latter methods are the simplicity of calculations, resulting from an approximate analytical solution to the problem of the relative motion of colliding particles. The Born and Glauber methods are applicable, as a rule, when one of the colliding particles is an electron or proton. The distorted-wave Born approximation was applied to compute the differential cross sections and angular dependency of the Stokes parameters under collisional excitation of hydrogen and noble gases by electrons.[111–113] The Glauber approximation has been described elsewhere.[44, 45, 206] The polarization characteristics of an atomic ensemble excited by electron impact have been calculated by these techniques.[75, 76, 116, 118])

These techniques are important in computations of the polarization characteristics of an ensemble created by particle impact, and theoretical methods for the calculation of the scattering amplitude for collisions of two heavy particles will be examined in this chapter. In the next section, Born resolution for the scattering amplitude will be reviewed and some experimental situations in which one can restrict oneself to the first order of this resolution will be pointed out. The third section will be devoted to the eikonal equation and the Glauber approximation will be derived from this equation. The equations pertaining to the impact parameter method itself, as well as symmetry conditions for the collisional relaxation matrix and the matrix of the cross sections of impact ordering moments will be considered in the penultimate section.

4.2. Born Expansion of the Scattering Amplitude

Let an atom A, before collision with a charged particle, be at a state with wave function Ψ_α and momentum $\mathbf{p}_\alpha = \mathbf{k}_\alpha \hbar$. After collision, the atom A will be in a state with wave function Ψ_β and the projectile will have impulse $\mathbf{p}_\beta = \mathbf{k}_\beta \hbar$ (Figure 24).

Figure 24. Collision of a light charged particle with a heavy scattering center.

The main problem of collision theory is to determine the scattering amplitude

$$f_{\alpha\beta}(\theta) = \frac{m}{2\pi\hbar^2} T_{\alpha\beta} \tag{4.2.1}$$

where $T_{\alpha\beta}$ is a transition matrix that depends on the interaction potential $V(rr')$:

$$T_{\alpha\beta} = \langle \Phi_\alpha | V(rr') | \Psi_\beta^{(+)} \rangle \tag{4.2.2}$$

Here

$$\Phi_\alpha = \Psi_\alpha e^{i(\mathbf{k}_\alpha \mathbf{r}')}$$

and $\Psi_\beta^{(+)}$ are solutions of the Schrödinger equation

$$[H + V(r, r')]\Psi_\beta^{(+)} = E\Psi_\beta^{(+)} \tag{4.2.3}$$

which has an asymptotic solution as a diverging wave.

The solution of this differential equation may be reduced to the solution of an integral equation of type

$$\Psi_\beta^{(+)} = \Phi_\beta + (H - E + i\varepsilon)^{-1} V \Psi_\beta^{(+)} \tag{4.2.4}$$

where

$$\Phi_\beta = \Psi_\beta e^{i(\mathbf{k}_\beta, \mathbf{r})}$$

The iteration method, and a zero-approximation choice of function Φ_β yields

$$\Psi_\beta^{(+)(0)} = \Phi_\beta \tag{4.2.5a}$$

and

$$\Psi_\beta^{(+)(1)} = \Phi_\beta + (H - E + i\varepsilon)^{-1} V \Phi_\beta \tag{4.2.5b}$$

if the following conditions are fulfilled:

$$Ka \gg 1 \qquad \text{and} \qquad \frac{m\bar{V}a}{\hbar^2 K} \ll 1 \tag{4.2.6}$$

where a quantity of the order of an atomic size and $\bar{V}$ is the mean interaction energy.

The scattering amplitude corresponding to the first Born approximation then takes the form

$$f^{(1)}_{\alpha\beta}(\theta) = \frac{m}{2\pi\hbar^2} \langle \Phi_\alpha | V | \Phi_\beta \rangle \tag{4.2.7}$$

For Coulomb interaction, integration over the projectile coordinates can be performed in the last expression. Then relation (4.2.7) may be rewritten in the following form for a one-electron atom:

$$f^{(1)}_{\alpha\beta}(\theta) = \frac{2e^2 m}{\hbar q^2} [F_{\alpha\beta}(q) - Z\delta_\alpha \beta] \tag{4.2.8}$$

where $q = |\mathbf{k}_\alpha - \mathbf{k}_\beta|$ is the transmitted impulse and

$$F_{\alpha\beta}(q) = \int d\mathbf{r} \psi_\alpha(r) e^{i(\overline{qr})} \psi_\beta(r) \tag{4.2.9}$$

A detailed description of accurate methods for computing the scattering amplitude appears elsewhere.[37] Here, we only point out the possibility of obtaining a more accurate solution accounting for the next term of the iteration process for an estimated representation of Ψ^+_β. By substituting Eq. (4.2.5b) into the right hand side of Eq. (4.2.2), one obtains the scattering amplitude of the second Born approximation in the form

$$\begin{aligned} f^2_{\alpha\beta}(\theta) &= \langle \phi_\alpha \| V (H - E + i\varepsilon)^{-1} V \| \phi_\beta \rangle \\ &= \frac{m}{\hbar^2 (2\pi)^4} \sum_\gamma \int d\mathbf{x} \frac{\langle k_\alpha \alpha \| V \| x\gamma \rangle \langle x\gamma \| V \| k_\beta \beta \rangle}{E_{k\alpha} - E_{x\gamma} + i\varepsilon} \end{aligned} \tag{4.2.10}$$

where

$$E_{k\alpha} = E_\alpha + \frac{\hbar^2 k_\alpha^2}{2m}$$

while κ and γ are the intermediate wave and quantum numbers, respectively.

A detailed examination of the second Born approximation[38] showed that for small scattering angles

$$f^2_{\alpha\beta} \sim \frac{f^1_{\alpha\beta}}{k}$$

but for large scattering angles the quantity $f^2_{\alpha\beta}$ is not small (see Tables 6 and 7).

The first Born approximation for the amplitude of exchange scattering may be written in the form[36, 38, 40, 65]

$$g^{(1)}_{\alpha\beta} = \frac{m}{2\pi\hbar^2}\int e^{-i(\mathbf{k}_\alpha,\mathbf{r}_1)}\psi_\alpha(r_1)\frac{1}{|\bar{r}_1 - \bar{r}_2|}\psi_\beta(r_2)e^{i(\mathbf{k}_\beta,\mathbf{r}_2)}d\mathbf{r}_1 d\mathbf{r}_2 \qquad (4.2.11)$$

It is convenient to expand this expression in terms of the inverse powers of the velocity:

$$g^{(1)}_{\alpha\beta} = g^{(1)}_{1\alpha\beta} - g^{(1)}_{2\alpha\beta} \qquad (4.2.12)$$

in which case

$$g^{(1)}_{1\alpha\beta} = 2\frac{F_{\alpha\beta}(q)}{k^2} \quad \text{and} \quad g^{(1)}_{2\alpha\beta} = \frac{4i}{k^2}\int d\mathbf{r}\psi_\beta(\mathbf{r})e^{iqr}\frac{\partial}{\partial z}\psi_\alpha(\mathbf{r}) \qquad (4.2.13)$$

One can see from Tables 6 and 7 that for intermediate angles the term $g^{(1)}_{1\alpha\beta}$ predominates. For scattering angles the order of $g^{(1)}_{2\alpha\beta}$ is the same as or greater than the second-order Born exchange term $g^{(2)}_{\alpha\beta}$.

The physical meaning of the real and imaginary parts of the amplitude for the second Born approximation was examined by Grien.[39] The real part of the amplitude within this approximation is responsible for the polarization interaction of the target with the projectile, while the imaginary represents the possibility of

TABLE 6
First- and Second-Order Scattering ($f^{(1),(2)}_{\alpha\beta}$) and Exchange Scattering ($g^{(1),(2)}_{1,2\alpha\beta}$) Born Amplitudes with Different Parameters of the Collision Experiment, for $l_\beta = l_\alpha \pm 1$

q	$f^{(1)}_{\alpha\beta}$	$f^{(2)}_{\alpha\beta}$	$g^{(1)}_{1\alpha\beta}$	$g^{(1)}_{2\alpha\beta}$	$g^{(2)}_{\alpha\beta}$
$q \ll 1$	kq^{-1}	k^{-1}	$k^{-2}q$	k^{-3}	k^{-3}
$q \sim 1$	1	k^{-1}	k^{-2}	k^{-3}	k^{-3}
$q \gg 1$	q^{-7}	$k^{-1}q^{-2}$	$k^{-2}q^{-5}$	$k^{-3}q^{-4}$	$k^{-3}q^{-2}$

TABLE 7
Same as Table 6, for $l_\beta \neq l_\alpha \pm 1$

q	$f^{(1)}_{\alpha\beta}$	$f^{(2)}_{\alpha\beta}$	$g^{(1)}_{1\alpha\beta}$	$g^{(1)}_{2\alpha\beta}$	$g^{(2)}_{\alpha\beta}$
$q \ll 1$	1	k^{-1}	$k^{-2}q^2$	$k^{-3}q$	k^{-3}
$q \sim 1$	1	k^{-1}	k^{-2}	k^{-3}	k^{-3}
$q \gg 1$	q^{-6}	$k^{-1}q^{-2}$	$k^{-2}q^{-4}$	$k^{-2}q^{-3}$	$k^{-3}q^{-2}$

interference between different channels. The real part for the second Born approximation must decrease with velocity, because the target does not have time for polarization at large velocities.

The first Born approximation is used extensively to compute the cross sections of elastic and nonelastic scattering processes for systems with small numbers of electrons.[37–40, 43] Two rather complicated problems arise for multielectron systems. The first is the construction of a many-electron wave function[41] and the second is connected to the summation over the intermediate states. The pseudo-level method[42] significantly simplifies the summation problem for light targets.

4.3. The Glauber Approximation

Suppose a charged particle with coordinate r' collides with a one-electron atom in state $\Psi_\alpha(r)$. The problem consists in establishing a wave function $\Psi_\gamma(r)$ as a solution of the Schrödinger equation (4.2.3) with interaction potential

$$V(rr') = \frac{e^2}{|\mathbf{r}-\mathbf{r}'|} - \frac{Ze^2}{|\mathbf{r}|}, \qquad H = -\frac{\hbar^2}{2m}\Delta_{r'} + \mathscr{H} \tag{4.3.1}$$

where $\mathscr{H}$ is the Hamiltonian of the target particle:

$$\mathscr{H}\psi_\gamma = \mathscr{E}_\gamma \psi_\gamma$$

The solution of Eq. (4.2.3) is sought in the form

$$\Psi_\gamma = \psi_\gamma(r) A_\gamma(r, r') e^{iS_\gamma(r,r')} \tag{4.3.2}$$

where A_γ and S_γ are real functions.

Substitution of the latter expression into Eq. (4.2.3) with

$$U(r, r') = \frac{2m}{\hbar^2} V(r, r')$$

yields the system of equations

$$\Delta_{r'} A_\gamma - A_\gamma(\Delta_{r'} S_\gamma)^2 + (k_\gamma^2 - U)A_\gamma = 0$$

$$2(\Delta_{r'} A_\gamma)(\Delta_{r'} S_\gamma) + A_\gamma(\Delta_{r'} S_\gamma) = 0 \tag{4.3.3}$$

If the first equation of system (4.3.3) is used with the function

$$W_\gamma(r, r') = U(r, r') - \frac{\Delta_{r'} A_\gamma}{A_\gamma}$$

then one can rewrite wave function (4.3.2) in the following form:

$$\Psi_\gamma = \psi_\gamma(r) A_\gamma(rr') \exp(i \int_{r'_0}^{r'} dl \sqrt{k_\gamma^2 - W_\gamma})$$

where r' is some point of the trajectory.

If we treat a straight line projectile path, function $A(r, r')$ will not depend on coordinate r' and the following relation will be valid in this case:

$$W_\gamma(r, r') = U(r, r')$$

Further on, we can write

$$\Psi_\gamma = \psi_\gamma(r) \exp(i \int_l dl \sqrt{k_2 - U}) \tag{4.3.4}$$

where l is the integration path.

If the interaction is weak (weak-coupling approximation), the integral in Eq. (4.3.4) may be expanded in terms of U/k_γ^2 and only the first two terms need be taken into account. Then the solution is given by

$$\Psi_\gamma = \psi_\gamma(r) \exp(ik_\gamma r') \exp\left(\frac{i}{2} k_\gamma \int_{-\infty}^{r'} U dl\right) \tag{4.3.5}$$

By introducing cylindrical coordinates (ρ, z, φ) in a frame of reference with the OZ axis oriented in the direction of the projectile velocity vector (Figure 25), the projections of the transmitted impulse in this case ($k_\alpha \approx k_\beta$) become

$$q_\rho = k \sin\theta \qquad \text{and} \qquad q_z = 2k \sin^2 \tfrac{1}{2}\theta$$

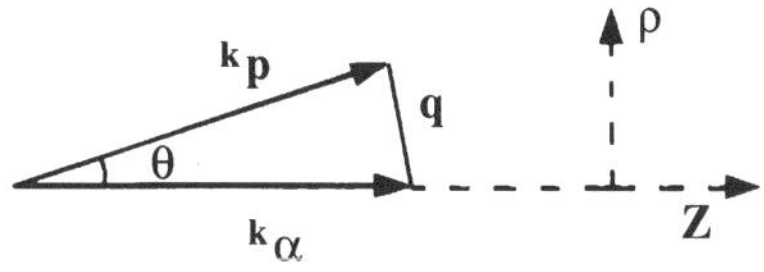

Figure 25. Estimate of the transmitted impulse during collision in a cylindrical system of coordinates.

For small scattering angles ($q_z \ll q_p$), the transmitted impulse is approximately orthogonal to the projectile velocity and integration over z in this case easily yields

$$f_{\alpha\beta}(\theta) = \frac{k}{2\pi i}\int d\mathbf{r}\int_0^\infty d\rho\rho e^{iq\rho}(1 - e^{2i\chi(\rho)})\psi_\alpha^*(\mathbf{r})\psi_\beta(\mathbf{r}) \qquad (4.3.6)$$

where

$$\chi(\rho) = -\frac{1}{\hbar v}\int_{-\infty}^{\infty} V(\rho, z)dz \qquad (4.3.7)$$

This approximation has wide application in computing elastic and elastic collisional cross sections for light charged particles with an atomic target, especially for high relative velocities.[39, 44] The validity of the Glauber approximation for inelastic scattering at low velocities has been specially examined.[39, 45, 116–121]

The scattering amplitude for the (e–H) reaction may be represented as a linear combination of hypergeometric functions.[46, 136] The asymptotic behavior of these functions show that the amplitude, even for $1s$–ns scattering, diverges at the point $q = 0$. This results in a violation of the optical theorem. On expanding Eq. (4.3.6) in terms of a power series in χ, one obtains an infinite sum consisting of alternate real and pure imaginary terms:

$$f_{\alpha\beta} = \sum_n \frac{1}{n!} f_{\alpha\beta}^{(n)}$$

where

$$f_{\alpha\beta}^{(n)} = \frac{ki^{n-1}}{2\pi}\int_0^\infty d\rho\rho e^{i(q,\rho)}\langle\psi_\beta|\chi^n(\rho)|\psi_\alpha\rangle \qquad (4.3.8)$$

Each term of this series incorporates an interaction of some power which corresponds to a particular term of the Born expansion. The first term in Eq. (4.3.8), ($f_{\alpha\beta}^{(1)}$), is the amplitude for the first Born approximation. The terms in expansion (4.3.4) of order higher than $(U/k)^2$ will contribute to amplitudes $f_{\alpha\beta}^{(n)}$ ($n \geq 3$). Hence, the divergence may originate from terms of expansion (4.3.8) containing some interaction with powers less than or equal to 2. As mentioned above, the first term of expansion (4.3.8) is real and the second term is purely imaginary. The reason for the equality $\mathrm{Re} f_{\alpha\beta}^2 = 0$ is the straight line trajectory approximation, which leads to the independence of function $A(r, r')$ on the projectile coordinates.

The real part of the second Born approximation describes the polarization interaction of the charged projectile and the target particle. In order to exclude diverging terms and describe the polarization interaction in more detail, a modified Glauber approximation was proposed.[39, 44] Within this approximation, the scattering amplitude is presented as

$$f_{\alpha\beta}^{(mod)} = f_{\alpha\beta} - f_{\alpha\beta}^{(2)} + f_{\alpha\beta}^{(B_2)}$$

where $f_{\alpha\beta}^{(B_2)}$ is the amplitude of the second Born approximation.

The cross section for $1s \rightarrow 1s$ scattering of an electron by a hydrogen atom was estimated as $1.05a_0^2$. The first Born and usual Glauber approximations give coefficients 0.51 and 0.64, respectively, while the experimental value is 1.20.

4.4. The System of Impact Parameter Equations

The system of differential equations for calculating scattering matrix elements in the impact parameter approximation and straight line trajectory will be introduced in this section. Such an approximation gives good results for collisions when long-range interaction prevails and the trajectory distortion is small. On the other hand, collisions with predominantly short-range interaction and high trajectory distortion may be described using the quasiclassic approximation.[47–49]

The trajectory concept may be used in this case if uncertainty in the scattering angle $\Delta\theta$ and in the impulse projection $\Delta p_\perp$ orthogonal to the propagation direction are much less than θ and $p_\perp$, respectively. In other words, the classical trajectory approximation is valid if

$$\frac{m\bar{V}a}{\hbar k^2} \gg 1$$

where $\bar{V}$ is the mean interaction potential and a is the interatomic force radius.

Two systems of reference are conventionally used to describe an impact process: the collisional system and a second system explained below. The $O\tilde{\tilde{Y}}$ axis of the latter system coincides with the $O\tilde{Y}$ axis of the collisional system and the $O\tilde{\tilde{Z}}$ axis is directed to a particle B moving with velocity v relative to atom A along the straight line trajectory, defined by impact parameter b (Figure 26). Later on, the parameters referred to in the second system will be denoted by the symbol ($\approx$).

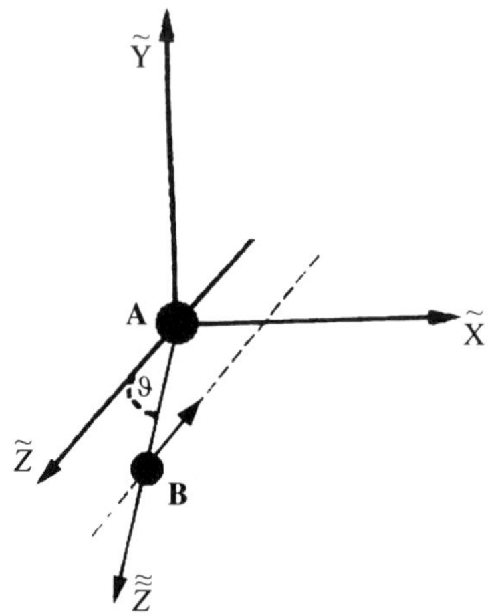

Figure 26. Systems of reference for describing collisions.

The wave function of atom A is the solution of the Schrödinger equation

$$i\hbar \frac{\partial}{\partial t}\tilde{\tilde{\Psi}} = (\hat{H}_0 + \hat{V})\tilde{\tilde{\Psi}} \tag{4.4.1}$$

which may be solved in the second system using the eigenfunctions of the $\hat{j}^2$ and $\hat{j}_z$ operators.

The ordinary expansion procedure for the wave function takes the form

$$\tilde{\tilde{\Psi}}(r, t) = \sum_{jm} \tilde{\tilde{C}}_{jm}(t)\tilde{\tilde{\psi}}_{jm}(r, t) \exp\left(\frac{iE_j t}{\hbar}\right) \tag{4.4.2}$$

The system of expansion coefficients $\tilde{\tilde{C}}_{jm}(t)$ will be time-dependent owing to the rotation of the second system. Substituting Eq. (4.4.2) into (4.5.1), and taking into account the commutation condition for operators $\hat{j}^2$ and $\hat{j}_z$, one has

$$i\hbar \frac{d}{dt}\tilde{\tilde{C}}_{jm} + i\hbar \sum_{m'} \tilde{\tilde{C}}_{j'm'} \langle jm| \frac{d}{dt} |j'm'\rangle - \sum_{j'} V_{jj'}(m)\tilde{\tilde{C}}_{j'm} = 0 \tag{4.4.3}$$

It is convenient to introduce a new variable ϑ, which is the angle between the $O\tilde{Z}$ and $O\tilde{\tilde{Z}}$ axes (cotan $\vartheta = vt/b$). The relation

$$\frac{d}{dt} = \frac{i}{\hbar}\frac{d\vartheta}{dt}\hat{j}_y$$

enables Eq. (4.4.3) to be written in the form

$$\frac{d}{d\vartheta}\tilde{\tilde{C}}_{jm} = -\frac{i}{\hbar}\left(\frac{d\vartheta}{dt}\right)^{-1}\sum_{j'} V_{jj'}^{(m)}(R)\tilde{\tilde{C}}_{j'm}\exp\left[\frac{i(E_j - E_{j'})t}{\hbar}\right]$$
$$+\frac{\delta_{jj'}}{2}[\sqrt{(j-m)(j+m+1)}\tilde{\tilde{C}}_{jm+1}$$
$$-\sqrt{(j+m)(j-m+1)}\tilde{\tilde{C}}_{jm-1}] \tag{4.4.4}$$

We note that from the standpoint of the general theory, transfer to the second reference system is not indispensable at all. The equations equivalent to (4.4.4) could be obtained in the collisional system as well. However, from the computational point of view, transfer to the second system of reference is preferential in case of low relative velocities ($v \leq 10^6$ cm/s). This follows from the fact that for small relative velocities, the time derivative $d\vartheta/dt$ is small.

System (4.4.4), and the system obtained after substitution of an expansion similar to (4.4.2), written in the collision reference system, into Eq. (4.4.1), has a large parameter on the right hand side. This parameter provides rapid oscillation of the solution for both systems. Oscillation of the solution of system (4.4.4) is much slower than for the system written in the collisional system. The later solution will comprise, first, harmonics of higher orders that stem from the nondiagonal parts of the interaction operator in a representation of the magnetic quantum numbers within the collisional reference system. Second, the spherical functions of these nondiagonal parts of the interaction depend on the angles fixing the position of the second system with respect to the collisional frame of reference.

A fundamental solution of system (4.4.4) is provided by the scattering matrix connecting expansion coefficients (4.4.2) before (i) and after (f) a collision:

$$\tilde{\tilde{C}}_{jm}^{(f)} = \sum_{j'm'} \tilde{\tilde{S}}_{mm'}^{jj'} \tilde{\tilde{C}}_{j'm'}^{(i)} \tag{4.4.5}$$

We shall now consider the symmetry properties of the scattering matrix, which may be created as a fundamental solution of the last system. The first symmetry property of that matrix follows directly from Eq. (4.4.5) and corresponds to the reflectional transformation of function Ψ_{jm} in the collision plane:

$$\tilde{\tilde{S}}_{mm'}^{jj'} = (-1)^{m-m'}\tilde{\tilde{S}}_{-m-m'}^{jj'} \tag{4.4.6}$$

If $U(\vartheta)$ is the matrix representing the fundamental solutions of system (4.4.4), the solution for an arbitrary value of angle ϑ may be written in the form

$$\tilde{\tilde{C}}_{jm}(\vartheta = \sum_{j'm'} U^{jj'}_{mm'}(\vartheta)\tilde{\tilde{C}}_{j'm'}(0) \tag{4.4.7}$$

By taking into account the symmetry of system (2.3.5) with respect to the time reflection operation $t \to -t$; $\vartheta \to \pi - \vartheta$, one has

$$\tilde{\tilde{C}}_{jm}(\pi - \vartheta) = \sum_{j'm'} U^{jj'*}_{mm'}(\vartheta)\tilde{\tilde{C}}_{j'm'}(\pi) \tag{4.4.8}$$

When $\vartheta = \pi/2$, the last relation yields

$$\tilde{\tilde{S}} = U^T(\pi/2)U(\pi/2) \tag{4.4.9}$$

where subscript "T" denotes the transposed matrix.

We now introduce a new variable $\vartheta' = \pi - \vartheta$, replace m by $-m$, and form the complex conjugate of the right and left hand sides of Eq. (4.4.4). We obtain that coefficients $\tilde{C}_{jm}(\vartheta)$ and $\tilde{C}^*_{jm}(\pi - \vartheta)$ satisfy the same system of equations. As a result, we can write

$$\tilde{\tilde{C}}^*_{j-m}(\pi - \vartheta) = \sum_{j'm'} U^{jj'}_{mm'}(\vartheta)\tilde{\tilde{C}}^*_{j'-m'}(\pi) \tag{4.4.10}$$

Now substitute $\vartheta = \pi/2$ in formulas (4.4.10) and (4.4.7) to obtain an expression for the scattering matrix within the second system, as follows:

$$\tilde{\tilde{S}}^{jj'}_{mm'} = \sum_{j''m''} U^{j''j}_{m''-m} U^{j''j}_{-m''m'} \tag{4.4.11}$$

The symmetry property of the scattering matrix in the second system follows from the previous formula:

$$\tilde{\tilde{S}}^{jj'}_{mm'} = \tilde{\tilde{S}}^{j'j}_{-m'-m} \tag{4.4.12}$$

The latter relation and the connection between the scattering matrices in the collisional and second systems yield the symmetry property of the scattering matrix in the collisional system of reference:

$$\tilde{S}^{jj'}_{mm'} = (-1)^{j-m+j'-m'}\tilde{S}^{j'j}_{m'm} \tag{4.4.13}$$

The symmetry property (4.4.9) based on the time reflection operation has a general character, but condition (4.4.12), which is strictly related to Eq. (4.4.5), is valid only for the scattering matrix calculated within the impact parameter and straight line trajectory approximation.

To conclude this section we make two remarks.

1. In case of an isolated level, when the dependence of the interaction anisotropy on the internuclear distance may be approximated by the power law

$$V_{jj'}^{(m)} \sim \frac{\Delta_j}{R^n} m^2 \delta_{jj'}$$

system (4.4.4) can be transformed to the form

$$\frac{d}{d\vartheta}\tilde{\tilde{C}}_{jm} = i\eta m^2 \sin^{n-2}\vartheta \tilde{\tilde{C}}_{jm} + \frac{1}{2}\Bigg[\sqrt{(j-m)(j+m+1)}\tilde{\tilde{C}}_{j,m+1} - \sqrt{(j+m)(j-m+1)}\tilde{\tilde{C}}_{j,m-1}\Bigg] \tag{4.4.14}$$

where $\eta = \Delta_j/\hbar v b^{n-1}$ is a dimensionless quantity known as the Massey parameter which, apart from velocity v and impact parameter b, depends on the anisotropy of interaction Δ_j.

In such a way, the cross section $\sigma_q^{kk'}$ on the right hand side of formula (3.8.6) [see also Eq. (3.8.3)] may be presented in the form

$$\tilde{\sigma}_q^{kk'}(jj';j_1j_1'|v) = 2\pi/(n-1)\cdot(\Delta_j/\hbar v)^{2/(n-1)}\cdot \sum_{mm'm_1m_1'}(-1)^{m'-m_1'}\begin{bmatrix} j & j' & k \\ m & -m' & q \end{bmatrix}\begin{bmatrix} j_1 & j_1' & k' \\ m_1 & -m_1' & q' \end{bmatrix}\cdot \int_0^\infty d\eta\,\eta^{-\frac{(n+1)}{(n-1)}}(\tilde{S}_{mm_1}^{jj_1}\tilde{S}_{m'm_1'}^{j'j_1'*} - \delta_{mm_1}\delta_{m'm_1'}) \tag{4.4.15}$$

All the specific features of the colliding particles are incorporated in the factor

$$s = \left(\frac{\Delta_j}{\hbar v}\right)^{2/(n-1)} \tag{4.4.16}$$

and the function $\tilde{\sigma}_q^{kk'}/s$ becomes a universal function which depends only on the angular momentum and the power of interaction anisotropy.

2. Let $\psi_1(r)$ and $\psi_2(r)$ be the wave functions of an atom before and after collision, where the impulses of a projectile are $p = k\hbar$ and $p' = k'\hbar$. In this case the solution of Eq. (4.4.4) may be written as

$$\psi(r, R) = C_1\psi_1(r)e^{i(\mathbf{k},\mathbf{R})} + C_2\psi_2(r)e^{i(\mathbf{k},\mathbf{R})} \qquad (4.4.17)$$

If the parameter $e^2/\hbar v$ is small, first-order perturbation theory may be applied to solve Eq. (4.4.4). The transition probability, which is proportional to $|C_2|^2$, is given by a relation similar to that of the first Born approximation.

4.5. Concluding Remarks

Some basic introductory information about quantum mechanical methods of atomic collisions, especially for collisions of a heavy atomic particle with a light charged projectile, is given in this chapter. These methods and approaches are very effective in practical problems of impact spectropolarimetric sensing of optically active entities in the solar and terrestrial atmospheres.

5

Theory of Anisotropic Collisional Relaxation

5.1. Introduction

Polarization effects in ensembles of atomic particles (atoms or ions), excited by an external process and undergoing collisions with slow atomic perturbers having no nuclear spin and no hyperfine structure of the quantum states, will be examined in this chapter. Using a kinetic approach, the effect of collisions in this case can be qualitatively interpreted as the transformation of an ordering of the distribution of the relative velocity vectors into an ordering of momenta of the ensemble of excited atoms.[10]

For an isotropic distribution of the relative velocity vectors (isotropic collisions), no collisional ordering of momenta can be created and the role of collisions is reduced only to depolarization, which is characterized by a one-exponent decay law. In this case an ensemble of excited particles is described by a set of $(2j+1)$ relaxation constants, each of which determines the collisional relaxation of the polarization moment of particular rank.[99, 122, 128]

The opposite situation takes place for collisions with anisotropic distribution of the relative velocity vectors (anisotropic collisions). In such a case collisions may create a new type of ordering of momenta of the excited atomic ensemble, and the type of this ordering reflects a symmetry of the collisional process. In the case of an axially symmetric excitation process, which takes place under beam excitation conditions, or sometimes, in an ordinary gas discharged plasma, transformation of aligned into oriented states, and vice versa,[52–55] as well as the collisional creation of longitudinal alignment[56, 57] are possible. For lower symmetry of a collisional process, other types of ordering of momenta may be created. For example, if the interaction operator is invariant under reflections in the collision plane, an orientation could arise. The last situation could be implemented experimentally, for example, with the help of a magnetic field tilted at some angle to the axis of collisions.[52–54]

The same type of symmetry is realized when the Stokes parameters of an optical emission are recorded only from particles scattered at a given angle or with the aid of a coincidence technique for the polarized photon and scattered particles. Such experiments are described in detail in review papers.(86, 108, 109) Most such experiments have been carried out with electrons, for which significant deflection angles could be realized,(104–106) but some experiments of this type with heavy low energy projectiles have been also implemented.(107, 110)

In the present chapter we restrict ourselves only to axially symmetric collisional processes. For a description of slow collisions of heavy atomic particles with axial symmetry, when long-range interaction plays a principal role, one can employ multipole resolution, which simplifies calculations and enables one to obtain more general results. The general symmetry properties of the matrix of cross sections of the collisional relaxation of polarization moments and the internal symmetry properties of the quantum states of colliding particles (occasional degeneration) will be considered. The matrix elements of the rate constants of collisional relaxation will be introduced in the third section and the special role of the quadrupole moment of the relative velocity distribution function will be pointed out. Collisional relaxation of polarization moments in the case of fine or hyperfine structure will be analyzed in the fifth section. Sections 6.7 and 6.8 will be devoted to a substantial description of relaxation processes of fine structure doublet states and a spectropolarimetric method for determining the drift velocity of ions in a plasma.

5.2. Symmetry of the Collisional Relaxation Cross Section Matrix

The evolution of polarization moments caused by population mixing during an axially symmetric process in the laboratory frame of reference may be expressed in the form

$$\frac{\partial}{\partial t}\rho_q^k = n_0 \sum_{k'} \langle v\sigma_q^{kk'} \rangle \rho_q^{k'} \tag{5.2.1}$$

where n_0 is the foreign gas density, $\langle v\sigma^{kkq'} \rangle$ is the matrix of the rate constants of collisional relaxation, comprised from collisional relaxation cross sections for the extreme anisotropic case of counterpropagating beams (3.8.7), or $\sigma_q^{kk'}$.

First, let us consider the symmetry properties of the matrix of rate constants of collisional relaxation. According to Eqs. (5.2.1) and (2.3.11) it is evident that

$$\sigma_q^{kk'*} = (-1)^{k+k'} \sigma_{-q}^{kk'} \tag{5.2.2}$$

Consequently, the elements of this matrix should be real and pure imaginary. There exists a relation between the symmetry of the collisional process and the reflection operation in the collisional plane, so

$$\sigma_q^{kk'} = (-1)^{k+k'} \sigma_{-q}^{kk'} \tag{5.2.3}$$

Within the framework of the impact parameter method with straight line trajectories, the cross section matrix elements are symmetrical if the upper subscripts are replaced:

$$\sigma_q^{kk_1} = \sigma_q^{k_1 k} \tag{5.2.4}$$

Following relations (5.2.1)–(5.2.4), it is clear that the collisional relaxation cross section matrix for a counterpropagating beam model has a block structure, and each block is connected with some q-value. Therefore, anisotropic collisions can mix polarization moments of different ranks. Within the block with $q = 1$, the relation between moments $\rho_1^{(2)}$ and $\rho_1^{(1)}$ leads to the possibility of mutual transformation of orientation and alignment.[52–55] Within the block with $q = 0$, the elements of the cross section $\sigma_0^{kk'} \neq 0$ only if $k + k'$ is even, therefore this block is split into two subblocks, each mixing zero components of polarization moments of a certain parity.

The combination of polarization moments ρ_0^0 and ρ_0^2 under relaxation leads to the possibility of the creation of collisional alignment under thermal conditions, for which the mean energy of particles is usually too small for direct impact excitation. The alignment creation process due to intermultiplet mixing at thermal relative velocities has been proposed[56, 57] and was later observed experimentally.[58, 59]

Taking into account only the collisional evolution of polarization moments, one can see that the population of an isolated level j

$$n_j = \sqrt{2j+1}\,\rho_0^0(j)$$

is time-dependent:

$$\frac{\partial}{\partial t} n_j = n_0 v \sum_k \sqrt{2j+1}\,\sigma_0^{0k}(j)\rho_0^k(j) = 0 \tag{5.2.5}$$

Due to the linear independence of polarization moments and taking into account Eq. (5.2.4), one has

$$\sigma_0^{0k}(j) = \sigma_0^{k0}(j) = 0$$

This means that the creation of alignment as well as polarization moments of higher rank is impossible under anisotropic collisional relaxation at an isolated level.

It should be emphasized that the symmetry property (5.2.4) was obtained under the approximation of the impact parameter method with straight line trajectories, which is valid mainly for collisions of heavy particles with predominant long-range interaction. Alignment may also arise under collisions with light atoms, when short-range interaction predominates. The mechanism responsible for that phenomenon is described in published papers.[47–49]

5.3. Rate Constants of Collisional Relaxation in the Case of Partial Anisotropy

Expression for the rate constants of a collisional relaxation process for any axially symmetric distribution of relative velocities have been derived in Chaper 3 [see Eq. (3.8.6)]. Those rate constants are linear combinations of cross sections of the collisional relaxation of polarization moments in the extreme anisotropic case of counterpropagating beams.

It was pointed out earlier [see Eq. (4.4.15)] that if the anisotropic interaction of colliding particles depends on the internuclear distance as a power law function $V \sim R^{-n}$, the cross section of collisional relaxation of electron polarization moments is represented also as a power law function with respect to the relative velocity, $\sigma_q^{kk'} \sim v^{-2/(n-1)}$. If we set

$$\xi \equiv \frac{n-3}{n-1}$$

then expression (3.8.10) can be transformed to the form

$$\langle v\sigma_q^{kk'}\rangle = \bar{v}\sum_{lq''}(-1)^{q+q'}\sigma_{q''}^{kk'}(\bar{v})\begin{bmatrix} k & k' & l \\ q & -q & 0 \end{bmatrix}\begin{bmatrix} k & k' & l \\ q'' & -q'' & 0 \end{bmatrix}\frac{4\pi}{2l+1}\frac{1}{\bar{v}^{\xi}} \times \int_0^{\infty} v^{\xi+2} f_l(v)\,dv \tag{5.3.1}$$

where $\bar{v}$ is a constant of velocity dimension, introduced in order to ensure that the integral on the right hand side is dimensionless; $\sigma_q^{kk'}(\bar{v})$ is the cross section of collisional relaxation in counterpropagating beams, calculated for velocity v, and

$f_l(v)$ is the multipole moment of the relative velocity distribution function $f(v)$ given by Eq. (3.8.8):

$$f_l(v) = \frac{2l+1}{4\pi} \int d\Omega f(v) P_l(\cos\theta)$$

The last formula and the symmetry properties of the relaxation matrix show that only terms with even l are not zero on the right hand side of Eq. (5.3.1):

$$\langle v\sigma_q^{kk'} \rangle = \bar{v} \sum_{lq''} (-1)^{q+q''} \sigma_q^{kk'}(\bar{v}) \begin{bmatrix} k & k' & 2l \\ q & -q & 0 \end{bmatrix} \begin{bmatrix} k & k' & 2l \\ q'' & -q'' & 0 \end{bmatrix} F_{2l}^{(\xi)} \tag{5.3.2}$$

The quantities $F_{2l}^{(\xi)}$ are called the dynamic multipole moments of the relative velocity distribution function, while the multipole moments $f_l(v)$ in Eq. (3.8.9) at fixed l, depending only on the relative velocity, are the corresponding kinetic quantities. The dynamic multipole moments are functions of the anisotropy factor of the interaction law:

$$\xi = \frac{n-1}{n-3}$$

These values are expressed as

$$F_l^{(\xi)} = \frac{4\pi}{2l+1} \frac{1}{\bar{v}^{\xi}} \int_0^{\infty} v^{\xi+2} f_l(v) dv \tag{5.3.3}$$

The triangle condition for the Clebsh–Gordan coefficient in formula (5.3.2) introduces the following restriction on the summation index l:

$$|k - k'| \leq 2l \leq k + k' \tag{5.3.4}$$

The ranks of polarization moments k and k' for the electronic state of an atomic particle A with atomic momentum l can take values between 0 and $2l$. Therefore, taking (5.3.4) into consideration, the collision relaxation process is governed only by a small number of dynamic multipole moments $F_{2l}^{(\xi)}$ of even rank $2l$. So for the electron angular moment $l = 1$, collisional relaxation is completely determined by the dynamic multipole moments with rank $2l = 0, 2$. When $l = 2$, it is determined by five dynamic multipole moments with ranks $2l = 0, 2, 4, 6$, and 8.

If all even dynamic multipole moments $F_{2l}^{(\xi)}$ (except the zero one) are zero and all odd moments take arbitrary values, anisotropic collisional relaxation is reduced to isotropic. Indeed, in such a case formula (5.3.2) contains only one term ($l = 0$) and, following the properties of the Clebsh–Gordan coefficients, one has the selection rule $k = k'$. Consequently, the matrix of collisional relaxation rate constants will be diagonal with respect to the ranks of the polarization moments and the elements of this matrix will not be q-dependent. As a result, relaxation will turn out to be isotropic and the rate constants in this case will take the form

$$\gamma^k = n_0 \bar{v} \frac{F_0^{(\xi)}}{2k+1} \sum_q \sigma_q^{kk}(\bar{v}) \tag{5.3.5}$$

The collisional transformation of polarization moments ρ_q^k and $\rho_q^{k'}$ is possible only if rate constants $\langle v\sigma_q^{kk'} \rangle$, given by formula (5.3.2), are not zero and the ranks of the dynamic multiple moments of the velocity distribution $F_{2l}^{(\xi)}$ are determined by relation (5.3.4). This situation is illustrated by Table 8 for all polarization moments with ranks between 0 and 8.

Polarization moments with $k = 0, 1, 2$ are closely connected to the polarization characeristics of the emitted optical radiation. The higher-order moments ($k = 3, 4, \ldots$) do not determine directly the polarization properties of radiation, but may affect indirectly the optical emission of an ensemble by being involved in the process of mutual collisional relaxation of higher-order moments with orientation ($k = 1$) and alignment ($k = 2$).

Table 8 shows that the second-order dynamic multiple moment of the velocity distribution function has a special place, being responsible for the most important process of the collisional generation of alignment from the population ($k = 2, k' = 0$) and the transformation of alignment from the orientation ($k = 1, k' = 0$).

TABLE 8
Dynamic Multipole Moments of the Relative Velocity Distribution Function of Atomic Particles, Responsible for Processes of Relaxation and Mutual Transformation of Polarization Moments of Excited Atomic Particles[a]

	$k_1 = 0$	$k_1 = 1$	$k_1 = 2$	$k_1 = 3$	$k_1 = 4$
$k = 0$	F_0	—	F_2	—	F_4
$k = 1$	—	$F_{0'}F_2$	F_2	$F_{2'}F_4$	F_4
$k = 2$	F_2	F_2	$F_{0'}F_{2'}F_4$	$F_{2'}F_4$	$F_{2'}F_{4'}F_6$
$k = 3$	—	$F_{2'}F_4$	$F_{2'}F_4$	$F_{0'}F_{2'}F_{4'}F_6$	$F'_2F_{4'}F_6$
$k = 4$	F_4	F_4	$F_{2'}F_{4'}F_6$	$F_{2'}F_{4'}F_6$	$F_{0'}F_{2'}F_{4'}F_{6'}F_8$

[a] Diagonal positions ($\kappa = \kappa_1$) represent dynamic multipole moments governing collisional decay of polarization moments of rank κ; nondiagonal positions ($\kappa \neq \kappa_1$) represent multipole moments governing mutual collisional transformation of polarization moments of ranks κ and κ_1.

5.4. Distribution Functions of Relative Velocities

Let us analyze a common problem of the collision of free particles in an ionized gas within a reference frame for which atomic particles of sort B with Maxwell velocity distribution

$$f_b(v_b) = \left(\frac{\beta}{\hbar}\right)^{3/2} e^{-\beta|v_b|^2} \tag{5.4.1}$$

collide with an ensemble of ions or atoms of sort A with preferential direction of motion specified by the Maxwell distribution function with drift

$$f_a(v_a) = \left(\frac{\alpha}{\hbar}\right)^{3/2} e^{-\alpha|v_a+v_0\mathbf{e}_z|^2} \tag{5.4.2}$$

where v_0 is the drift velocity while

$$\alpha = \frac{m_a}{2kT_a} \qquad \text{and} \qquad \beta = \frac{m_b}{2kT_b}$$

If we introduce the new variables

$$\mathbf{u} = \frac{\alpha v_a + \beta v_b}{\alpha + \beta} \qquad \text{and} \qquad v = v_a - v_b$$

then the relative velocity distribution may be obtained by integrating the product of functions (5.4.1) and (5.4.2) over the variable u:

$$f(v) = (\gamma/\pi)^{3/2} e^{-\gamma(v_0^2 - 2v_0 v \cos\theta + v^2)} \tag{5.4.3}$$

where

$$\gamma = \frac{\alpha \cdot \beta}{\alpha + \beta}$$

and θ is the angle between the relative velocity and the drift ion velocity directions.

The distribution of the relative velocity vectors of ions with respect to atoms represents in itself a Maxwell distribution with shift along the *OZ* axis. The difference from Eq. (5.4.2) is in the sign of the drift velocity v_0, and the parameter γ substituted in place of α. In this case, the surfaces of equal collisional probability in the space of relative velocities are spheres centered on the *OZ* axis at the point v_0.

At zero drift velocity, the spheres of equal collision probability have center at the coordinate origin while the distribution is transformed into an isotropic one. On increasing the drift velocity, the anisotropy also increases. A quantitative characteristic of this anisotropy is described by the ratio of the drift velocity to the most probable relative velocity given by the Maxwellian distribution or the λ parameter:

$$\lambda = v_0\sqrt{\gamma} \tag{5.4.4}$$

When λ increases, the surfaces of equal collisional probability shift along the OZ axis, and when $\lambda \to \infty$ the relative velocity distribution becomes the same as that for counterpropagating monokinetic beams (Figure 27).

Substitution of Eq. (5.4.3) into (5.4.2) yields the multipole moment of the distribution function:

$$f_l(v, \lambda) = \frac{2l+1}{2\pi}\frac{\lambda^2}{v_0^3} e^{-\lambda^2[1+(v/v_0)^2]}\sqrt{\frac{v_0}{v}} I_{l+\frac{1}{2}}\left(2\lambda^2\frac{v}{v_0}\right) \tag{5.4.5}$$

where $I_{l+\frac{1}{2}}(2\lambda^2 v/v_0)$ is the Bessel function of a purely imaginary argument.

If the latter expression is substituted into Eq. (5.3.3) and we set

$$\bar{v} = \gamma^{-1/2} \tag{5.4.6}$$

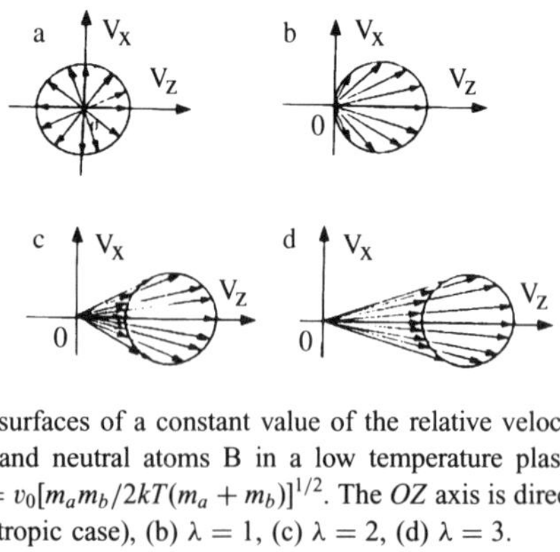

Figure 27. Profiles of surfaces of a constant value of the relative velocity distribution function of drifting excited ions A and neutral atoms B in a low temperature plasma for different values of anisotropy parameter $\lambda = v_0[m_a m_b/2kT(m_a + m_b)]^{1/2}$. The OZ axis is directed along the drift velocity vector $\mathbf{v}_0$: (a) $\lambda = 0$ (isotropic case), (b) $\lambda = 1$, (c) $\lambda = 2$, (d) $\lambda = 3$.

then the resulting expression could be regarded as a representation of the dynamical multiple moments in terms of the degenerate hypergeometric function:

$$F_l^{\xi}(\lambda) = \lambda^l e^{-\lambda^2} \frac{\Gamma\left(\dfrac{1+\xi+3}{2}\right)}{\Gamma\left(l+\dfrac{3}{2}\right)} \Phi\left(\frac{l+\xi+3}{2}, l+\frac{3}{2}; \lambda^2\right) \tag{5.4.7}$$

Let us now study the asymptotic behavior of functions (5.4.5) and (5.4.7). When $\lambda \to 0$, the properties of the Bessel and degenerate hypergeometric functions enable us to write

$$f_l(v, \lambda) \sim \frac{2l+1}{2\pi\Gamma(l+3/2)} \gamma^{3/2} \lambda^{2l} \left(\frac{v}{v_0}\right)^l \exp\left\{-\lambda^2\left[1+\left(\frac{v}{v_0}\right)^2\right]\right\} \tag{5.4.8}$$

and

$$F_l^{\xi}(\lambda) \sim \lambda^l e^{-\lambda^2} \frac{\Gamma[(l+\xi+3)/2]}{\Gamma(l+3/2)} \tag{5.4.9}$$

Hence it is evident that $f_l(v, \lambda)$ and $F_l^{\xi}(\lambda)$ decrease to zero if $l \neq 0$; or, physically speaking, anisotropic relaxation is transformed into isotropic relaxation.

In the opposite case of large λ

$$f_l(v, \lambda) \sim \frac{2l+1}{\sqrt{8\pi^3}} \frac{\lambda}{v_0^2 v} \exp\left[-\lambda^2\left(1-\frac{v}{v_0}\right)^2\right] \tag{5.4.10}$$

and

$$F_l^{\xi}(\lambda) \sim \lambda^{\xi} \tag{5.4.11}$$

These two formulas show that, when $\lambda \to \infty$, the relations describe collisions in counterpropagating beams. The discrepancies of the dynamical multiple moment for $l = 0, 2, 4$ and $\xi = 0.6, 0, -1$ on parameter λ are presented in Figure 28.

These values describe anisotropic collisions, characterized by a power law interaction with the potential function depending on the interatomic distance with exponent $n = 6, 3, 2$. The power law with $n = 6$ describes the interaction of an excited neutral atom with a structure-free particle. The curve with $n = 3$ descirbes the interaction of an excited atom with an ion and that with $n = 2$ describes collisional relaxation of the occasionally degenerate levels of the hydrogen atom.

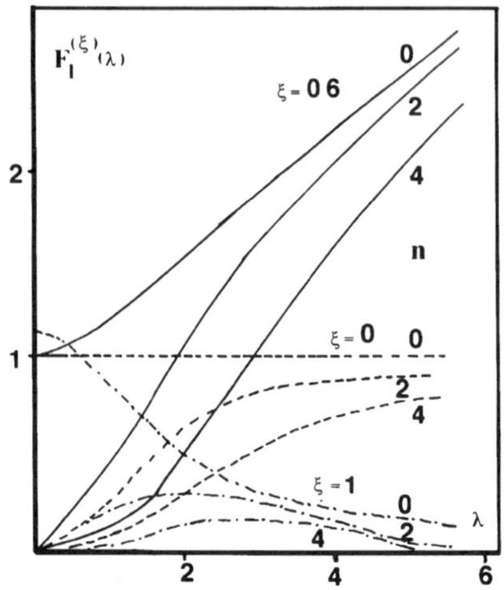

Figure 28. Dynamic multipole moment $F_l^{\xi}(\lambda)$ for $l = 0, 2, 4$ as a function of anisotropy parameter λ for $\xi = 0.6$ (solid line), $\xi = 0$ (dashed line), $\xi = -1$ (dash-dot line).

While discussing the distribution functions of relative velocities it is worth mentioning that, apart from distribution function (5.4.3) which is commonly used to compute spectropolarimetric effects of atomic particles undergoing slow collisions in a low temperature plasma, other expressions for this function relevant to different physical problems are used. For example, to calculate impact polarization effects created by precipitation of fast charged particles into the upper layers of the solar chromosphere, where an optically active medium itself comprises a low ionized gas, the distribution function was proposed[(51)] in the form

$$f(v) \sim \left[\frac{\cos\theta}{v^4\cos\theta + \Phi(v)}\right]^{5/2}, \qquad \theta \in \left[0, \frac{1}{2}\pi\right] \tag{5.4.12}$$

where

$$\Phi(v) = \frac{16\pi e^4 N_0}{m_e m_p}\ln B \tag{5.4.13}$$

In this expression N_0 is the particle density of a medium within the layer where the H_α emission line core is formed, and $\ln B$ is the Coulomb logarithm characterizing energy losses of a particle while passing through an ionized medium.

For quantity B the following expression[172] is used:

$$B = \frac{2mv^2}{\hbar\bar{\omega}} \tag{5.4.14}$$

where $\bar{\omega}$ is the averaged value of the frequency.

Suppose we take the typical value $N_0 = 2.6 \times 10^{18}\ \text{cm}^{-1}$ from Hénoux *et al.*[51] Also, we approximate quantity $\hbar\omega$ by an energy gap between levels $n = 1$ and $n = 3$ of the atomic hydrogen that mainly form the solar atmosphere at this level, and introduce dimensionless parameter $\lambda = v/v_0$ ($v_0 = e^2/\hbar$ is the atomic velocity). Expression (5.4.12) then assumes the form

$$f(v) \sim \left(\frac{\cos\theta}{\lambda^4 \cos\theta + 2t}\right)^{5/2} \tag{5.4.15}$$

where $t = \ln 36 + 2\ln\lambda$.

The multipole moments of the relative velocity distribution function, indispensable for computing spectropolarimetric effects, have been calculated for function (5.4.15) with a view to astrophysical applications. Functions $f_k(v)/f_0(v)$ for $k = 1, 2, 3$ are presented in Figure 29. The absolute value of this ratio decreases rapidly with increasing order of the multipole moment.

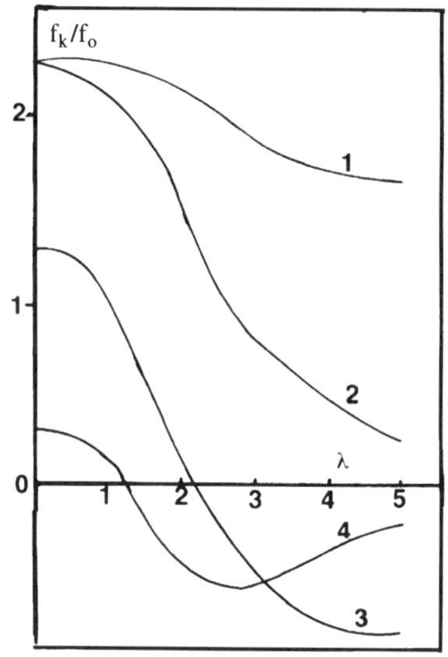

Figure 29. Ratio f_k/f_0 ($k = 1, 2, 3, 4$) as a function of dimensionless parameter $\lambda = v/v_0$, ($v_0 = e^2\hbar$).

There are also other treatments for a velocity distribution of charged particles bombarding the upper part of the solar chromosphere and producing observable optical effects. This distribution was calculated numerically[198] using a computer simulation of proton kinetics in a magnetized plasma of the solar flare loop. This approach was also utilized to calculate impact spectropolarimetric effects accompany energy transport processes during solar flares or other nonthermal phenomena in a solar plasma. General ideas on the kinetics of charged particles were also applied to studies and quantitative spectropolarimetric sensing of different optical emissions in the upper atmosphere of the Earth.

5.5. Collisional Relaxation of Polarization Moments under Intermultiplet Mixing

When populations of quantum states differ from equilibrium, the population mixing process caused by anisotropic collisions could provoke the collisional creation of longitudinal components of polarization moments of even rank, including the collisional formation of alignment from the population of atomic states.

In order to calculate spectropolarimetric observables resulting from this process, one should obtain the rate constants of the collisional relaxation of polarization moments (5.3.1), describing different types of ordering of angular moments $J = L + S$ for fine structure levels J (S is the electron spin). In the case of a narrow fine-structure multiplet, it is possible to express algebraically these rate constants $\langle v\sigma_q^{kk'}(l)\rangle$.[61–63]

Indeed, suppose a splitting ΔE of an L–S multiplet is much less than the kinetic energy of the colliding particles:

$$\frac{\|\Delta E|}{\hbar} \ll \frac{v}{R} \tag{5.5.1}$$

In other words, this condition means that the frequency corresponding to multiplet splitting at distance R, related to the effective cross section σ of the process under study by the relation $R \sim \sqrt{\sigma/\pi}$, is much less than the angular frequency of the motion of particle B with respect to excited atom A.

Subject to a reliable approximation, one can assume that L–S coupling is violated during the collision and reorientation of the momentum vector $\mathbf{L}$ takes place. If the energy of interaction depends only on the value of the orbital moment and the spin vector $\mathbf{S}$ does not change its orientation during a collision (the spin inertia principle[64]), then only the orbital moment $\mathbf{L}$ will participate in the relaxation process, which is described by formula (3.8.6) (see Figure 30).

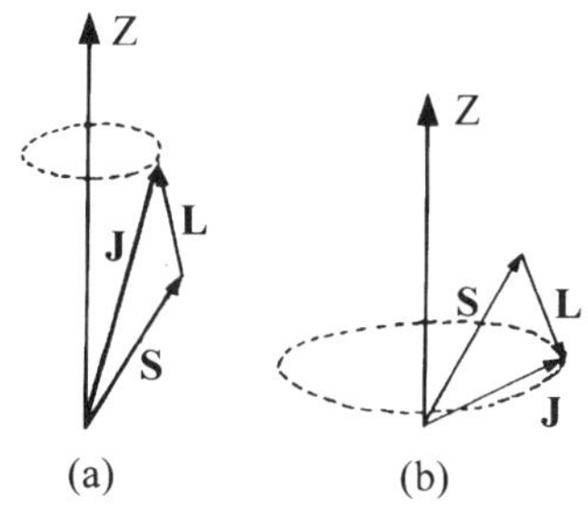

Figure 30. Reorganization of an $L-S$ coupling scheme during a collision: (a) before collision, (b) after collision.

After a collision, coupling of the L and S values is reestablished and the polarization moments of the fine structure of the multiplet form again from the orbital and spin polarization moments.

Collisional violation and recoupling of the binding between vectors **L** and **S** may be described by well-known methods of angular moment algebra.[27] These general considerations enable an expression for the rate constants of anisotropic collisional relaxation of the j components of a narrow LS fine-structure multiplet to be written in the form

$$\begin{aligned}\langle v\sigma_Q^{KK_1}(JJ_1)\rangle = &\sum_{\kappa_1,\kappa,l,p,q}(-1)^{k+k_1+Q+p+q+l}(2J+1)(2J_1+1)(2l+1).\\ &\times[(2\kappa+1)(2\kappa_1+1)(2K+1)(2K_1+1)]^{1/2}\\ &\times\begin{Bmatrix} S & S & l\\ L & L & \kappa\\ J & J & K\end{Bmatrix}\begin{Bmatrix} S & S & l\\ L & L & \kappa_1\\ J_1 & J_1 & K_1\end{Bmatrix}\begin{Bmatrix} \kappa_1 & \kappa & p\\ K & K_1 & l\end{Bmatrix}\\ &\times\begin{bmatrix} \kappa_1 & \kappa & p\\ q & q & 0\end{bmatrix}\begin{bmatrix} K_1 & K & P\\ Q & -Q & 0\end{bmatrix}\langle v\sigma_q^{\kappa\kappa_1}(L)\rangle\end{aligned} \tag{5.5.2}$$

where

$$\begin{Bmatrix} S & S & l\\ L & L & \kappa\\ J & J & K\end{Bmatrix}$$

is the $9j$ Wigner symbol.[27]

In this case, the differential equation governing the collisional relaxation of the fine structure polarization moments takes the form

$$\frac{d}{dt}\rho_Q^k(J) = -n_0 \sum_{k_1, J_1} \langle v\sigma_Q^{kk_1}\rangle(J, J_1)\rho_Q^{k-1}(J_1) \tag{5.5.3}$$

This approximation and the above relations are valid only if collisional mixing with other fine structure multiplets may be neglected, or, in other words, if the collision kinetic energy is much less than the energy separation between the centers of different multiplets.

Because of the independence of the even dynamical multipole moments (5.4.3) on the sign of the relative velocity vectors **v**, the equations analyzed in this section are applicable also to anisotropic relaxation of polarization moments within a narrow multplet of an excited neutral atom colliding with ions in a plasma. In the general case of a low ionized gas, the situation is more complicated and these fine collisional polarization effects coexist with ordinary well-known collisional phenomena. Usually, the number of neutral particles in a low temperature gas discharge plasma is much higher than that for ions, and threfore the anisotropic relaxation is accompanied by the isotropic process proceeding during collision between excited and ground state neutral atoms.

5.6. *Anisotropic Collisional Alignment of a Narrow Multiplet*

Let us examine the specific spectroscopic problem of collisional alignment of the $j = 3/2$ level of a narrow $^2P_{1/2,3/2}$ doublet of an excited ion under its drift in an ionized gas (Figure 31). According to Eq. (2.2.1), the populations of the $j = 1/2, 3/2$ levels are expressed in terms of the polarization moment of zero rank as

$$n_{1/2} = \sqrt{2}\rho_0^0(1/2) \qquad \text{and} \qquad n_{3/2} = 2\,\rho_0^0(3/2)$$

The longitudinal alignment of the $j = 3/2$ level is represented by the second-order polarization moments $a_{3/2} = 2\rho_0^2(3/2)$.

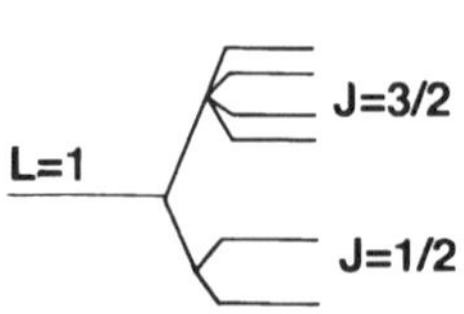

Figure 31. Energy diagram of a $^2P_{1/2,3/2}$ doublet.

If we extract from system (3.5.3) the block of equations with $q = 0$, which describe the complex collisional relaxation of $j = 1/2$ and $j = 3/2$ populations and the longitudinal alignment of the $j = 3/2$ level, then we have the following system:

$$\begin{aligned}\frac{d}{dt}n_{1/2} &= n_B[-\langle v\sigma_p\rangle n_{1/2} + \tfrac{1}{2}\langle v\sigma_p\rangle n_{3/2} + \tfrac{1}{2}\langle v\sigma_a\rangle a_{3/2}] - \Gamma_0 n_{1/2} + N_{1/2} \\ \frac{d}{dt}n_{3/2} &= n_B[\langle v\sigma_p\rangle n_{1/2} - \tfrac{1}{2}\langle v\sigma_p\rangle n_{3/2} - \tfrac{1}{2}\langle v\sigma_a\rangle a_{3/2}] - \Gamma_0 n_{3/2} + N_{3/2} \\ \frac{d}{dt}a_{3/2} &= n_B[\langle v\sigma_a\rangle n_{1/2} - \tfrac{1}{2}\langle v\sigma_a\rangle n_{3/2} - \langle v\sigma_d\rangle a_{3/2}] - \Gamma_0 a_{3/2}\end{aligned} \tag{5.6.1}$$

where Γ_0 is the natural damping constant, $N_{1/2}$ and $N_{3/2}$ are the rates of excitation of these levels, $\langle v\sigma_p\rangle$ is the rate constant for collisional population mixing between the $j = 1/2$ and $j = 3/2$ levels, while $\langle v\sigma_a\rangle$ and $\langle v\sigma_d\rangle$ are positive coefficients related to the rate constants for the creation of the alignment of the $j = 3/2$ state and the collisional alignment destruction (depolarization), respectively.

These positive coefficients could be expressed in terms of the corresponding rate constants of collisional relaxation processes as follows:

$$\begin{aligned}\langle v\sigma_p\rangle &= -\sqrt{2}\langle v\sigma_0^{00}(1/2, 3/2)\rangle = -\sqrt{2}\langle v\sigma_0^{00}(3/2, 1/2)\rangle \\ \langle v\sigma_a\rangle &= -\sqrt{2}\langle v\sigma_0^{20}(3/2, 1/2)\rangle = -\sqrt{2}\langle v\sigma_0^{02}(1/2, 3/2)\rangle \\ \langle v\sigma_d\rangle &= \langle v_0^{22}(3/2, 3/2)\rangle\end{aligned} \tag{5.6.2}$$

In analyzing this result we note that the independence of the ratio of the collisional population transfer rate for direct and inverse processes on the multiplet splitting ΔE is a drawback of the oversimplified model of the breakdown of L–S coupling during a collision. Indeed, following the detailed balance principle,[65] collisionally induced transitions between levels with increased energy are more difficult than those with the opposite transitions. This situation is accounted for by the factor

$$Y = \exp\left(-\frac{|\Delta E|}{kT_{\text{eff}}}\right) \tag{5.6.3}$$

where T_{eff} is the effective temperature of the atomic ensemble determined by the mean square of the relative velocity.

For velocity distribution (5.4.3), this mean velocity exceeds the corresponding pure Maxwellian one by the value of the drift velocity. Therefore, the effective temperature is higher than the mean temperature of atomic particles in a plasma. It is expressed by the relation

$$T_{\text{eff}} = (1 + \tfrac{3}{2}\lambda^2)T \tag{5.6.4}$$

In the case of a normal fine structure, the $j = 3/2$ level lies higher than the $j = 1/2$ level and the factor Y, given by expression (5.6.3), should be introduced into system of equations (5.5.1) for the terms corresponding to transitions $j = 1/2 \to j = 3/2$ with energy increase. As a result

$$\begin{aligned}
\frac{d}{dt}n_{1/2} &= n_B[-Y\langle v\sigma_p\rangle n_{1/2} + \tfrac{1}{2}\langle v\sigma_p\rangle n_{3/2} + \tfrac{1}{2}\langle v\sigma_a\rangle a_{3/2}] - \Gamma_0 n_{1/2} + F_{1/2} \\
\frac{d}{dt}n_{3/2} &= n_B[Y\langle v\sigma_p\rangle n_{1/2} = \tfrac{1}{2}\langle v\sigma_p\rangle n_{3/2} - \tfrac{1}{2}\langle v\sigma_a\rangle a_{3/2}] - \Gamma_0 n_{3/2} + F_{3/2} \\
\frac{d}{dt}a_{3/2} &= n_B[Y\langle v\sigma_a\rangle n_{1/2} - \tfrac{1}{2}\langle v\sigma_a\rangle n_{3/2} - \langle v\sigma_d\rangle a_{3/2}] - \Gamma_0 a_{3/2}
\end{aligned} \tag{5.6.5}$$

For the thermal equilibrium populations of doublet levels, with account for statistical weights and Boltzmann factors, these equations satisfy

$$Yn_{1/2} - \tfrac{1}{2}Yn_{3/2} = 0$$

Therefore, the terms describing collisional population transfer between the $j = 1/2$ and $j = 3/2$ levels turn out to be zero. This conclusion, resulting formally from Eqs. (5.5.5), is in accordance with the detailed balance principle.

The first two terms on the right hand side of the third equation describe the process of alignment creation from populations undergoing anisotropic collisions. As they possess the same rate constants $\langle v\sigma_a\rangle$, the efficiency of the alignment creation is proportional to the quantity

$$\Delta n = Yn_{1/2} - (1/2)n_{3/2}$$

or, in other words, to the deviation of the doublet level populations from its equilibrium values. By these means, anisotropic collisions may generate alignment only from nonequilibrium populations of the fine multiplet levels. The efficiency of this process increases with growth in the degree of deviation of populations from its equilibrium value.

The third term within square brackets on the right hand side of the first two equations (5.6.5) describe the effect of the alignment of angular momentum $j = 3/2$ on the rate constants of the collisionally induced population mixing. This effect for the general case of anisotropic collisions was predicted theoretically.[56] According to published results[66] for extreme anisotropic collisions in counterpropagating beams, the efficiency of collisional population transitions between the fine structure levels of a multiplet depends on the magnitude and sign of the $j = 3/2$ level alignment, changing its value by a factor of 5–10.

Rate constants $\langle v\sigma_p \rangle$, $\langle v\sigma_a \rangle$, and $\langle \sigma_d \rangle$ in system (5.6.5) were calculated using formula (5.6.2) and relation (5.4.2) for the dynamic multipole moments of the relative velocity distribution function (5.4.3) for excited ions drifting in an atmosphere of neutral atoms. Calculations were conducted for two internuclear distances: $n = 6$, which describes the interaction of an excited ion with a neutral particle; and $n = 2$, corresponding to the interaction of a charged particle with an excited atom at the degenerate state.

5.7. Anisotropic Relaxation and Polarization of Light

Formula (3.6.3) and relations for the reduced matrix elements[27] are used to obtain the following expression for the light intensity emitted during the transition $(JL) \rightarrow (J_0 L_0)$:

$$I_{\mathbf{e}_\lambda} = (-1)^{J+J_0} C(L, L_0)(2J+1)(2J_0+1) \times \begin{Bmatrix} L_0 & J_0 & S \\ J & L & 1 \end{Bmatrix}^2 \sum_{kq} \begin{Bmatrix} 1 & 1 & k \\ J & J & J_0 \end{Bmatrix} \Phi_{kq}(\mathbf{e}_\lambda) \rho_q^k(J) \tag{5.7.1}$$

where $\mathbf{e}_\lambda$ is the polarization orth and $\Phi_{kq}(\mathbf{e}_\lambda)$ is defined by formula (3.6.4).

For optical transitions from ${}_2P_J$ doublet states ($L = 1$, $J = 3/2$, $J = 1/2$), one has the following expressions for the observable polarization characteristics:

$$(I_Y - I_Z)_{J=3/2} = C(1, L_0)\sqrt{2}(-1)^{J_0+3/2}(2J_0+1) \times \begin{Bmatrix} 3/2 & 3/2 & 2 \\ 1 & 1 & J_0 \end{Bmatrix} \begin{Bmatrix} L_0 & J_0 & 1/2 \\ 3/2 & 1 & 1 \end{Bmatrix}^2 a_{3/2} \tag{5.7.2a}$$

$$(I_Y + I_Z/2)_{J=3/2} = C(1, L_0)(-1)^{J_0+3/2}(2J_0+1) \times \begin{Bmatrix} 3/2 & 3/2 & 0 \\ 1 & 1 & J_0 \end{Bmatrix} \begin{Bmatrix} L_0 & J_0 & 1/2 \\ 3/2 & 1 & 1 \end{Bmatrix}^2 n_{3/2} \tag{5.7.2b}$$

$$(I_Y + I_Z/2)_{J=1/2} = C(1, L_0)(-1)^{J_0+1/2}(2J_0+1) \times \begin{Bmatrix} 1/2 & 1/2 & 0 \\ 1 & 1 & J_0 \end{Bmatrix} \begin{Bmatrix} L_0 & J_0 & 1/2 \\ 1/2 & 1 & 1 \end{Bmatrix}^2 n_{1/2} \tag{5.7.2c}$$

Under stationary conditions we obtain from Eq. (5.6.5)

$$\frac{a_{3/2}}{Yn_{1/2} - (1/2)n_{3/2}} = \frac{\langle v\sigma_a \rangle p}{\langle v\sigma_d \rangle p + \Gamma_0} \tag{5.7.3}$$

It should be emphasized that this relation is valid regardless of the rates of excitation of levels $J = 1/2$ and $J = 3/2$, or $N_{1/2}$ and $N_{3/2}$ which, as a rule, are unknown in the plasma. Equation (3.7.2) can be employed to express quantities $a_{3/2}$, $n_{1/2}$, and $n_{3/2}$ in terms of the polarization characteristics of the light I_X, I_Y, and I_Z. For transitions from an *ArII* doublet $4p^2P_J \to 4p^2D_{3/2}$ and $4p_2P_J \to 4s^2P_{3/2}$ we have

$$\frac{\langle v\sigma_a \rangle p}{\langle v\sigma_d \rangle p + kT\Gamma_0} = \frac{12.5(I_Y - I_Z)_{3/2}}{5(I_Y + 0.5I_Z)_{3/2} - Y(I_Y + 0.5I_Z)_{1/2}} \tag{5.7.4a}$$

and

$$\frac{\langle v\sigma_a \rangle p}{\langle v\sigma_d \rangle p + kT\Gamma_0} = \frac{(I_Y - I_Z)_{3/2}}{(I_Y + 0.5I_Z)_{3/2} - 2Y(I_Y + 0.5I_Z)_{1/2}} \tag{5.7.4b}$$

For given temperature, gas pressure p, and radiative damping constant Γ_0 the left hand side of these relations depends only on the rate constants $\langle v\sigma_a \rangle$ and $\langle v\sigma_d \rangle$ (see Tables 9–11) and, as a result, on the anisotropy parameter λ. Therefore, measurement of the polarization characteristics enables one to determine parameter λ and, consequently, the drift velocity.[93]

5.8. *Depolarizing Collisions with Charged Particles*

One successful application of the impact parameter method is a theoretical description of collisions with charged particles destroying the ordering of momenta of an excited ensemble. This type of collision is of major significance in studies of ionized gases by the spectropolarmetric method.

Suppose a charged particle B moves with constant velocity v past an excited atom A* along a rectiliner trajectory with impact parameter b, determined by the distance between the nuclei (see Figure 32). The instantaneous axis $O\tilde{\tilde{Z}}$ connecting the nuclei A* and B, and the flight direction of particle B (the $O \sim \tilde{Z}$ axis) form an angle

$$\vartheta = \frac{b}{vt} \tag{5.8.1}$$

TABLE 9
Rate Constants for Collisional Alignment Creation $\langle v\sigma_a \rangle$, Destruction $\langle v\sigma_d \rangle$, and Population Transfer $\langle v\sigma_p \rangle$ as a Function of Anisotropy Parameter λ with Interaction Anisotropy Dependent on Internuclear Distance, $\Delta \sim R^{-n}$: $n = 6$, $\dot{\xi} = 0.6$[a]

λ	$\langle v\sigma_a \rangle$	$\langle v\sigma_d \rangle$	$\langle v\sigma_p \rangle$
0	0	2.867	1.998
0.5	0.141	3.074	2.094
1.0	0.492	3.596	2.349
1.5	0.916	4.245	2.690
2.0	1.318	4.893	3.058
2.5	1.669	5.501	3.422
3.0	1.974	6.070	3.774
3.5	2.246	6.606	4.111
4.0	2.491	7.117	4.433
4.5	2.718	7.606	4.743
5.0	2.929	8.077	5.041
5.5	3.129	8.532	5.329
6.0	3.318	8.973	5.607
6.5	3.500	9.400	5.878
7.0	3.769	9.737	6.110

[a] Rate constants are expressed in units of $(\Delta c/\hbar)2/(n-1)\bar{v}(n-3)/(n-1)$.

TABLE 10
Same as Table 9, for $n = 3$, $\dot{\xi} = 0$

λ	$\langle v\sigma_a \rangle$	$\langle v\sigma_d \rangle$	$\langle v\sigma_p \rangle$
0	0	2.728	1.901
0.5	0.109	2.781	1.901
1.0	0.360	2.893	1.901
1.5	0.614	2.988	1.901
2.0	0.799	3.038	1.901
2.5	0.916	3.055	1.901
3.0	0.989	3.057	1.901
3.5	1.035	3.055	1.901
4.0	1.066	3.052	1.901
4.5	1.088	3.049	1.901
5.0	1.104	3.046	1.901
5.5	1.116	3.044	1.901
6.0	1.125	3.042	1.901
6.5	1.132	3.040	1.901
7.0	1.173	3.030	1.901

TABLE 11
Same as Table 9, for $n = 2$, $\xi = -1$

λ	$\langle v\sigma_a \rangle$	$\langle v\sigma_d \rangle$	$\langle v\sigma_p \rangle$
0	0	3.078	2.145
0.5	0.079	2.878	1.979
1.0	0.236	2.409	1.602
1.5	0.345	1.906	1.224
2.0	0.374	1.505	0.946
2.5	0.357	1.220	0.760
3.0	0.326	1.019	0.634
3.5	0.294	0.873	0.543
4.0	0.266	0.763	0.475
4.5	0.241	0.678	0.422
5.0	0.220	0.609	0.380
5.5	0.202	0.553	0.346
6.0	0.187	0.507	0.317
6.5	0.174	0.468	0.292
7.0	0.168	0.433	0.272

where t is the time varying from ∞ to $-\infty$ during the collision, and $t = 0$ at the moment of closest approach of the particles.

The atom A is considered to be at a degenerate energy level with nonzero electron angular moment $J = L$. In the course of the collision this energy level splits as a result of the interaction between A* and B. If we take the principal term in the multipole expansion of the potential energy of system A*B, then the interaction energy of the quadrupole moment Qe of the shell of the neutral atom A given by

$$Qe = \langle L, 0| \sum_{j=1}^{N} (3z_j^2 - r_j^2)/2 |L, 0\rangle \tag{5.8.2}$$

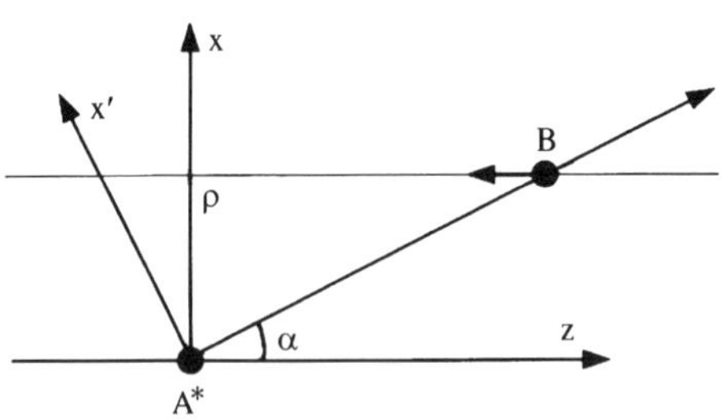

Figure 32. Scheme of collisional interaction of an excited atomic particle A* with a light charged projective B in the straight trajectory approximation.

with the charge Ze of the precipitating particle B, is given by the formula[62, 173]

$$V_{LL}^{(M)}(R) = E_{LM}(R) = E_L - ZQe^2 \frac{L(+1) - 3M^2}{L(L+1)R^3} \tag{5.8.3}$$

This represents an energy splitting of the sublevels of the excited atom A* in the field of particle B.

Here, E_L is the energy of the nonperturbed atom A*, $R = \sqrt{b^2 + v^2t^2}$ is the instantaneous distance between A* and B, and M is the projection of the electron moment of atom A* on the OZ axis, which changes direction during the collision.

The general quantum mechanical description of a system of colliding atomic particles is incorporated in the Schrödinger equation for the wave function Ψ of system A*B:

$$i\hbar \frac{\partial \Psi}{\partial t} = H\Psi \tag{5.8.4}$$

where $\hat{H}$ is the Hamiltonian of the system.

An approximate solution of the Schrödinger equation can be represented in the form of a superposition of eigenfunctions ψ_{LM} of atom A that refer to its energy sublevels (5.8.3) split in the field of particle B:

$$\Psi = \sum_{LM} C_{LM}(t)\psi_{LM}(r, t) \exp\left(-i\frac{E_L t}{\hbar}\right) \tag{5.8.5}$$

The expansion coefficients C_{LM} and basis functions $\psi_{LM}(r, t)$ in this expression are time-dependent, because they account for the instantaneous orientation of the OZ' axis connecting the colliding particles. The interaction in the studied case is given by a power law dependency, so the system of equations (4.5.14) for coefficients C_{LM} will take the following form after substitution of Eq. (5.8.5) into (5.8.4):

$$\begin{aligned} i\frac{\partial C_{LM}}{\partial \vartheta} &= \eta \sin\vartheta \frac{L(L+1) - 3M^2}{3L^2} C_{LM} \\ &+ \frac{i}{2}[(L+M+1)^{1/2}(L-M)^{1/2} \\ &\times C_{L,M+1} - (L-M+1)^{1/2}(L+M)^{1/2} C_{L,M-1}] \end{aligned} \tag{5.8.6}$$

where $M = -L, L+1, \ldots, L-1, L$.

It is important that these equations and their solutions depend on the characteristics of every collision event of particles A and B only via the dimensionless Massey parameter η given by

$$\eta = \frac{3ZQe^2L}{(L+1)\hbar v \rho^2} \tag{5.8.7}$$

In this case the system of differential equation (5.8.6) operates as a unitary matrix, transforming initial values of C_{LM} specified before a collision (at $\vartheta = 0$) to their final values after a collision (at $\vartheta = \pi$). This relation can be written in a form making use of elements of matrix S:

$$C_{LM}(\pi) = \sum_{M'} S_{M,M'}(\eta) C_{LM'}(0) \tag{5.8.8}$$

These elements depend only on the Massey parameter η, which characterizes every given collision.

It was shown in Chapter 3 that the polarizaton of light emitted by an ensemble of atoms A from the state with angular moment L is determined by the polarization moments of (3.4.11) of this ensemble:

$$\rho_q^{(k)} = \sum_{m'} (-1)^{L-M'} \begin{bmatrix} L & L & k \\ M & -M' & q \end{bmatrix} \rho_{MM'} \tag{5.8.9}$$

Here $\rho_{MM'}$ are the density matrix elements for this ensemble in the laboratory reference frame with fixed orientation of axes (OX, OY, OZ), independent of the directions of individual collisions.

The polarization moments $\rho_q^{(k)}$ of various ranks k are responsible for the different types of polarizaton of the light emitted by the atoms. In particular, linear polarization of the emission of an atomic ensemble consisting of atoms A undergoing depolarization collisions with particles B is governed by the second-rank polarization moment, or the alignment tensor $\rho_q^{(2)}$ with five components ($q = -2, -1, 0, 1, 2$).

As before, let us first consider the ultimate anisotropic monokinetic case when all collisions occur at the same relative velocity v along the same direction. In this case, all the vectors are parallel to the OZ axis of the laboratory frame of reference and the projections q of the polarization moments onto this axis are conserved under collisions. In terms of the relaxation matrix elements, the collisional relaxation of polarization moments is given by the equations

$$\frac{\partial \tilde{\rho}_q^{(k)}}{\partial t} = -\sum_{k'} \tilde{R}_q^{kk'} \tilde{\rho}_q^{k'} \tag{5.8.10}$$

where the sign $\tilde{}$ indicates that the corresponding value refers to the collision system of reference while matrix $\tilde{R}_q^{kk'}$ is defined by relation (3.8.3).

On substituting the integration variables in expression (3.8.3) and using relation (5.8.7) for the impact parameter b in terms of the Massey parameter η, one can derive an expression for the relaxation matrix as

$$R_q^{kk'} = \frac{3\pi Z Q e^2 L}{(L+1)\hbar} n_0 \sum_{MM'M_1,M_1'} (-1)^{M_1'-M'+1} \begin{bmatrix} L & L & k \\ M & -M' & q \end{bmatrix} \begin{bmatrix} L & L & k' \\ M_1 & -M_1' & q \end{bmatrix}.$$
$$\times \int \frac{d\eta}{\eta^2} [S_{MM_1}(B) S_{M'M_1'}(B) - \delta_{MM_1}\delta_{M'M_1'}] \qquad (5.8.11)$$

These matrix elements are essentially independent of the relative velocity v of colliding particles. This is a specific feature of the charge–quadrupole interaction, the power law for which Eq. (5.7.3) is reciprocal to the third power of the interatomic distance. Therefore, in this case it is insignificant whether or not the distribution of relative velocities for collisions of excited atoms with charged particles is monokinetic: the spread of velocities does not affect the collisional relaxation of the polarization moments. The possible spread of the distribution over the directions of the relative velocity vector v is more important.

In the general case of a partially anisotropic, axially symmetric distribution of relative velocities of colliding particles, $f(v, \theta)$, the relaxation matrix elements can be calculated if the real, partially anisotropic relaxation is considered as a superposition of the infinite number of ultimately anisotropic processes created by elementary beams inclined at different angles θ to the OZ axis of the laboratory frame of reference. An elementary process was examined before. The contribution of every elementary process could be accounted for by a corresponding statistical weight. This methodological approach leads to the following formula for the elements of the relaxation matrix for a partly anisotropic colloidal process:

$$\langle R_q^{kk'} \rangle = \sum_{q_1 q_2} \int D_{qq_1}^k(\varphi\theta\Psi) D_{q_2 q_1}^{k'*}(\varphi\theta\Psi) f(v,\theta) R_{q_1}^{kk'} v^2 \sin\theta d\theta d\varphi d\Psi dv \qquad (5.8.12)$$

This expression contains integrals of products of Wigner D-functions and the real velocity distribution function. The values of $\langle R_q^{kk'} \rangle$, diagonal with respect to k, are the probabilities of collisional destruction of polarization moments $\rho_q^{kk'}$. They are related to the rate constants $\langle v\sigma_q^k(v)\rangle$ of the corresponding process by

$$\langle R_q^{kk'} \rangle = n_0 \langle v\sigma_q^{(k)}(v) \rangle \qquad (5.8.13)$$

In the isotropic case of equiprobable directions of all collisions, the distribution function f is independent of θ and integral (5.8.12) can be easily calculated. The values $\langle R_q^{(kk')}\rangle$, which are independent of the projections of polarization moments q onto the symmetry axis and their magnitudes, can then be obtained by arithmetic averaging of the corresponding matrix elements in the case of ultimate anisotropic relaxation. These values could be used to calculate the collisional rate constants:

$$\gamma^k = \sum_q \frac{R_q^{kk}}{(2k+1)} \tag{5.8.14}$$

It is essential that for the anisotropic axially symmetric distribution of relative velocities, the projections q of the polarization moments are conserved during collisions, but the relaxation matrix $\langle R_q^{kk'}\rangle$ given by Eq. (5.8.12) may be nondiagonal with respect to the ranks of polarization moments k and k'. This may result in a specific mutual collisional transformation of polarization moments of different ranks with corresponding polarization effects in the linear emission spectrum. However, in the isotropic case, the relaxation matrix is diagonal with respect to both k and q values and, consequently, no polarization moment transformation and spectropolarimetric effects could occur.

Here, we are interested in the processes underlying the collisional destruction of alignment tensor $\rho_q^{(2)}$. According to calculations[60] based on numerical integration of system (5.8.6), the relaxation matrix elements describing the destruction of the alignment of the electronic state with $L = 2$ in the course of ultimate anisotropic collisions with charged particles are

$$R_0^{22} = 3.12s, \qquad R_1^{22} = 3.98s, \qquad R_2^{22} = 2.23s, \qquad \gamma^2 = 3.11s \tag{5.8.15}$$

where $s = 2|ZQ|e^2 n_0/\hbar$.

This theory was applied to experiments on the depolarization of radiation of He atoms resulting from transitions from the upper state 4^1D_2 in a low temperature beam plasma discharge, which could qualitatively simulate a number of astrophysical situations.[60] The quadrupole moment of this state was calculated under the assumption that the helium 4d-function is like the corresponding hydrogen one. The result is $Q = 144$ au.

The electron density $n^{(e)}$ in the plasma under study is composed of concentrations of thermal electrons n_1 with a virtually isotropic velocity direction distribution and rapid beam electrons n_2 flying from cathode to anode, exciting He atoms, and inducing the alignment of their electron angular momenta. Measurements of these concentrations showed that n_2 is 10–15% of the value of n_1. As the elements of the collisional relaxation matrix (5.8.10) with various q are rather close to each other (their difference from $\gamma^{(2)}$ is below 30%), an approximation of the completely isotropic relaxation was exploited[60] in order to interpret the experimental results.

As a consequence of the quasi-neutrality of the plasma, the concentrations of free electrons and ions are equal. As collisional relaxation of polarization moments under the charge–quadrupole interaction approximation is independent of the velocity of the colliding particles, the depolarizing interaction with ions is the same as for collisions with electrons. Within the framework of the applied approximation of isotropic collisions, the relaxatiòn of alignment is then determined by the value $\gamma^{(2)} = 2n^{(e)}\, 0.51 \times 10^{-5}\mathrm{s}^{1}$. This value is derived from system (5.8.15) at $Z = 1$ after substituting into it the value of the quadrupole moment of the electron shell for the excited He atom.

For example, the value of the rate constant for the destruction of the alignment of helium atoms in the 4^1D_2 state, caused by helium collisions with charged particles in the plasma beam discharge, was computed[60] to be $\langle v\sigma^{(2)}(v)\rangle = 0.51 \times 10^{-5}\ \mathrm{cm}^{-3}\mathrm{s}^{-1}$.

5.9. Concluding Remarks

Polarization of atomic ensembles resulting from slow atomic collisions, due to collisional relaxation in a different system of closely spaced levels, is a relatively new phenomenon which was observed and anlayzed theoretically in connection with the low temperature plasma of a gas discharge.[14–17] Diagnostic possibilities of this phenomenon are very promising, especially for sensing slow relative velocities of colliding partners and boundary gradients of parameters arising under anisotropic conditions in different natural and artificial ionized entities.

6

Theory of Charge Exchange Polarization

6.1. Introduction

The determination of polarization characteristics of the optical emission of an atomic ensemble excited by collisions in the presence of a charge exchange process is very significant owing to practical problems of plasma diagnostics under various natural and laboratory conditions. The ordering of momenta of colliding particles during a charge exchange process is described by the equation

$$\mathrm{A} + \mathrm{B}^{q+} \rightarrow \mathrm{A}^{m+} + \mathrm{B}^{(q-m)+} \tag{6.1.1}$$

and will be treated in this chapter.

An ion B colliding with an atom A, captures one or several electrons of the latter, thus changing its electric charge. The multielectron charge exchange process involves considerable computational difficulties typical of two-center multielectron problems, so we restrict ourselves in this chapter to the one-electron charge exchange process ($m = 1$).

Interest in astrophysics, atomic physics, theory of atomic collisions, and in laboratory collisional experiments with reactions of type (6.1.1) is growing continuously. Nevertheless, theoretical and experimental studies have been restricted, as a rule, to a determination of the charge exchange cross section. For example, the charge exchange cross sections of N^{q+} ions ($q = 2, 4, 5$) interacting with atoms of hydrogen and noble gases have been obtained.[129–132] Spectropolarimetric effects have usually been beyond the scope of these studies. However, spectropolarimetric methods and the polarization spectroscopy of the products of reaction (6.1.1) can provide valuable data not only on local parameters (such as the relative velocity of colliding partners), but also on anisotropies (temperature gradients, field orientation, etc.), representing the structure of an ionized entity. This main possibility is connected to the fact that the Stokes parameters of the emitted radiation are dependent on the quadrupole ordering of velocity vectors, or to the quadrupole kinetic characteristics in the velocity space of the colliding partners.

The amount of work conducted on spectropolarimetric studies of emission of recharged ions is quite small. We mention only work[77, 78] on the degree of polarization of emission of ioins Ne^{q+} after charge exchange with neutral sodium, and work[166, 167] on polarization observation in the spectra of Ar^{7+} and Kr^{7+} ions during charge exchange collisions. Experimental and theoretical analyses of polarization effects in the emission of Ar^{7+} ions resulting from charge exchange of Ar^{8+} ions on lithium have been conducted.[168, 181]

One of the first and obvious examples of a prospective practical implementation of spectropolarimetric methods during charge exchange for sensing an artificially sustained hot plasma in a generator could involve studies[67, 68] in which a beam of hydrogen atoms passing through a plasma is utilized as a diagnostic tool. Electrons from the neutral hydrogen atom were captured by the impurity atoms. By examining the emission spectrum of ions as a result of such a process, it is possible to probe the parameters of a high temperature plasma.

Several theoretical methods of computing the spectropolarimetric characteristics of recharged ions subjected to collisions will be treated in the present chapter. Theoretical approaches to an approximate solution of the three-body problem with Coulomb interaction will be considered.[70] Approximate methods, such as the distorted wave technique, will be analyzed.[85] Some examples of numerical calculations will be undertaken on the basis of the pseudolevel technique. A one-electron charge exchange process for $q \gg 1$ should be mentioned. It is an important case because, usually, the ionization potential of the atom A is much less than the ionization potential of the ion B^{q+} and the recharged $B^{(q-1)+}$ ion will be in the highly excited, near-Rydberg state, which allows one to neglect the fine structure of the recharged levels.

6.2. Theory of an Electron in the Field of Two Coulomb Centers

In order to outline the ways of posing and solving the problem of electron behavior in the field of two Coulomb centers, we shall follow the general theory published elsewhere.[82, 83]

An electron mass is much smaller than an atomic mass. Hence the Schrödinger equation describing the motion of an electron in the field of two Coulomb centers with electric charges eZ_a and eZ_b positioned distance R apart can be written as

$$\left(-\frac{\hbar^2}{2\mu}\Delta_r - \frac{e^2 Z_a}{r_a} - \frac{e^2 Z_b}{r_b}\right)\Psi(\mathbf{r}, R) = E\Psi(\mathbf{r}, R) \tag{6.2.1}$$

where r_a and r_b are the distances between the electron and the centers A and B, and $\mathbf{r}$ is the vector directed from the center of mass of the system of particles A, B to the electron location (Figure 33).

In terms of spheroidal coordinates

$$\xi = \frac{r_a + r_b}{R}, \qquad \eta = \frac{r_a + r_b}{R}, \qquad \varphi = \arctan\frac{y}{x} \tag{6.2.2}$$

the variables split so that a solution of this equation for the wave function $\Psi(\mathbf{r}, R)$ can be sought in the form

$$\Psi(\mathbf{r}, R) = N_{\kappa q m}(p, a, b)\Pi_{\kappa m}(p, a, \xi)\Xi_{qm}(p, b, \eta)\frac{e^{im\varphi}}{\sqrt{2\pi}} \tag{6.2.3}$$

Functions $\Pi_{km}(p, a, \xi)$ and $\Xi_{qm}(p, b, \eta)$, called the radial and angular spheroidal Coulomb functions, satisfy the equations

$$\frac{d}{d\xi}(\xi^2 - 1)\frac{d}{d\xi}\Pi_{\kappa m}(p, a, \xi) + \left[-\lambda_{\kappa m}^{(\xi)} - p^2(\xi^2 - 1) + a\xi - \frac{m^2}{\xi^2 - 1}\right]\Pi_{\kappa m}(p, a; \xi) = 0 \tag{6.2.4a}$$

and

$$\frac{d}{d\eta}(1 - \eta^2)\frac{d}{d\eta}\Xi_{qm}(p, b, \eta) + \left[\lambda_{qm}^{(\eta)} - p^2(1 - \eta^2) + b\eta - \frac{m^2}{1 - \eta^2}\right]\Xi_{gm}(p, b; \eta) = 0 \tag{6.2.4b}$$

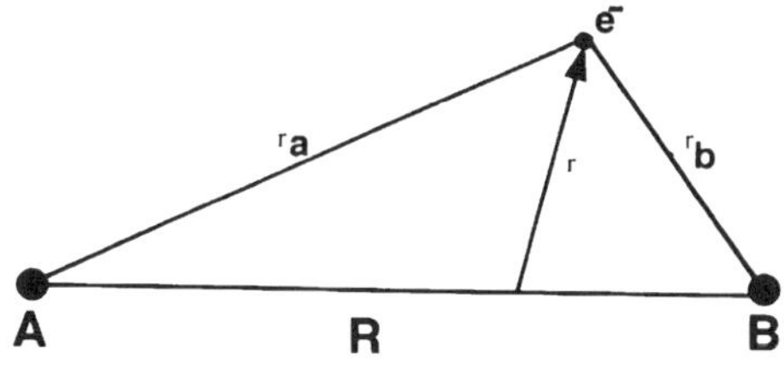

Figure 33. Geometrical scheme for the problem of electron motion in the field of two Coulomb centers.

where

$$p = \frac{R\sqrt{-2E\mu}}{2\hbar}, \qquad a = \frac{e^2 R\mu(Z_a + Z_b)}{\hbar^2}, \qquad b = \frac{e^2 R\mu(Z_b - Z_a)}{\hbar^2}$$

Here m is the magnetic quantum nuber, μ is a reduced mass, k and q define the numbers of zeros of functions $\Pi_{km}(p, a, \xi)$ and $\Xi_{qm}(p, b, \eta)$, while $\lambda_{km}^{(\xi)}$ and $\lambda_{qm}^{(\eta)}$ are the splitting constants.

Equations (6.2.4a) and (6.2.4b) are equivalent to the initial equation (6.7.1) only if the following condition is valid:

$$\lambda_{\kappa m}^{(\xi)}(p, a) = \lambda_{qm}^{(\eta)}(p, b) \tag{6.2.5}$$

This condition can be regarded as the equation for determining p. By solving Eq. (6.2.5) we obtain the value p, which depends, in general, on k, q, and m, finally yielding the energy spectrum.

$$E = E_{\kappa qm}(R, Z_a, Z_b) \tag{6.2.6}$$

Function $E_{kqm}(R)$ is called the energy term, or simply the term. The characteristics of the term do not involve the quantities k, q, m themselves but their combinations. Hence the quantity $n = k + q + m + 1 (n = 1, 2, 3, \ldots)$ is called the principal quantum number and the sum $l = q + m$ is called the orbital quantum number. The orbital quantum number varies from 0 to $n - 1$ and, finally, the magnetic (azimuthal) quantum number m takes values between 0 and l.

These terms are usually designated according to quantum chemistry notation. A particular term as a whole is identified by the quantum number n. Terms with $l = 0, 1, 2, \ldots$ are designated by letters $s, p, d, \ldots$ respectively, while greek letters $\sigma, \pi, \delta, \ldots$ denote the magnetic quantum numbers $m = 0, 1, 2, \ldots$. Hence for the term $1s\sigma$ the quantum numbers are $n = 1$, $l = 0$, $m = 0$ while for the term $2d\pi$ they are $n = 2$, $l = 2$, $m = 1$. When the collisional process is symmetric with respect to reflection in the plane, orthogonal to the line connecting centers A and B, and crossing it at the midpoint between A and B, the terms are characterized by the special parity with respect to reflection in this plane. Even terms are designated by index g (gerade) and odd terms by u (ungerade).

When $R \to 0$, the terms are reduced to the levels of the energy structure of a hydrogen-like atom with charge $Z_a + Z_b$:

$$E_{nlm}(R = 0) = -\frac{\mu e^4 (Z_a + Z_b)^2}{2\hbar^2 n^2} \tag{6.2.7}$$

The wave function in this case is given by

$$\Psi(\mathbf{r}, 0) = R_{nl}(r)Y_{lm}(\theta, \varphi)$$

When $R \to \infty$, the problem is split into two one-center problems and an electron possesses a probability to be found near the first or near the second center. In this respect all terms can be divided in two types: a eZ_a term, which is reduced at $R \to \infty$ to the level of a hydrogen-like system with charge Z_a; and a eZ_b term, transformed to the same system of levels of a hydrogen-like center B with charge eZ_b.

The spheroidal coordinates when $R \to \infty$ are reduced to parabolic coordinates. The wave function $\Psi(\mathbf{r}, R)$ for large internuclear distances is transferred into the hydrogen function, expressed in terms of parabolic coordinates and localized at the corresponding center.

For the eZ_a term

$$\Psi(\mathbf{r}, R) = f_{n_1 m}(\zeta) f_{n_2 m}(\tau) e^{im\varphi}$$

$$R(\xi - 1)|_{R\to\infty} \to \zeta, \qquad R(1 + \eta)|_{R\to\infty} \to \tau, \qquad n = n_1 + n_2 + m + 1 \tag{6.2.8a}$$

and for the eZ_b term

$$\Psi(\mathbf{r}, R) = f_{n'_1 m}(\zeta) f_{n'_2 m}(\tau) e^{im\varphi}$$

$$R(\xi - 1)|_{R\to\infty} \to \zeta, \qquad R(1 - \eta)|_{R\to\infty} \to \tau, \qquad n' = n'_1 + n'_2 + m + 1 \tag{6.2.8b}$$

where n and n' are principal quantum numbers related to the energy of the eZ_a and eZ_b terms when $R \to \infty$ by a relation similar to Eq. (6.2.7).

Sets of quantum numbers (n, n_1, n_2, m), (n', n'_1, n'_2, m) and (k, q, m) are connected to one another by so-called correlation diagrams. For radial Coulomb spheroidal functions we have simply

$$\kappa = n_1 = n'_1 \tag{6.2.9}$$

while for angular functions we have much more complicated relations.[82–84]

After these introductory remarks and initial definitions we can discuss an algorithm pertaining to the solution of Eqs. (6.2.4), in order to find an algorithm to compute the molecular terms of system AeB.

Suppose $W(x)$ is the solution of a boundary problem of type (6.2.4a) or (6.2.4b). The conditions at points $x = \infty, \pm 1$ can be obtained by substituting

$$W(x) = (1 - x^2)^{m/2} \exp[-p(1 \pm x)] f(x) \tag{6.2.10}$$

and seeking the solution of function $f(x)$ as an expansion in terms of a system of functions:

$$f(x) = \sum_{s=0}^{\infty} a_s(p, \lambda) U_s(x) \tag{6.2.11}$$

These functions can be chosen such that the recurrent relation for coefficients $a_s(p, \lambda)$ consists of three terms:

$$A_s(p, \lambda) a_{s+1} + B_s(p, \lambda) a_s + C_s(p, \lambda) a_{s-1} = 0 \tag{6.2.12}$$

By representing this system in matrix form, it is possible to obtain an infinite system of equations with which to determine coefficients a:

$$Q\mathbf{a} = \begin{pmatrix} B_0 & A_0 & 0 & 0 & . & . & . \\ C_1 & B_1 & A_1 & 0 & . & . & . \\ 0 & C_2 & B_2 & A_2 & . & . & . \\ . & . & . & . & . & . & . \\ . & . & . & . & . & . & . \\ . & . & . & . & . & . & . \\ . & . & . & . & . & . & . \end{pmatrix} \begin{pmatrix} a_0 \\ a_1 \\ a_2 \\ . \\ . \\ . \\ . \end{pmatrix} = 0 \tag{6.2.13}$$

The condition for a nonzero solution of this system is that the determinant of this system is zero:

$$\det Q = \phi(\lambda, p) = 0 \tag{6.2.14}$$

If the coefficients a_s satisfy the three-term recurrent condition, the determinant of matrix Q can be expressed in terms of chain fractions. The method of summation of these chain fractions are well-developed[(84)] and it enables one to find the roots of equation (6.2.14) rather easily.

Returning to Eq. (6.2.4), after obtaining the conditions at the particular points ∞ and ± 1, functions $\Pi_m(p, a, \xi)$ and $\Xi_{qm}(p, b, \eta)$ can be expanded in terms of the system of functions to provide the three-term recurrent condition for the expansion coefficients.[(82)] The coefficient p, related to the energy by relation

(6.2.5), and the splitting constant λ can be found by solving a system of type (6.2.14):

$$\phi^{(\xi)}(p, \lambda) = 0 \qquad \text{and} \qquad \phi^{(\eta)}(p, \lambda) = 0 \tag{6.2.15}$$

where $\phi^{(\xi)}$ and $\phi^{(\eta)}$ are determinants of infinite matrices of type (6.2.15) for Eqs. (6.2.4a) and (6.2.4b), respectively.

The energy as a function of parameters R, a, and b can be found by formula (6.2.15) after computing the value of p. In order to distinguish all the solutions of system (6.2.15), namely, those parameters p and λ which correspond to the special set of parameters (k, q, m), system (6.2.15) is solved initially subject to the condition $R \ll 1$ using asymptotics for $p_{kqm}(R)$ and $\lambda_{kqm}(R)$ as $R \to 0$. Then, by increasing R, it is possible to find p and λ at different values of R.[82]

In order to describe theoretically the charge exchange process, we select the origin of the system of reference at the center of mass of a system consisting of kernals A, B, and an electron. The Schrödinger equation takes the form

$$\left(-\frac{\hbar^2}{2\mu}\Delta_R - \frac{\hbar^2}{2M}\Delta_r - \frac{e^2 Z_a}{r_a} - \frac{e^2 Z_b}{r_b} + \frac{e^2 Z_a Z_b}{R}\right)\Psi(\mathbf{r}, \mathbf{R}) = E\Psi(\mathbf{r}, \mathbf{R}) \tag{6.2.16}$$

where

$$\mu = \frac{m_a m_b}{m_a + m_b} \qquad \text{and} \qquad M = \frac{m_e(m_a + m_b + m_e)}{m_a + m_b}$$

The wave function $\Psi(\mathbf{r}, \mathbf{R})$ is sought in the form of an expansion in terms of the eigenfunctions of Eq. (6.2.1):

$$\Psi(\mathbf{r}, \mathbf{R}) = \sum_{\kappa q m} f_{\kappa q m}(\mathbf{R})\psi_{\kappa q m}(\mathbf{r}, R) \tag{6.2.17}$$

Substitution of this equation into (6.2.16) yields the following system of equations with which to determine the coefficients $f_{kqm}(\mathbf{R})$:

$$\left[-\frac{\hbar^2}{2\mu}\Delta_R + \varepsilon_s(R) + \frac{e^2 Z_a Z_b}{R}\right] f_s(\mathbf{R}) - \frac{\hbar^2}{2\mu}\sum_{s'}[\langle\psi_s|\Delta_R|\psi_{s'}\rangle + 2\langle\psi_s|\nabla_R|\psi_{s'}\rangle\nabla_R] f_{s'}(\mathbf{R})$$
$$= E f_s(\mathbf{R}) \tag{6.2.18}$$

where s denotes the set of quantum numbers (k, q, m), angle brackets refer to averaging over the electron coordinates, and $\varepsilon_s(R)$ is the eigenvalue of (5.2.1) for the electron problem.

The condition

$$\nabla_R\langle\psi_1|\nabla_R\psi_2\rangle = \langle\nabla_R\psi_1|\nabla_R\psi_2\rangle + \langle\psi_1|\Delta_R\psi_2\rangle \tag{6.2.19}$$

applied to Eq. (6.2.18) gives

$$\left(-\frac{\hbar^2}{2\mu}\Delta_R + \varepsilon_s + \frac{e^2 Z_a Z_b}{R}\right) f_s - \frac{\hbar^2}{2\mu} \times \sum_{s'}[-\langle\nabla_R\psi_s|\nabla_R\psi_{s'}\rangle + \nabla_R\langle\psi_s|\nabla_R\psi_{s'}\rangle - \langle\psi_s|\nabla_R\psi_{s'}\rangle\nabla_R] f_{s'} = E f_s \tag{6.2.20}$$

Separation of the angular variables in this equation allows it to be transformed to the form

$$\left[-\frac{1}{R}\frac{\partial}{\partial R}R^2\frac{\partial}{\partial R} + K_s^2 + \frac{\alpha}{R} + \frac{L(L+1)}{R^2}\right] G_s^{(L)}(R) + \sum_{s'L'} W_{ss'}^{(LL')} G_{s'}^{(L')}(R) = 0 \tag{6.2.21}$$

where L is the total angular moment and $G_s^{(L)}$ is the radial part of function f_s.

We further assume that the electron was located before the collision at the center A and the system was described by the eZ_a term, while after the collision the electron is located at the center B and the system is in a state described by the eZ_b term. Using the correlation diagram[82] and calculating the matrix elements $W_{SS'}^{(LL')}$, it is possible to determine an indispensable number of functions $f_S(\mathbf{R})$ by resolving Eq. (6.2.20) and consequently composing the products $f_S(\mathbf{R}) f_{S'}(\mathbf{R})|_{R=\infty}$. It is finally possible to derive the cross section for the creation of the momenta ordering of corresponding rank.

This method of calculating the charge exchange cross section of OIX, CVII, and NeXI ions on hydrogen atom was exploited elsewhere[79–81, 174] All the terms correlated at high internuclear distances with a two-center system, one center being the hydrogen ion H^+ and the other the term corresponding to an absence of recharging.

To conclude this section, attention should be drawn to an important point connected with the representation of the probing function as a sum of type (6.2.17). To simply illustrate this point, let us use the two-state approximation

$$\Psi(\mathbf{r}, \mathbf{R}) = f_\alpha(\mathbf{R})\psi_\alpha(\mathbf{r}, \mathbf{R}) + f_\beta(\mathbf{R})\psi_\beta(\mathbf{r}, \mathbf{R}) \tag{6.2.22}$$

where α and β characterize a set of quantum numbers k, q, and m referring to the selected terms.

Suppose further that term α belongs to the eZ_a type. This means that as $R \to \infty$, function $\psi_\alpha(\mathbf{r}, R)$ is transformed to wave function $\psi_\alpha(\mathbf{r}_a)$ describing an electron localized at center A in quantum state $|\alpha\rangle$. Also, term β belongs to the

eZ_b type, so as $R \to \infty$ it describes an electron positioned at center B in state β with wave function $\psi_\beta(\mathbf{r}_B)$.

For large distances R, the first term of Eq. (6.2.22) corresponds to a system consisting of an electron positioned at center A and the center B with impulse k_b. So as $R \to \infty$ the first term has asymptotic

$$\psi_\alpha(r_\alpha)\exp[i(\boldsymbol{\kappa}_b, \mathbf{r}_b)] \tag{6.2.23}$$

where vectors R_b (see Figure 33) can be represented as

$$\begin{aligned}\mathbf{R}_b &= \mathbf{R} - \frac{m_e}{m_a + m_e}\mathbf{r}_a \sim \mathbf{R} - \frac{m_e}{m_a}\mathbf{r}_a \\ &= \left[1 - \frac{m_e m_b}{(m_a + m_b)m_a}\right]\mathbf{R} - \frac{m_e}{m_a}\mathbf{r}\end{aligned} \tag{6.2.24}$$

and, in general, is not equal to R.

Therefore the right asymptotic behavior of Eq. (6.2.22) is possible only if the first term of the expression is multiplied by the factor

$$\exp[i(k_b, \frac{m_e}{m_a}\mathbf{r})]$$

and the second term by the factor

$$\exp\left[i(k_a, \frac{m_e}{m_a}\mathbf{r})\right]$$

Expression (6.2.22) then assumes the form

$$\Psi(\mathbf{r}, \mathbf{R}) = \tilde{f}_\alpha(\mathbf{R})\psi_\alpha(\mathbf{r}, R)\exp\left[i\left(\boldsymbol{\kappa}_b, \frac{m_e}{m_a}\mathbf{r}\right)\right] + \tilde{f}_\beta(\mathbf{R})\psi_\beta(\mathbf{r}, R)\exp\left[i\left(\boldsymbol{\kappa}_a, \frac{m_e}{m_b}\mathbf{r}\right)\right] \tag{6.2.25}$$

where functions $\tilde{f}_\alpha(\mathbf{R})$ and $\tilde{f}_\beta(\mathbf{R})$ differ from functions $f_\alpha(\mathbf{R})$ and $f_\beta(\mathbf{R})$ by the R-dependent phase term originating from Eq. (6.2.24).

The exponential factors in Eq. (6.2.25) provide rather complicated computations, but neglect of these exponents could lead to large errors at high velocities.

6.3. Distorted Wave Approximation Theory

In a frame of reference related to the center of mass, the Schrödinger equation for the system $A + B + e$ consisting of two centers A and B and an electron, initially connected to center A, can be written in terms of variables $\mathbf{R}_b$ and $\mathbf{R}_a$ as

$$\left[-\frac{\hbar^2}{2\mu_a}\Delta_{r_a} - \frac{\hbar^2}{2M_b}\Delta_{R_b} + V_a(r_a) + V_b(r_b) + U(R)\right]\Psi = E\Psi \tag{6.3.1a}$$

where

$$\mu_a = \frac{m_e m_a}{m_e + m_a} \quad \text{and} \quad M_b = \frac{m_b(m_a + m_b + m_e)}{m_a + m_e}$$

Also, Δ_{r_a} and Δ_{R_b} are Laplace operators operating on the coordinates of vectors $\mathbf{r}_a$ and R_b, respectively (see Figure 34), while functions $V_a(r_a)$ and $V_b(r_b)$ and $U(R)$ describe, respectively, the interaction of cores A and B with the electron and the interaction of the cores with each other.

In terms of variables $\mathbf{R}_a$ and $\mathbf{r}_b$ it is possible to write the equation analogously to Eq. (6.3.1a):

$$\left[-\frac{\hbar_2}{2\mu_b}\Delta_{r_b} - \frac{\hbar^2}{2M_a}\Delta_{R_a} + V_a(r_a) + V_b(r_b) + U(R)\right]\Psi = E\Psi \tag{6.3.1b}$$

where

$$\mu_b = \frac{m_e m_b}{m_e + m_b} \quad \text{and} \quad M_a = \frac{m_a(m_a + m_b + m_e)}{m_b + m_e}$$

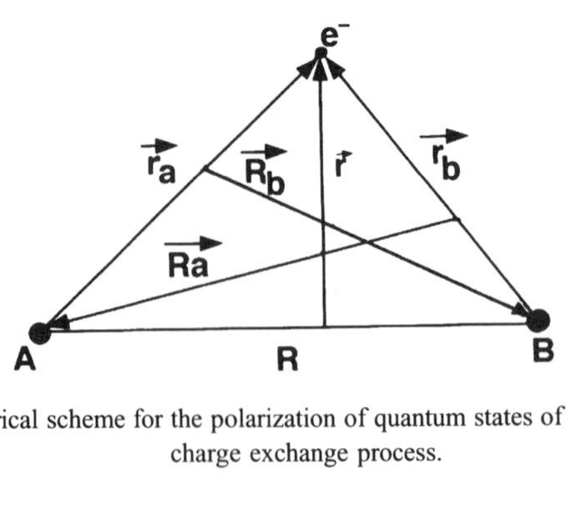

Figure 34. Geometrical scheme for the polarization of quantum states of atomic particles under the charge exchange process.

The wave function of system $\mathrm{A}+\mathrm{B}+e$ is sought in the following form:

$$\Psi=\sum_{\alpha}\phi_{\alpha}(\mathbf{R}_{a})\varphi_{\alpha}(\mathbf{r}_{a})+\sum_{\beta}\phi_{\beta}(\mathbf{R}_{b})\varphi_{\beta}(\mathbf{r}_{b}) \tag{6.3.2}$$

where functions $\varphi_{\alpha}(\mathbf{r}_{a})$ and $\varphi_{\beta}(\mathbf{r}_{b})$ refer to systems $(\mathrm{A}+e)$ and $(\mathrm{B}+e)$ in states $|\alpha\rangle$ and $|\beta\rangle$ with energy ε_{α} and ε_{β}, respectively:

$$\begin{aligned}\left[-\frac{\hbar^{2}}{2\mu_{a}}\Delta_{r_{a}}+V_{a}(r_{a})-\varepsilon_{\alpha}\right]\varphi_{\alpha}(r_{a})=0\\ \left[-\frac{\hbar^{2}}{2\mu_{b}}\Delta_{r_{b}}+V_{b}(r_{b})-\varepsilon_{\beta}\right]\varphi_{\beta}(r_{b})=0\end{aligned} \tag{6.3.3}$$

The coefficients $\phi_{\alpha}(\mathbf{R}_{b})$ and $\phi_{\beta}(\mathbf{R}_{a})$ characterize the probability for an electron to stay at center A or to undergo transition to center B as a result of a collision.

Suppose the electron before the collision was at center A in state $|\alpha_{0}\rangle$. Substitution of (6.3.2) into (6.3.1) yields a system of linked equations:

$$\begin{aligned}&\left[-\frac{\hbar^{2}}{2M_{b}}(\Delta_{R_{b}}+_{K}\beta^{2})+\langle\varphi_{\alpha}|V_{b}(r_{b})+U(R)|\varphi_{\alpha}\rangle\right]\phi_{\alpha}(\mathbf{R}_{b})\\ &\quad+\sum_{\beta}\langle\varphi_{a}|-\frac{\hbar^{2}}{2M_{a}}(\Delta_{R_{a}}+K_{\alpha}^{2})+V_{a}(r_{a})+U(R)|\varphi_{\beta}\rangle\phi_{\beta}(\mathbf{R}_{a})=0\end{aligned} \tag{6.3.4a}$$

$$\begin{aligned}&\left[-\frac{\hbar^{2}}{2M_{a}}(\Delta_{R_{a}}+K_{\alpha}^{2})+\langle\varphi_{\beta}|V_{a}(r_{a})+U(R)|\varphi_{\beta}\rangle\right]\phi_{\beta}(\mathbf{R}_{a})\\ &\quad+\sum_{\alpha}\langle\varphi_{\beta}|-\frac{\hbar^{2}}{2M_{b}}(\Delta_{R_{b}}+K_{\beta}^{2})+V_{b}(r_{b})+U(R)|\varphi_{\alpha}\rangle\phi_{\alpha}(\mathbf{R}_{b})=0,\end{aligned} \tag{6.3.4b}$$

An approximate solution of these equations can be derived by means of the distorted wave technique.[175] Starting with the first equation, we see that the second term is determined by the overlapping integral of functions φ_{α} and φ_{β} and consequently it decreases exponentially with growth of R. Therefore, neglecting the second term in Eq. (6.3.4a), function $\phi_{\alpha}(\mathbf{R}_{b})$ satisfies

$$\left[-\frac{\hbar^{2}}{2M_{b}}(\Delta_{R_{b}}+K_{\beta}^{2})+\langle\varphi_{\alpha}|V_{\beta}(r_{b})+U(R)|\varphi_{\alpha}\rangle\right]\phi_{\alpha}(\mathbf{R}_{b})=0 \tag{6.3.5}$$

and the solution of this equation should possess asymptotic behavior

$$\phi_\alpha(\mathbf{R}_b) \sim \delta_{\alpha\alpha_0} e^{(\mathbf{K}_b,\mathbf{R}_b)} + f_{\alpha\alpha_0}(\theta_b \varphi_b) \frac{e^{iK_b R_b}}{R_b} \tag{6.3.6}$$

After this solution has been obtained, it is substituted into the right hand side of Eq. (6.3.4b) to yield the solution $\phi_\beta(R_a)$ with asymptotic behavior

$$\phi_\beta(\mathbf{R}_a) \sim f_{\alpha_0\beta}(\theta_a \varphi_a) \frac{e^{iK_a R_a}}{R_a} \tag{6.3.7}$$

For a known quantity $f_{\alpha_0\beta}$, it is possible to compose the combination $f_{\alpha_0\beta} f^*_{\alpha_0\beta'}$ and, after averaging over the angular variables, one finally obtains the cross section of ordering of momenta.

6.4. Pseudolevel Technique for Computing the Polarization of Ions

Suppose a heavy atom A with core charge Z_a and one valent electron collides with an ion B, possessing charge Z_b. As a result of the collision, particle A, moving along the linear trajectory, looses the valent electron, which is transferred to center B. This process can be formally described as

$$\mathrm{A} + \mathrm{B}^{Z_b^+} \rightarrow \mathrm{A}^+ + \mathrm{B}^{(Z_b-1)^+} \tag{6.4.1}$$

In order to compute the cross sections of collisional ordering of momenta of the excited states of an ensemble of recharged ions $\mathrm{B}^{(Z_b-1)+}$, we apply the pseudolevel technique, which was first proposed by Vainstein *et al.*[65] and later exploited for computing spectroscopic observables for recharged ion emission.[126,168,181] The idea behind this method is that the wave function of the electron in a two-center field is constructed in a form of an expansion in terms of the full system of functions of the $\mathrm{B}^{(Z_b-1)+}$ ion, $\psi_m(\mathbf{r}_b)$, as follows:

$$\Psi(\mathbf{r}_b t) = \sum_m C_m(t) \psi_m(\mathbf{r}_b) \exp(i\varepsilon_m^{(b)} t) \tag{6.4.2}$$

where m is the set of quantum numbers specifying the state of ion $\mathrm{B}^{(Z_b-1)+}$, and $\varepsilon_m^{(b)} = E_m^{(b)}/\hbar$.

When $t \to -\infty$, function (6.4.2) transfers to the wave function of the electron at the center A:

$$\psi_a(\mathbf{r}_a) \exp\left(-i\frac{E_0^{(a)}t}{\hbar}\right)$$

A consequence of this condition is the following asymptotic behavior of the coefficients:

$$C_m(t)|_{t\to-\infty} \sim \exp(-i\omega_m t)S_m(t) \tag{6.4.3}$$

where

$$S_m(t) = \int d\mathbf{r}\psi_m\left(\mathbf{r} + \frac{m_a}{m_a + m_b}\mathbf{R}\right) \exp\left[im\frac{(\mathbf{v}, \mathbf{r})}{\hbar}\right]\psi_a\left(\mathbf{r} - \frac{m_b}{m_a + m_b}\mathbf{R}\right) \tag{6.4.4}$$

and

$$\omega_m = \varepsilon_m^{(b)} - \frac{E_0^{(a)}}{\hbar}$$

$E_0^{(a)}$ being the total energy of the electron at center A, including the kinetic energy associated with the relative motion A–B.

Let us choose a particular state with $m = n$ and rewrite expression (6.4.2) in the form

$$\Psi(\mathbf{r}_b, t) = \sum_{m\neq n} C_m(t)\psi_m(\mathbf{r}_b) \exp(-i\varepsilon_m^{(b)}t) + C_n(t)\psi_n(\mathbf{r}_b) \exp(-i\varepsilon_n^{(b)}t) \tag{6.4.5}$$

If the first term of Eq. (6.4.5) is expressed as the product

$$b_n(t)\phi_n(\mathbf{r}_b, \mathbf{R}) \exp(-\varepsilon_a t), \qquad \langle\phi_n|\phi_n\rangle = 1$$

then this expression can be transformed to

$$\Psi(\mathbf{r}_b, t) = b_n(t)\phi_n(\mathbf{r}_b, \mathbf{R}) \exp(-i\varepsilon_a t) + C_n(t)\psi_n(\mathbf{r}_b) \exp(-i\varepsilon_n^{(b)}t) \tag{6.4.6}$$

This means that a multilevel problem is reduced to a two-level problem or to a particular level $|n\rangle$ with wave function $\psi_n(\mathbf{r}_b)$ and a pseudolevel with wave

Expression (6.4.13) is written in a reference system connected to an ion beam (collision system), while calculation of the matrix elements is facilitated within a system centered at the argon ion with OZ axis directed to the lithium atom moving with constant velocity v near the argon ion along the linear trajectory characterized by impact parameter s. Transition to this reference system is implemented with the help of three-dimensional rotation matrices. With this transformation, matrix elements S_n, such as $\langle\psi_a|V_a|\psi_n\rangle$, $\langle\psi_a|V_b|\psi_a\rangle$, and $\langle\psi_n|V_a|\psi_a\rangle$, have been calculated analytically in elliptic coordinates with the aid of the REDUCE language. The matrix element $\langle\psi_n|V_a|\psi_n\rangle$ was calculated, making use of the expansion of

$$\frac{1}{r_a} = \frac{1}{|\mathbf{R} - \mathbf{r}_b|}$$

in terms of Legendre polynomials.

An additional relation for spherical functions yields

$$\langle\psi_n|V_a|\psi_n\rangle = (-1)^m N_{nl}^2 \frac{2l+1}{\sqrt{4\pi}} \sum_L \begin{bmatrix} l & l & L \\ m & -m & 0 \end{bmatrix} \begin{bmatrix} l & l & L \\ 0 & 0 & 0 \end{bmatrix} \cdot \frac{P_L(\cos\vartheta)}{(2L+1)} \int_0^\infty B_L(r_b, R) R_{nl}^2(r_b) r_b^2 dr_b \qquad (6.4.14)$$

where

$$B_L(r, r') = \frac{r_<^L}{r_<^{L+1}}$$

Quantities $\tau>$ and $\tau<$ are values bigger and smaller than r_b and R; R_{nl} is the radial part of Eq. (6.4.13) while ϑ is the angle between the OZ axis of the subsequent system with respect to the collision system.

For fixed values of n and l, computation of the last expression can be easily undertaken analytically with the help of the REDUCE language by neglecting the transmitted impulse in the course of calculating $S_m(r)$ in expression (6.4.4). This means that the overlapping integral of functions $\psi_m(\mathbf{r}_b)$ and $\psi_a(\mathbf{r}_b - \mathbf{R})$ was taken as the quantity $S_m(t)$.

As a result of solving system (6.4.10) $l+1$ times, a set of coefficients C_m and hence combinations $C_m C_m^*$ were obtained. These combinations were then averaged over the impact parameter b to yield the following expression for the polarization moment under stationary conditions which characterize an ordering of the angular momenta induced in the ensemble of ions ArVIII as a result of charge exchange collisions:

$$\rho_0^{\kappa}(l) = \frac{n_0 v \sigma_0^k(l, v)}{\gamma_0} \tag{6.4.15}$$

Here, the cross section of angular momenta ordering $\sigma_0^k(l, v)$ is determined by

$$\sigma_0^k(l, v) = \sum_m (-1)^{l-m} \begin{bmatrix} l & l & k \\ m & -m & 0 \end{bmatrix} \int_0^\infty b\, db C_m C_m^* \tag{6.4.16}$$

where n_0, v, and γ are the lithium atom number density, relative velocity, and radiative damping constant, respectively.

Equation (3.6.7) can be employed to compute the following expression for the degree of polarization of radiation of a recharged ion, corresponding to the transition $l \rightarrow l_0$:

$$\frac{I_z - I_y}{I_z + I_y} = \frac{3\rho_0^2(l)}{2\sqrt{2}t\rho_0^0(l) - \rho_0^2(l)}$$

where t is the ratio of $6j$ symbols:

$$t = \frac{\begin{Bmatrix} 1 & 1 & 0 \\ l & l & l_0 \end{Bmatrix}}{\begin{Bmatrix} 1 & 1 & 2 \\ l & l & l_0 \end{Bmatrix}}$$

Numerical calculations were conducted for the particular case of transitions $8k \rightarrow 7i$, $8g \rightarrow 7f$, and $8p \rightarrow 7d$ of ArVIII ions resulting from charge exchange collisions in the energy range $60\,\text{keV} < E < 120\,\text{keV}$. Figure 35 represents the dependence of the degree of polarization of recharged ions ArVIII for the above-mentioned transitions. Solid and dashed lines describe the experimental and theoretical dependencies, respectively. Taking into consideration that the experimental error was around 40%, agreement between the theoretical and experimental results can be considered rather satisfactory. Experimental data (Figure 35) show that the degree of polarization for transitions from upper levels, having higher momenta, is considerably higher than for upper levels with small l. Theoretical computation does not provide the same monotonic decrease in the degree of polarization.

The general characteristics of the velocity dependence of the cross section were examined in detail for a light projectile in collision with a neutral target.[(69–72)] For example, in the case of proton–hydrogen atom collisions $\sigma \sim v^{-12}$, so the

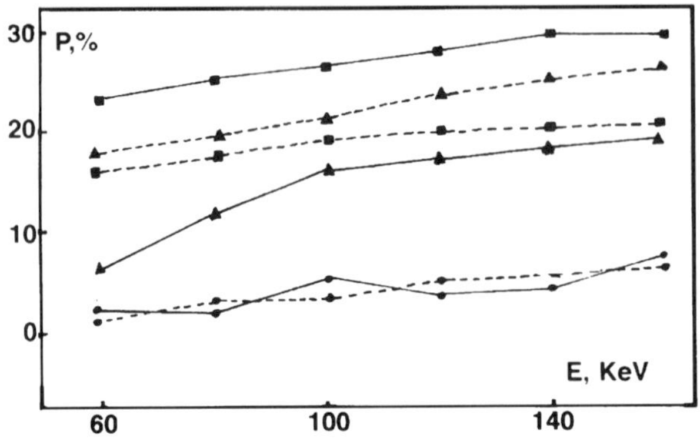

Figure 35. Experimental dependencies for the degree of polarization of ArVIII ions after the charge exchange process for transitions $8k \to 7i$ (■), $8g \to 7f$ (▲) and $8p \to 7d$ (●) on the energy of the incident ionic beam.[168]

charge exchange cross section decreases more rapidly than for excitation and ionization.[73]

A generalization of the Brinckman–Kramers formula for velocities $v \sim v_0$ was proposed,[73] as well as a simplified formula for estimating the charge exchange cross section for a stripped ion and atom. When the target has several atomic shells, the possibility of electron capture from the inner shell arises with increase in velocity.[65, 73] This capture is a resonance process. It means that electron capture from the inner shell has maximal probability when the ionization potential of that shell is equal to the kinetic energy of the projectile. Therefore, as the total cross section is the sum of contributions from all the shells, the charge exchange cross section decreases slowly as $\sigma \sim E^{-k}$ $(2.5 < k < 6)$, and the case $k = 2.5$ is realized for very heavy targets only.

Another method for computing the cross section for fast particle charge exchange was proposed[74] where approximation similar to the first Born approximation was used. Within this approximation, function ψ_b in the charge exchange amplitude was replaced by its asymptotic expression. The relations derived in this work first give identical results with and without approximations, and second, numerical calculations of the charge exchange for a completely stripped ion on an atom provided very good agreement with experiment. Analogous results, obtained with the Brinckman–Kramers formula, gave a small discrepancy with experiment. Moreover, an accurate comparison of calculations of charge exchange cross sections using the Brinckman–Kramers formula and the first Born approximation shows that these two techniques give close results only in the case of charge exchange between a hydrogen atom and a proton $(Z_a = Z_b = 1)$. The very different results of these two methods of calculations is typical for more complicated cases.

6.5. Concluding Remarks

Despite the fact that charge exchange collisions have been analyzed in many collisional, optical spectroscopic experiments and by different theoretical methods, polarization of atomic ensembles created by the charge exchange process between two atomic particles has not been studied well until now. A very effective pseudolevel technique for the theoretical description of polarization effects was presented in this chapter. The method could be used in many problems of spectropolarimetric sensing, especially for solar and terrestrial atmospheres and hot plasma.

7

Spectropolarimeters for Solar Problems

7.1. Introduction

The observational spectropolarimetric techniques discussed in this chapter and the features of the particular instrumental schemes are attributed to pragmatic research objectives of observation, analysis, and measurement of collision-induced linear polarization of different solar atmosphere emissions. The optically active ionized entities in the solar atmosphere are analyzed in this book as practical examples and real objects of impact spectropolarimetric sensing. The main idea of this chapter is not to provide a thorough and rigorous description of astrophysical instruments and practical spectropolarimetric techniques for solar atmosphere sensing, but rather to illustrate the features specific for polarization measurements and how they differ from well-known intensity spectroscopic techniques.

Solar flares are one of the brightest emitting entities in the solar atmosphere with highly expressed structural features, such as the presence of beams of precipitating particles, temperature gradients, and a complicated magnetic field topology. These flares represent a complicated astrophysical phenomenon involving many physical effects and having also an important geophysical significance in regard to the state of the upper atmosphere of the Earth. This explains the profound interest of astrophysicists in the comprehensive study of this phenomenon. The appearance and development of solar flares are intimately connected with the configuration, orientation, and dynamics of the magnetic field in solar active regions. Therefore, the wide range of observational methods includes the study of the magnetic field in the flare plasma and they are also based on spectropolarimetry. One of the most interesting optical effects of a solar flare is the pronounced linear polarization of emission lines. The polarization of the emission of spectral lines contains much new information, so the main observations and quantitative spectropolarimetric estimates of solar flare emissions have been performed independently by leading research groups in a number of observatories during solar activity events.

Presented here are instrumental techniques for linear polarization observation developed by four leading groups in this field at Meudon Observatory (France), Crimean Astrophysical Observatory (Ukraine), Astrophysical Observatories of the Institute of Solar and Terrestrial Physics (Russia), and Sacramento Peak Observatory (USA). Observations of the linear polarization of emissions of the solar atmosphere have been undertaken by these groups on different spectral lines and by means of various instrumental techniques. They provide reliable impact spectropolarimetric data in the sense that they reflect the collisional excitation of a medium and the polarization of atomic particles in the course of different astrophysical events.

Unfortunately, in almost all observations the influence of the instrumental polarization was not fully eliminated, although in all studies attempts were made to assess or reduce maximally the influence of this effect. At the same time much direct observational evidence (such as a change in the degree and azimuth of the plane of polarization in the process of observations, the rapid disappearance or appearance of polarization and the different degrees of polarization for locations adjacent to the object under study) confirms the reliability of the linear polarization observations during solar flares and other nonthermal entities in the solar atmosphere.

Historically, only the group from the Meudon Observatory (France), from the very beginning, focused specially on the search, observation, and investigation of linear polarization associated with collisional processes during solar flares.[(221)] The most extensive observations of linear polarization in emission features were undertaken by the Crimean group over a long period of time. They systematically and carefully put ever-increasing effort into obtaining reliable evidence for linearly polarized radiation in emission features, collecting the spectropolarimetric information without putting forward any specific concept as to the origin of the polarization of an ensemble of emitting particles in the solar atmosphere. The observational technique elaborated by the Americans is interesting and unique in the sense that their research goal was to verify the extremely high degree of polarization in some moustaches observed by the Crimean authors. The aim of the spectropolarimetric observations initiated by the Irkutsk group was to study the turbulent Stark effect. However, the form of the polarization profiles of hydrogen lines showed clearly the linear polarization at the line center, which led the observers to study collision polarization effects.

Regardless of the initial goal of observations, the type of astronomical instrument, the polarization optics, and the recording technique, all spectropolarimetric observations described in this chapter clearly indicate the impact nature of spectral line polarization during different event in the solar atmosphere.

The sufficient level of accuracy of the spectropolarimetric measurements achieved by all the observational techniques allows us to use the spectropolarimetric data for sensing the energy transport by accelerated particles precipitating to the studied locations from the primary source of energy.

7.2. Spectropolarimetric Techniques of Meudon Observatory (France)

Impact polarization during solar flares was studied[176] on *SI* line 1436.9 Å. The selection of this line was due to the following reasons. First, this chromospheric line with rather high excitation potential (8.63 eV) is mostly selective to energetic electrons. Second, the polarization at threshold is relatively high, so the 90° polarization fraction P_{90} with respect to the exciting beam could theoretically attain 100%.[102] This means that the degree of polarization could only be slightly dependent on the direction of observation.

If β is the angle between the beam direction and the line of sight, the degree of polarization is given by[6, 7, 178]

$$P = \frac{P_{90} \sin^2 \beta}{1 - P_{90} \cos^2 \beta}$$

and completely polarized light ($P = 1$) corresponds to $P_{90} = 1$, regardless of the value of β.

This line does not show any self-absorption in flares and could reasonably be assumed to be optically thin. Hence the line polarization will not be affected by scattering. In fact, the impact polarization of *SI* 1439.6 Å resulting from collisional excitation by a beam is not known accurately. Experimental results indicate that the degree of polarization near a level threshold is lower than that predicted by simple theoretical arguments. However, it is not necessary to know accurately the degree of impact polarization by a beam in order to reach at the first step some qualitative conclusions on anisotropy in the energetic electron velocity distribution function. The most important parameter to derive from this qualitative estimate is not the degree of polarization, but rather the azimuth or orientation of the polarization plane, which allows one to deduce the symmetry of anisotropy of the excitation mechanism and subsequently to come to a conclusion as to the physical nature of the excitation process.

In the course of observations, the Ultraviolet Spectrometer and Polarimeter (UVSP) on the Solar Maximum Mission (SMM) was used. This instrument has been thoroughly described elsewhere[177] and details of the polarimeter have been presented.[180] The telescope is an aplanatic Gregorian which focuses the Sun on the entrance slit of an Ebert–Fastie spectrometer. Magnesium fluoride wave plates were inserted behind the entrance slit and rotated in steps of 22.5°. A wave plate with a maximum retardance of 286° in the region from 1300 Å to 1600 Å was used. A linear polarizer following the rotating wave plate was required to serve as analyzer. The linear polarizer for the wave plate was the spectrometer grating.

The modulation of intensity as a function of the angle Θ of the fast axis of the wave plate resulting from the rotation of the plate is given by

$$I_M(\Theta) = I + PQ(\tilde{a} + \tilde{b}\cos 4\Theta) - P\tilde{b}U \sin 4\Theta + PV \sin\delta \sin 2\Theta \qquad (7.2.1)$$

where

$$P = \frac{I_y - I_x}{I_y + I_x}$$

is the polarization of the linear polarizer, δ is the retardation angle of the wave plate,

$$\tilde{a} = \frac{1 + \cos\delta}{2}, \qquad \tilde{b} = \frac{1 - \cos\delta}{2}$$

and I, U, Q, and V are Stokes parameters defined in the usual way relative to a set of X–Y axes.

The phase angle of the linear polarization was measured counterclockwise and the rotation angle Θ was measured clockwise when looking toward the source. Fourier analysis was used to evaluate the Stokes parameters. The coefficients of the component modulated at the 4ω ($\omega = d\Theta/dt$) frequency gave U and Q.

A part of the solar image was rastered by moving the secondary mirror of the Gregorian telescope. An enstrance slit of size $10'' \times 10''$ was used. The field of view of $50'' \times 50''$, including 25 pixels, was covered. An exit slit of width 0.5 Å was used in order to integrate the full line profile. Unlike the Zeeman effect, impact polarization affects the integrated profile. At each raster pixel, the intensity I_M was sampled at each of the 16 wave plate positions Θ, giving the polargram.

The gate time was 0.32 s and the time needed to obtain a $50'' \times 50''$ polarogram was 148 s. Up to 23 polarograms could be obtained per orbit. All these polarograms were centered at the brightest point of the active region under study. The brightest point was established at the beginning of each orbit using an experiment mode which located the brightest observed point in a $2' \times 2'$ raster.

In each polarogram point the Stokes parameters I, U, Q, and V were obtained by Fourier analysis of the signal. Equation (7.2.1) was corrected for the time dependence of I, assuming a linear variation in I with time and a constant value of Q/I and U/I. The function

$$I[\Theta(t)]\left[1 + \frac{PaQ}{I[\Theta(t)]}\right]$$

was obtained by regression analysis using the least-squares principle. The following function was defined as a result:

$$S_i = \frac{i_M(\Theta) - I[\Theta(t)][1 + PaQ/I(\Theta(t))]}{I[\Theta(t)][1 + PaQ/I(\Theta(t))]} \tag{7.2.2}$$

$$\cong Pb\frac{Q}{I}\cos 4\Theta - Pb\frac{U}{I}\sin 4\Theta + P\frac{V}{I}\sin\delta\sin 2\Theta \tag{7.2.3}$$

where i defines the polarogram point, and it was assumed that

$$\frac{PaQ}{I} \ll 1$$

A set of typical parameters for the 1427 Å spectral line is as follows: $P = 0.23 \pm 50\%$, $\sigma = 286°$, $b = 0.36$, $Pb = 0.83$. In this case the modulation was expected to be 8.3% for the initial polarization degree of 100%. The Fourier coefficients of the 4ω component gave parameters U/I and Q/I. The angle ϕ between the polarized electric vector and the OX axis was obtained by means of the relations

$$\frac{Q}{I} = P\cos 2\phi \qquad \text{and} \qquad \frac{U}{I} = P\sin 2\phi$$

where the amplitude of polarization is given by

$$P = \frac{\sqrt{U^2 + Q^2}}{I}$$

The OX axis is parallel to the spectrograph dispersion plane and perpendicular to the North–South direction. Taking into consideration the fact that the initial position of the wave plate corresponds to $\Theta = 45°$, the orientation of polarization relative to the North–South direction can be found.

The polarization was derived either for a single point in a single polargram, using S_i data, or by averaging S_i over the space or time, i.e.

$$S_s = \frac{1}{n_c}\sum_s S_i \qquad \text{or} \qquad S_t = \frac{1}{n_c}\sum_t S_t$$

An associated signal-to-noise ratio (S/N) derived from Poisson statistics was

$$\frac{S}{N} = 2\frac{A}{\sqrt{I}}$$

where A is the amplitude of the 4ω component and I is the mean intensity of counts, defined by

$$I = \frac{1}{n_c}\sum I_i$$

Here, n_c is the number of S_i functions contributing to S_t or to S_s.

Systematic errors due to instrumental effects were estimated by integrating all the sets of $S_t(\Theta)$ data over all points observed during the full period of observation. There were 312 polarograms used, representing 7800 points and 7800 $S_t(\Theta)$ curves. The residual amplitude of the 4ω component did not exceed 0.066% modulation, which is equivalent to 0.8% of the degree of polarization.

More extended observations have been undertaken in this group on the hydrogen H_α line[3, 51, 182, 184] An observational technique was chosen to study the center of the $H_\alpha$line, which is important for impact spectropolarimetric effects. The filter with a bandpass of 0.75 Å was used in combination with a flare patrol instrument. Even if H_α is not the best line for observing impact polarization, due to the strong turbulent width which could effectively reduce polarization by means of the Zeeman and Stark effects, estimates of these factors[3] showed that the contribution of these two effects to the polarization fraction does not exceed 0.5%.

The instrument used in these observations was an H_α flare patrol heliograph. The telescope was a 15 cm refractor. The axial symmetry of the optics ensures that our measurements are unaffected by the instrumental polarization. The monochromatic Lyot filter had a 0.75 Å bandpass. A rotating half-wave plate was inserted in front of the linear polarizer at the entrance of the filter. The half-wave plate was rotated through 22.5° per step, and one set of 16 H_α filtergrams was obtained per minute on the film. The time delay between two successive images was 2 seconds. The rotation of the half-wave plate introduced some modulation of the intensity $\mathcal{I}$ of the radiation that passed through the linear polarizer mounted after it

$$\mathcal{I} = \tfrac{1}{2}(I - U\cos 4\alpha - Q\sin 4\alpha) \tag{7.2.4}$$

where Q and U are linear Stockes parameters, and α is the angle between the direction of transmission of the polarizer and any axis of the half-wave plate.

The Stokes parameter V is not measured. The two entrance and exit sides of the half-wave plate were not strictly parallel. Therefore, the wave-plate rotation introduces some motion of the image. Consequently, a cross-correlation technique must be used to compensate for these motions. Cross-correlation in the course of these observations was made on digitized data, using an area unaffected by the flares. Then, by substracting the appropriate digitized images, the Stokes parameters were obtained. In order to reduce the effect of temporal variation in the flare intensity I, a set of three consecutive images was used. This gives

$$U_{2i+1} = (-1)^i[(\mathcal{I}_{2i} + \mathcal{I}_{2i+2}) - 2\mathcal{I}_{2i+1}]$$

and

$$Q_{2i} = -(-1)^i[(\mathcal{I}_{2i} + \mathcal{I}_{2i+2}) - 2\mathcal{I}_{2i+1}] \tag{7.2.5}$$

with

$$\alpha = \frac{\pi}{8} \times i$$

where i varies from one to seven.

The polarization fraction and the azimuth of the polarization were easily derived from the Stokes parameters.

7.3. *Polarimetric Observations in the Crimean Astrophysical Observatory (Ukraine)*

The basic instrument for spectropolarimetric observations in the Crimean Astrophysical Observatory was the Large Solar Vacuum Telescope BST-2 ($D/F = 45\,\text{cm}/21\,\text{m}$) and the Large Coronograph *CG*-2 ($D/F = 50\,\text{cm}/13\,\text{m}$ or 20 m). Spectra of the H_α line were recorded in the III and IV orders of the grating spectrographs (1.7 and 3.0 mm/Å dispersion) by Kopdak IIF photoplates and Izopanchrom-2 film.

The polarization analyzer was fixed at the exit of the spectrograph. Three different devices were used as polarization analyzer: a Wollaston prism, a polarizing mosaic, and a birefringent prism. The intensities of two orthogonally polarized light beams were equalized by fixing the axes of the analyzer at 45° to the entrance slit of the spectrograph.

Every polarization analyzer has been examined.[(185)] The Wollaston prism was 15 mm in diameter with deflection angle between the ordinary and extraordinary beams of about 5°. The drawback of this analyzer is the small image dimensions, which forced the observers to use the IVth spectral order instead of the IIIrd, and the small difference in image scale at two different polarizations (~1%).

The effective aperture of the polarizing mosaic is a parallelogram with linear dimensions of 17 × 24 mm and an angle of 70°. Narrow bands located one after another parallel to the smaller side of the parallelogram transmit light linearly polarized at an angle of +45° and −45° to the dispersion. There exist some drawbacks connected to spectropolarimetric observations with the aid of the polarizing mosaic. First, different parts of a spectral line are recorded in two orthogonal polarizations. Interpolation, used to reconstruct the real image, involves difficulties in determining the degree of polarization, especially in parts, where there exists spectral blending. Second, the image in one linear polarization is slightly shifted (at 0.015 mm) with respect to the image in the other one. Despite the fact that this shift could be accounted for during photoprocessing of images, the polarizing mosaic was used in this group only by Babin and Koval.[(185)]

The free aperture of the birefringent prism of 30 mm was higher than that of the Wollaston prism. The extraordinary beam passed the prism practically without deflection, and the ordinary one was deflected at 6°. As the optical axis of the prism was oriented at 45° to the refraction plane, two orthogonally polarized spectra were located one above the other and the influence of the instrumental polarization was minimal. Thanks to these advantages, the majority of polarization observations in this group have been implemented with the help of the birefringent prism.[(186–190)] Some instrumental modifications in the polarization analyzer were introduced;[(187)] a system of two polarizing prisms was utilized. This system allowed the distance between two orthogonally polarized spectra (from 2 mm to 12 mm) to be changed by shfting one prism with respect to the other.

Some improvements in polarization observations have been undertaken in this group[(190)] because the main drawback of early polarization measurements was that spectral observations were performed only for one position of the analyzer, fixed at an angle of 45° to the slit of the spectrograph. In fact, it did not allow one to carry out adequate determination of the degree of polarization and the azimuth of linear polarization. A technique for simultaneous determination of spectra of the same location on the Sun on the H_α line at four positions of the axes of the analyzer of linear polarization differing by 45° was proposed[(190)] and realized. The IVth-order spectrum was obtained in an ordinary manner, but for recording the IIIrd-order spectrum two additional plane mirrors were used. This scheme is presented in Figure 36. The light of the IIIrd-order spectrum behind differential grating G was reflected by mirror M1 to the other plane mirror M_2,

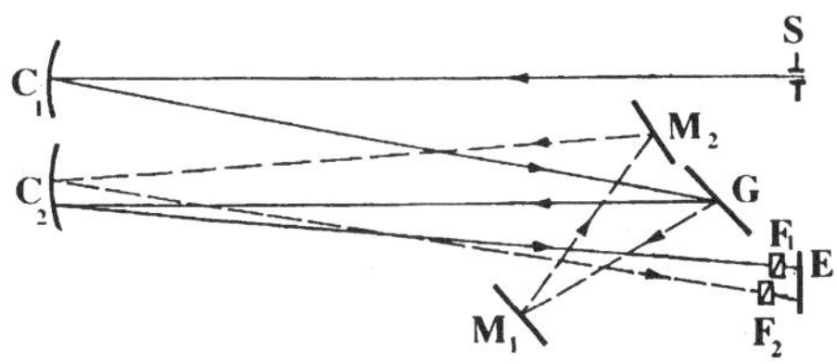

Figure 36. Optical scheme of CG-2 spectrograph for the study of linear polarization: S—entrance slit, C_1—collimator, G—diffraction grating, C_2—chamber mirror, $M_{1,2}$—plane mirrors, $F_{1,2}$—polarizing prisms, E—photoplate. The IVth-order spectrum is indicated by the solid line; the IIIrd-order spectrum is indicated by the dashed line after grating G, not to scale.

then to chamber mirror C_2, and finally to photoplate E. The analyzers of the linear polarization, or two birefringent prisms, were located near the focal plane of the spectrograph at a distance of 10–15 cm from the photoplate. The axes of the prism, mounted in IIIrd order, were inclined at 45° to the spectrograph slit; for the prism, in IVth order the angles were 0° and 90°.

Studies of the polarization dynamics of mustaches have been undertaken on the Large Coronograph CG-2 with a birefringent (Halle) filter for the H_α line.[191] Besides, the use of the filter allows one to obtain the image of all the active region, within which it is possible to observe several mustaches as a run. The disadvantage of the H_α filter is that the spectroscopic information is inferred only within a very narrow spectral range around the center of the H_α line. The analyzer of linear polarization in the observational scheme was a system consisting of the entrance polarizer of the H_α filter in combination with a rotating $\lambda/2$ plate. As the mustache emission is mostly concentrated in the wings of a line, the bandpass of the H_α filter (Halle filter) was shifted for ± 1 Å. The optical scheme of the coronograph CG-2 for studying the linear polarization is presented in Figure 37. The filter was mounted in front of the focus of the coronograph ($F = 20$ m) after the plane reflecting mirrors m_1 and m_2. The instrumental polarization of these mirrors was specially studied and was found to be near-linear. In the course of observations, the H_α filter was oriented around its optical axis in such a way that the axis of transmission of the first polaroid was parallel to the plane of the instrumental polarization.

The phase plate was fixed in the primary focus of the CG-2 instead of the artificial moon and it could rotate in its own plane on both sides at 5° per second. The phase plate in the primary focus was positioned conveniently such that the intensity of only the studied polarized radiation was modulated under its rotation at the entrance of the analyzer. A series of shots were made corresponding to five positions of the $\lambda/2$ phase plate $(0°, 22.5°, 45°, 67.5°, 90°)$ in the course of observations with the H_α filter.

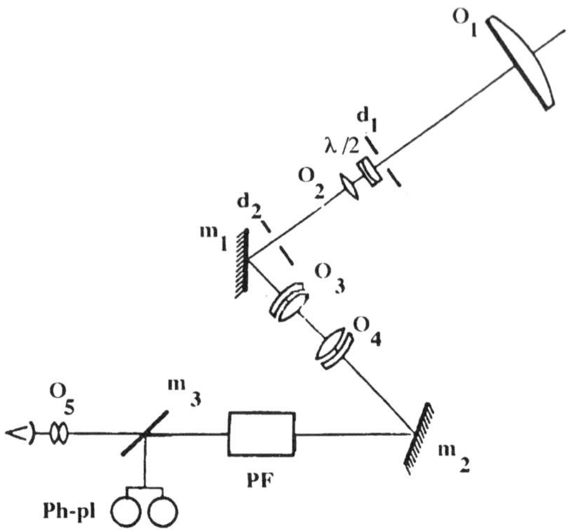

Figure 37. Optical scheme of the coronograph CG-2 and an installation for the study of linear polarization of emission: O_1—main objective, d_1—diaphragm in primary focus, O_2—optical field lens, d_2—Lio diaphragm, $m_{1,2}$—plane mirrors, $O_{3,4}$—objectives, PF—interferences polarizing filter, m_2—semitransparent mirror, Ph-pl—photochamber, O_5—lens.

7.4. The Spectropolarimeter of the Institute of Solar and Terrestrial Physics (Irkutsk, Russia)

Spectropolarimetric observations of solar flares and mustaches have been undertaken at the Institute of Solar and Terrestrial Physics. Two different instruments were employed in the course of these observations. The first was the Automatized Solar Telescope (AST) of the cellostate system of Sayany Observatory (2000 m altitude above sea level), which comprised the telescope itself, a spectrograph, a coordinatemeter with photoelectrical sensors for the position of the image, and a system to control the telescope orientation and monitor the image quality. Observations[192–196] were conducted with the help of this instrument and accompanying optics.

The main mirror of the telescope was 80 cm in diameter with a focus length of 20 m. The spectrograph of the telescope was assembled by the horizontal Eber–Fasty scheme. The collimating mirror of 20 cm diameter had a focus length of $F = 7$ m; for the chamber mirror, $D/F = 50\,\text{cm}/7\,\text{m}$. The diffraction grating with 600 grooves per millimeter and a working surface of 220×200 mm concentrated light in II and III spectral orders. Dispersion within the IIIrd and IVth orders was 1.65 and 2.75 mm/Å, respectively.

Published observations[50, 96, 197] were carried out with the Large Solar Vacuum Telescope. This instrument is situated on the shore of Lake

Baykal.[203, 224, 225]. One mirror tracking system (polar siderostate) assembled on a 25-m-high metal column reflected sunlight in the polar axis direction (latitude 52°) to a 760-mm aperture of a double objective lens with a focal length of 40 m. The metal tube of the telescope mounted between the main 1-m-diameter mirror close to ground level and the objective lens on top of the column was evacuated and sealed to the entrance window situated in the vicinity of the objective lens. A solar image 40 cm in diameter passed through the lower window, which was used as a part of the scanning and guiding systems. Two small sealed glass windows were used to illuminate the spectrograph and optical bench with narrow bandpass filters. This telescope has been operating since 1980 for different observations.

The spectrograph was assembled according to the horizontal Ebert design with two moving camera mirrors. Differential grating with 600 grooves per millimeter, 200 × 300 mm of working surface, and a blazing angle of 53° in the left hand orders were used. The efficiency of this optical system is optimal when the telescope and spectrograph have the same apertures. The optical scheme and dimensions of the spectrograph were determined by the aperture ratio of the telescope, which was 1 : 53, and by the vertical size of the grating.

A scheme of the spectrograph is presented in Figure 38. Mirrors were made of glass ceramic. Special railings were mounted on the floor of the spectrograph camera allowing one to displace mirrors within a range of 5 m. When two camera mirrors are used simultaneously, the spectrum from the second camera mirror is located in the cassette plane above the spectrum from the first one. The minimal angles of the beams on the mirrors and on the grating that enabled the optimal optical arrangement were $\alpha_1 = 1°05'$, $\alpha_2 = 1°21'$, and $Q = 2^2 2'$. The dispersion within the working IIIrd and IVth orders was 2.8 and 4.4 mm/Å, respectively.

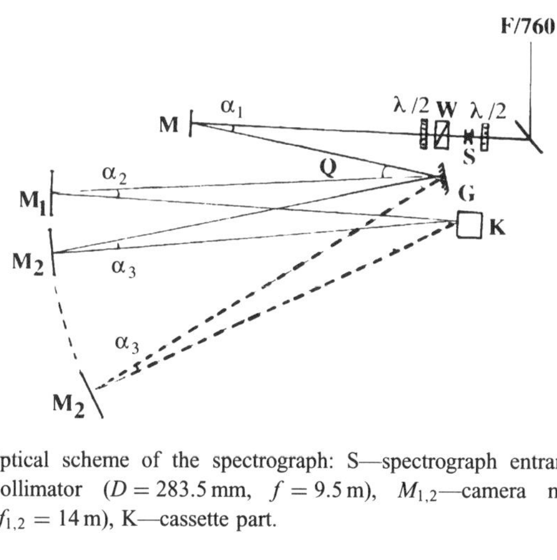

Figure 38. Optical scheme of the spectrograph: S—spectrograph entrance slit, G—diffraction grating, M—collimator ($D = 283.5$ mm, $f = 9.5$ m), $M_{1,2}$—camera mirrors ($D_1 = 598$ mm, $D_2 = 603$ mm, $f_{1,2} = 14$ m), K—cassette part.

A Wollaston prism was placed just behind the spectrograph slit. The prism angle and the distance from the slit were chosen such that both the ordinary and extraordinary beams provided the same contribution to the collimator. Within two spectral bands, the same spectral lines were located precisely at the same position, one below the other, and were adjusted by rotating the prism. The oscillations of the electric vector in the ordinary beam were parallel to the vertical axis or the direction of both the slit and the grooves of the differential grating; in the extraordinary beam they were orthogonal. In the course of observations, an achromatic $\lambda/2$ phase plate, with one of the axes either coinciding with the direction of the entrance slit or making an angle of 22.5° with it, was placed in front of the spectrograph. The intensities of the two spectral bands were equalized by means of a $\lambda/2$ plate fixed on the Wollaston prism holder. Intensity equalization for two orthogonal polarizations effectively decreased the role of instrumental polarization of the diffraction grating.

Linear polarization of H_α and H_β lines was observed simultaneously.[192–197] In order to make time-coincident pictures in the H_α and H_β lines, the grating was adjusted in such a way that, by observing the H_α line in the plane of the plateholder in the third order of the grating, the H_β line was observed in the fourth order. Light coming from the other orders was eliminated by means of filters, placed before the plateholder. Every shot contained four spectra of H_α and H_β lines in two orthogonal linear polarizations. A photoelectric guider was utilized to provide the exact orientation of the studied location with respect to the entrance slit. Spectrograms were processed photoelectrically[50, 192–197] and with the help of the CCD matrix.[96]

7.5. Solar Spectropolarimeter of Sacramento Peak Observatory (USA)

In order to simultaneously observe a number of mustaches or, as this phenomenon is also called, Ellerman bombs, a narrow band filter with bandpass of 6 Å was used together with a linear polarization analyzer.[200] The optical scheme is presented in Figure 39.

The Vacuum Tower Telescope F/72 representing a 50-cm solar disk image was used as main optical element after which a light beam was passed through the $\lambda/2$ phase plate and the field lens. A mask was used to fix a part of the image of about 50 arc sec, which was adjusted at the lens OI. At the prime focus the birefringent lithium niobate crystal wafer was mounted with the Ronchi ruling. The faces of the wafer were at 45° to the optical axis. Light passing through this system was transformed into a system of orthogonally polarized beams, as shown in Figure 40. The thickness and orientation of the crystal were chosen in such a

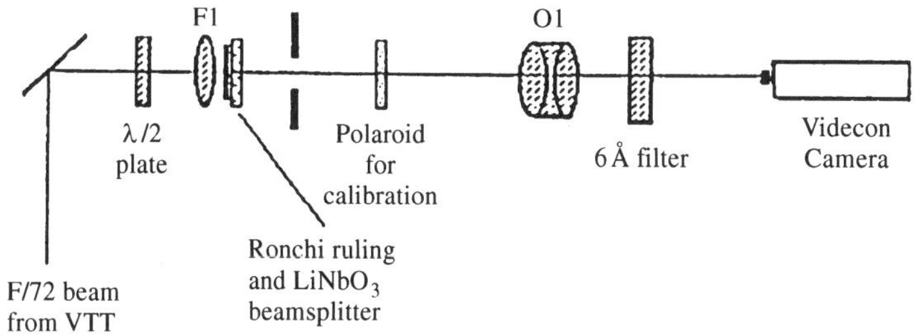

Figure 39. Optical scheme for spectropolarimetric measurements at Sacramento Peak Observatory. FI and OI (125 mm 1 : 1) are field and reimaging lenses, respectively.

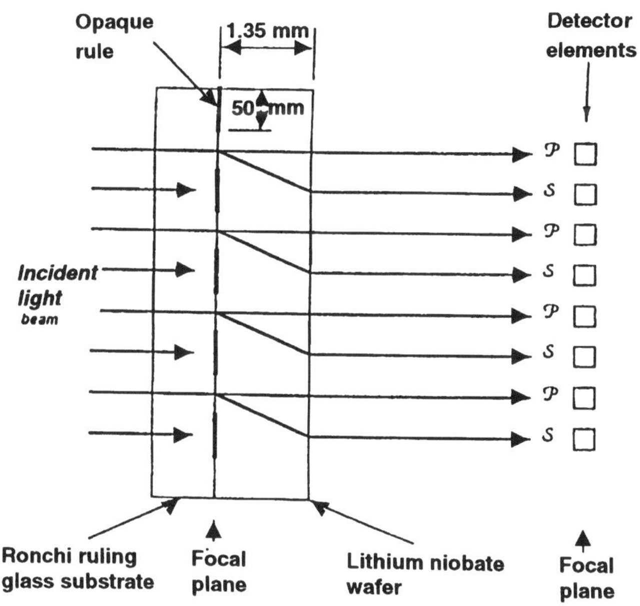

Figure 40. The polarization analyzer used in Sacramento Peak spectropolarimetric measurements of mustaches.

way that the displacement of the extraordinary beam matched the pitch of the rulings and effectively produced simultaneous interlaced polarized pictures.

Despite light losses in this original polarization analyzer, resulting from the opaque rulings, the optical paths of two orthogonally polarized light beams were practically the same and they were detected simultaneously by the same detector array. The picture was digitized by a videcon camera receiver into 500 × 500 pixel arrays with 8-bit precision. Every final video image consisted of two

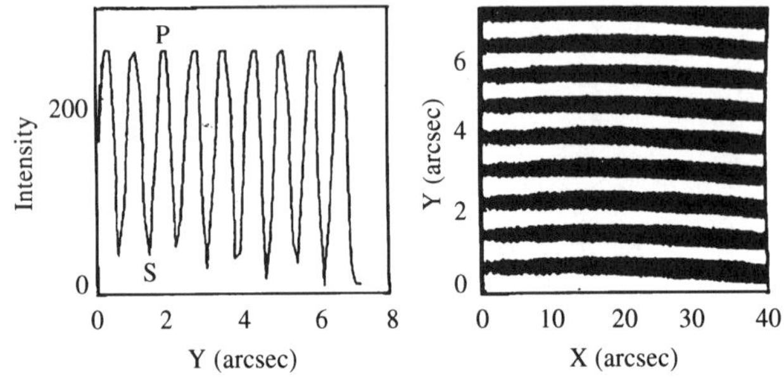

Figure 41. Calibration image in the detector plane for the polarization analyzer of the Sacramento Peak spectropolarimeter.

interlaced pictures. Calibration of the system was made by means of the linear polarized beam rotated through 90° and is illustrated in Figure 41.

In the course of observations, a half-wave plate was rotated at 45° and the final data were obtained from differences between the 0° and 45° orientations. The role of the $\lambda/2$ plate was to reverse the phase of signals $\mathcal{P}$ and $\mathcal{S}$. The degree of polarization was calculated by the formula

$$P = \frac{I_\mathcal{P} - I_\mathcal{S}}{I_\mathcal{P} + I_\mathcal{S}}$$

where $I_\mathcal{P} = I'_\mathcal{P} - I'_{ph}$, $I_\mathcal{S} - I'_\mathcal{S} - I'_{ph}$, and I'_{ph} is the background photospheric intensity.

The intensity of the studied object varied from 10% to 30%, exceeding the continuum level, and the degree of polarization was measured by this technique for emissions of more than 1.1 I'_{ph}. The effective instrumental polarization was never more than 1–2%. The azimuth of polarization was not estimated owing to rotation of the $\lambda/2$ phase plate.

7.6. Concluding Remarks

In conclusion, two main approaches can be specified for analyzing the features of all existing ground-based spectropolarimeters used to observe and estimate spectropolarimetric parameters on hydrogen Balmer lines emitted by ionized entities in the solar atmosphere. The first technique is based on the Lyot filter and the second on polarization optics. The Lyot filter provides the spatial distribution of linear polarization over a broad field of view.(184) Relative to Lyot filter measurements, spectropolarimetric observations based on polarization optics have some disadvantages and advantages:

1. The disadvantages are the limited number of locations in the active region that can be observed with the spectrograph slit at a given observation time and the fact that, due to image motion, the Stokes parameters Q and U may not correspond to the same location on the Sun, leading to erroneous values of the azimuth angle of the polarization vector. Image motion may affect also the derivation of the Stokes parameters from filtergrams. However, in this case an intensity cross-correlation technique may be used to eliminate the effects of motions.

2. The advantages are the possibility of using more than a single line, and additional significant information provided by measurement of the spectral dependence of linear polarization along the line profile.

Observations of line profiles bring additional information about the source of energy deposit at the origin of the observed polarization and allow one to avoid perturbations of the polarization signal due to the Doppler shift:

1. Line profiles can be affected by Doppler motions, shifting the central part of the lines out of the wavelength band transmitted by the filter and modifying the intensity and polarization of the light that passed through the filter. By recording the polarized line profile, no information will be lost even in the presence of strong Doppler shifts.

2. Measurement of the wavelength dependence of linear polarization along the line profile allows one to estimate the relative importance of other possible sources of linear polarization, such as the Zeeman effect or the Stark effect. Observations showed that these effects were not significant, either in H_α or H_β hydrogen lines.

3. In association with polarization measurements, recording the shape of the line spectrum provides complementary information on the mechanism of energy transport and deposit in the solar atmosphere. As an example, it has been shown[(219)] that electron and proton bombardments lead to rather different shapes of the H_α line profile.

Instruments and practical methods of ground-based spectropolarimetric measurements of nonthermal entities in the solar atmosphere, have been presented and compared in this chapter. They may be useful for extending the impact spectropolarimetric technique to prospective quantitative spectropolarimetric studies of the outer ionosphere of the Earth and other problems of spectropolarimetric remote sensing.

8

Solar Flare Observations at the Crimean Astrophysical Observatory (Ukraine)

8.1. Introduction

Spectropolarimetric observations and measurements at the Crimean Astrophysical Observatory (Ukraine) resulted from traditional Zeeman spectropolarimetry employed to measure the local magnetic field strength. By studying the magnetic field from polarization spectrograms in an H_α line obtained with a circular polarization analyzer in the mid-1970s, Koval[(226)] identified structural elements of different polarity in some flare knots. It would be logical to attribute this difference to the magnetic field. At the same time, these polarization data may well be distorted by other effects, such as the presence of linearly polarized emission caused by resonance scattering, the turbulent Stark effect, and anisotropic excitation of atoms by accelerated beams. This aspect of solar flare emission was quite poorly studied at that time. Therefore, the Crimean group during the 1970s and 1980s undertook systematic and careful observations of linear polarization in solar flare emissions.

This chapter presents observational results on linearly polarized emission of solar flares and other emission features obtained by the Crimean group in the 1980s and published in a series of works.[(185–188)]

8.2. Technique of Observation and Data Processing

The observational data at the Crimean Astrophysical Observatory were mostly obtained using the Large Coronograph ($D/F = 51\,\text{cm}/20\,\text{m}$ or $13\,\text{m}$) and the solar Tower telescope ($D/F = 45\,\text{cm}/21\,\text{m}$).[(185–188)] The spectra in the H_α

line were recorded in the third or fourth orders of a grating spectrograph with a dispersion of 1.7 mm/Å and 3.0 mm/Å, using photoplates and films. For Kodak plates the exposures were 0.06 and 0.2 s for the Coronograph and Tower telescope, respectively, while for the Izopanchrom film the exposure was 1/60 s. A birefringent prism located at the exit slit of the spectrograph or a set of two polarization prisms mounted behind the entrance slit were used as analyzers of linear polarization.

Continuous observations performed by this group enabled the emission to be recorded at the explosive phase of flare development. Short exposures permitted high quality spectrograms to be obtained with a spatial resolution of 0.6″ for solar flares and moustaches. Unfortunately, most of the spectropolarimetric observations of this group were undertaken only for a single position of the polarization analyzer with axes oriented at $\pm 45^\circ$ to the direction of the spectrograph slit.

The degree of polarization of a studied object was determined by comparing its intensity on two spectrograms, corresponding to orthogonal linear polarizations registered simultaneously. The degree of polarization was obtained to be some percents. The main errors in the photometric processing of spectra for different polarizations taken with a Wollaston prism and a polarization prism are related to uncertainties in fixing the studied location for two orthogonal linear polarizations. Therefore, each studied location was processed photometrically for several (up to seven) independent fixed coordinates. The profiles of intensities were averaged over all the measurements after which the degree of polarization for an emission was determined by the formula

$$P = \frac{I_\perp - I_\parallel}{I_\perp + I_\parallel}$$

The spectrum for a quiet location of the solar disk from the same spectrogram was used to determine the instrumental polarization. Special supplementary measurements showed that the instrumental polarization varied within the observed region along the direction of dispersion as well as in the perpendicular direction. Therefore, in order to obtain a more precise value for the instrumental polarization, two to four photometric cuttings were made corresponding to the quiet region of the photosphere on both sides of the studied emission.

The variation in the degree of polarization with wavelength was determined for every cutting and then averaged over all cuttings; the mean value was taken as the instrumental polarization for the studied emission. Later, the mean-square error of the mean degree of polarization σ was determined. As a rule, the mean-square error of the mean degree of polarization of the studied emission was of the same order of magnitude as the mean-square error of the instrumental polarization.

8.3. Degree of Polarization along the Hydrogen H_α *Line*

Polarization spectra have been observed and processed[(185)] for two solar flares of second grade. Both flares were observed during the period from the beginning of the flare until the post-maximal phase. The solar flare of grade 2B was observed on 27.09.1978 (27°N, 19°W). It started at 1428, attained its maximum at 1442, and finished at 1546 UT. Spectropolarimetric studies were undertaken during the interval 1435–1447 UT. Processing of the spectrograms showed that the degree of polarization of emission for the nodes of this flare either do not differ from the instrumental value, or the difference did not exceed 2σ.

The solar flare of grade 2N on 01.10.1978 (13°S, 57°E) commenced at 0653, attained its maximum at 0730, and ended at 0851 UT. The studied spectrograms were taken during 0717–0737 UT. Two emission nodes were studied on the spectrogram taken at 0717. The degree of polarization for these nodes within the error bar did not differ from the instrumental value. Figure 42 gives the wavelength dependence of the degree of polarization for cuttings along the emission and the wavelength variation in the instrumental polarization. Vertical lines indicate the mean-square error of the mean value of the degree of polarization. The upper part of the figure represents the intensity variation for the studied photometric cutting of the flare emission. Variations in emission polarization and instrumental polarization for the post-maximal phase of the flare (0737) are given in Figure 43. The variation in the degree of polarization for this event does not differ from the instrumental value, excluding a small part in the blue wing of the H_α line, where this difference, anyway, does not exceed 3σ.

Spectropolarimetric effects of four flares belonging to the first grade were also studied in this group: on 20.09.1979 (0803, 0805 UT), 23.06.1980 (0359 UT), 15.10.1980 (1248 UT), and 19.10.1980 (0902 UT), as shown in Figure 44. Polarization measurements of the emission as well as the emission eruptions at the solar disk are presented in Table 12 only for cases (10 out of 27) when emission polarization differed from instrumental emission more than for 3σ.

The first column of Table 12 gives the date of observation and time when the corresponding spectrogram was taken. The second column gives the maximal value $(P_e - P_i)$, the mean-square error of this value averaged over the given wavelength ranges, namely, $\sigma_{av} = \sqrt{\bar{\sigma}_e^2 + \bar{\sigma}_i^2}$, is in the third column, the fourth gives the absolute value of $(P_e - P_i)$ expressed in terms of σ_{av} with some remarks in the ninth column.

An example of a 2B grade (X 3.5) flare on 26.07.1981[(187)] within the tail of the extensive South–East group of spots ($\rho = 0.55$) is presented in Figures 45 and 46. This group was characterized by the complicated structure of the magnetic field and the existence of the δ configurations. The brightest node of the flare was located in the vicinity of the δ configuration. The positions of the

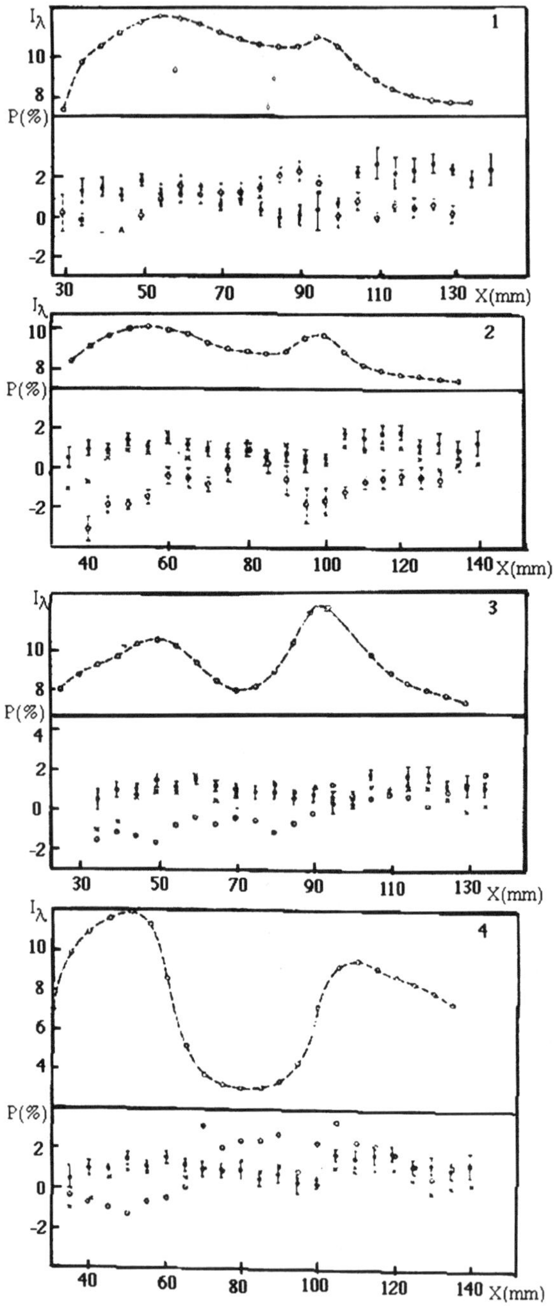

Figure 42. Dependency of the degree of polarization of the H_α hydrogen line on wavelength for emission nodes (1–4) of the flare on 01.10.1978 at 0718.5 UT (circles) and instrumental polarization (dots) and photometric cutting of the studied emission expressed in intensities (upper curve).

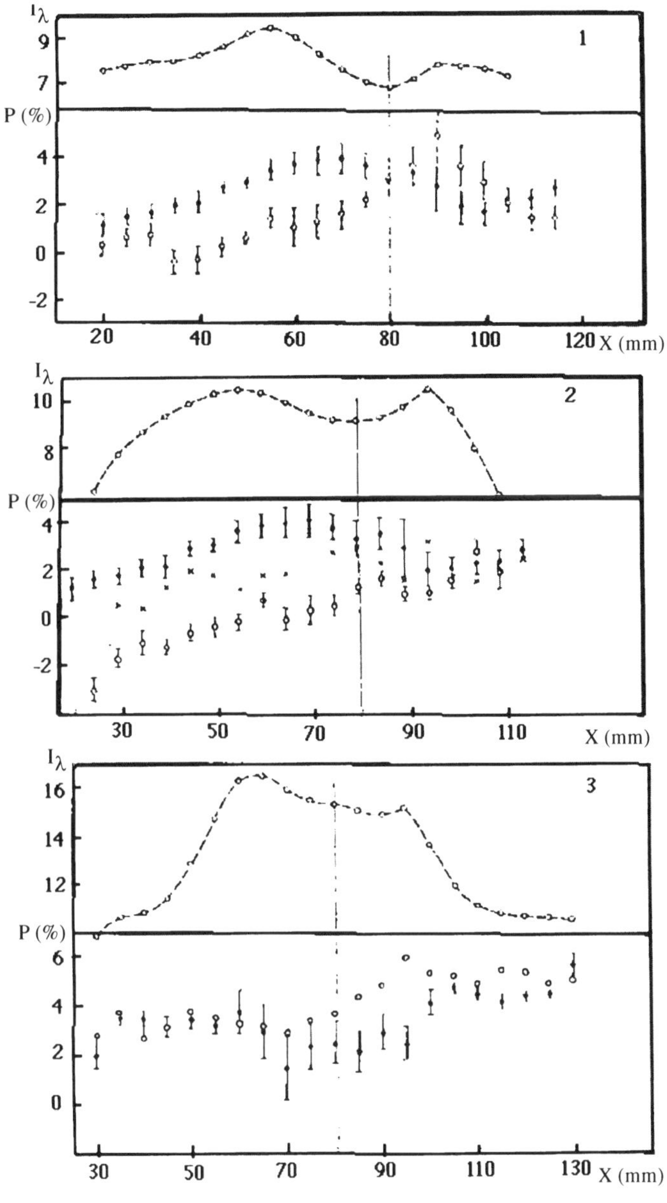

Figure 43. The same as Figure 42, for moments 0721 (1,2) and 0737 (3) UT.

TABLE 12
Polarization of Emission Detail for the Flares Observed in Ref. 185

Date, UT	$P_e - P_i$, %	σ_{av}, %	$\dfrac{\mid P_e - P_i \mid}{\sigma_{av}}$	Remarks
01.10.78, $07^h18.5^m$	2.3	0.65	3.5	Import.2
–	−2.1	0.68	3.1	–
–	−2.4	0.46	5.2	–
–	−2.5	0.55	4.5	–
–	−2.8	0.47	6.0	–
–	−2.5	0.42	6.0	–
–	1.7	0.54	3.1	–
–	−2.3	0.45	5.1	–
$07^h21.5^m$	−2.5	0.68	3.7	–
–	−2.2	0.6	3.7	–
20.09.79, 11^h05^m	2.6	0.47	5.5	Import.1
–	−2.3	0.46	4.9	–
23.06.80, 06^h59^m	2.7	0.56	4.8	–
11.07.80, 07^h40^m	3.3	0.76	4.3	–

spectrograph slit are indicated as I–III in Figures 45 and 46. According to solar geophysical data, this flare was accompanied by complicated pulses that started at 1349 UT. Moreover, the SMM satellite recorded pulses of γ radiation within the interval 1353.18–1354.40 UT. These data justify that pulsed acceleration of particles took place in the course of the studied flare.

Twenty emission contours of the H_α hydrogen line corresponding to different parts of the flare and different moments were studied during the flare on 26.07.1982. The degree of linear polarization exceeding the value 3σ was obtained for six of the emission contours. As regards the other contours, the degree of polarization within the bars of experimental error did not differ from the instrumental polarization.

Figure 47 demonstrates a flare of 1B importance on 14.06.1982 that appeared at the tail of the North–East group ($\rho = 0.87$), representing in itself a big spot with nuclei of different polarity located within its demishade. According to solar geophysical data, this flare was accompanied by pulsed emission of radiowaves, an X-ray burst (1–8 and 0.5–4 Å), and sudden ionospheric perturbation. Spectral observation of this flare was conducted for maximal and post-maximal phases of the pulsed burst.

Linear polarization of emission was detected and investigated for the flare on 14.06.1982 using five emission contours. Noticeable polarization was obtained for only one of these contours. One of the locations represented the emission spectrum during hard X-ray emission, but the degree of polarization was not significant. No one emission location represented any significant polarization during the maximum of soft X-ray emission.

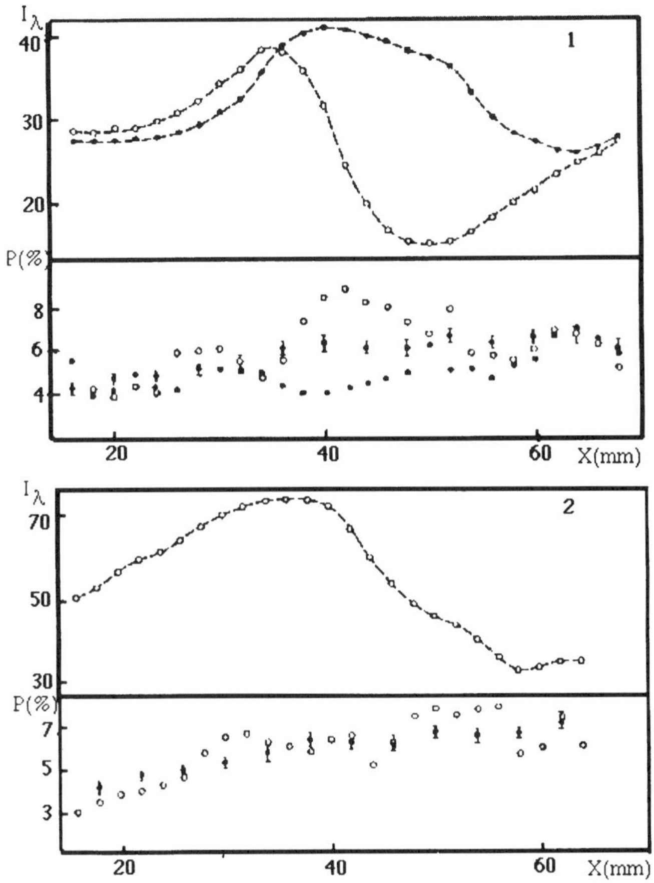

Figure 44. The same as Figure 44, for the flare on 20.09.1979: (1) at 0805 UT, (2) at 0803 UT.

Linear polarization was safely detected for six of twenty studied contours of the H_α emission line for the flare on 27.07.1981 (Figure 46) and for one of five contours of the flare on 14.06.1982 (Figure 47). Linear polarization of emission was observed either for every emission contour or only in the central part of the H_α line, in contrast to other work[(185)] where for the strongly asymmetric emissions the predominant polarization was observed in the red wing. A tendency toward an increase in the degree of polarization with time within this flare node was also noted.

It should be remarked that this set of observations for the degree of polarization is presumably underestimated systematically due to the experimental scheme used in which all observations were made for only one position of the

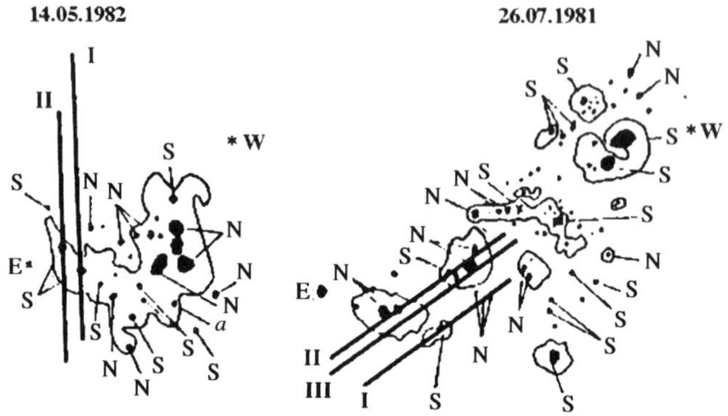

Figure 45. Drawings of a group of spots where flares were observed with indicated positions of the entrance slits of the spectrograph.

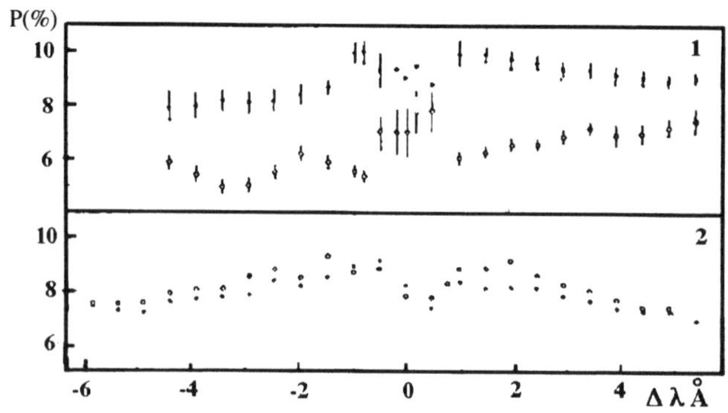

Figure 46. Wavelength dependence of the degree of polarization for the H_α hydrogen line of the emission core of a flare on 26.07.1981 at 1356.05 UT (1) and 1356.00 UT (2).

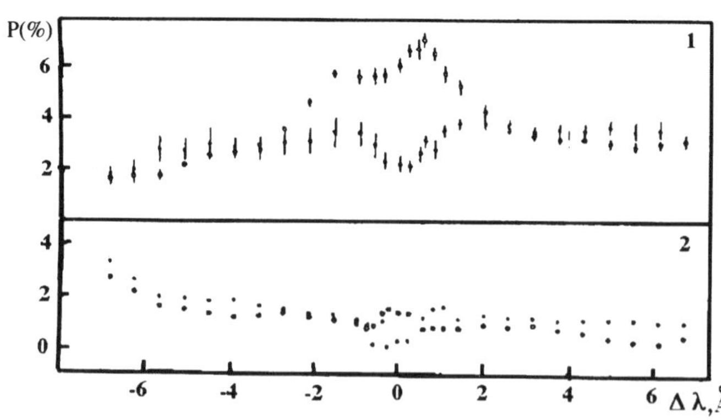

Figure 47. Wavelength dependence of the degree of polarization for the H_α hydrogen line of the emission core of a flare on 14.06.1982 at 0628.50 UT (1) and 0626.20 (UT) (2).

analyzer (the optical axes of the prism analyzer were fixed at 45° to the entrance slit of the spectrograph). For example, if the polarization plane of the studied radiation is parallel to the entrance slit of the spectrograph, polarization cannot be detected by the experimental technique employed.

8.4. Spatial Distribution of the Degree of Polarization

Visual study of spectrograms of the flares described in the previous section indicated that some emission nodes had a different structure in orthogonal linear polarizations. In some cases a crucial difference in structure was detected, in other cases some small emission details (such as mustaches or continuous emission bands) indicated a noticeable difference in brightness.[(186)]

Photometric processing of spectrograms was performed in order to reveal the structural details of the emission nodes of flares and to estimate the degree of polarization. Photometric cuttings were taken in a direction perpendicular to that of the dispersion for different parts of the spectrum, more often within the red wing of the hydrogen H_α, between atmospheric lines λ 6563.53 and λ 6564.22 Å. The latter were used as reference lines for fixing the same wavelengths at the spectrograms taken in orthogonal linear polarizations.

When the emission did not have any extended wings, the photometric cuttings were taken at the center of the H_α line. The intensities were normalized to unity at those locations where the cutting crossed a nonperturbed solar atmosphere. The distributions of intensities in two orthogonal polarizations obtained in such a way were compared with the aid of a reference image of thread fixed across the entrance slit of spectrograph. Results of the photometric processing are given in Figure 48.

The degree of polarization for the emission details was calculated by the formula

$$P = \frac{I_0 - I_e}{I_0 + I_e}$$

where I_0 and I_e are the intensities for ordinary and extraordinary light beams at some place counting out from the level of the nonperturbed photosphere.

Some features can be pointed out on analyzing these data. Three emission details are seen on the spectrogram taken at 550.53 UT for the 2n grade flare on 19.07.1982 (Figure 48a). Noticeable differences in the intensity distribution for orthogonal linear polarizations across the dispersion were found only for the middle node. Within the one linear polarization band a sharp single emission structure of dimension $\sim 3.4''$ was detected. Within another polarization, a double

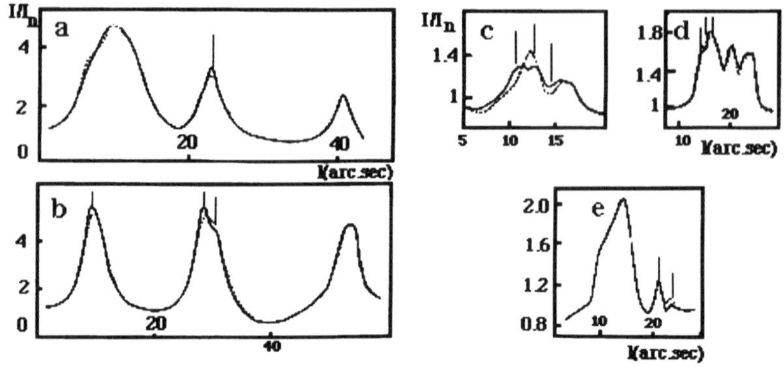

Figure 48. Photometric cutting of flare nodes perpendicular to the direction of dipers ion (solid line, the ordinary beam; dashed line, the extraordinary beam): (a) 2n flare on 19.07.1982 at 0550.53 UT; (b) the same at 0551.30 UT; (c) sn flare at 1136.06 UT; (d) sb flare on 20.07.1982 at 1256.30 UT; (e) sn flare on 19.07.1982 at 0920 UT.

emission structure with a distance between maxima of $\sim 0.8''$ was observed. The degree of polarization for the central part of the emission was about 9%.

The spectrogram taken at 551.30 UT for the other part of the flare shows a difference for two nodes (Figure 48b). For node I it is the same as for the previous case with a distance between peaks of $\sim 0.6''$ and degree of polarization about 3.5%. In the central part of node II the degree of polarization varies between 5% and -3% with a change of sign.

The polarization spectrogram for the SN flare dated 19.07.1982 ($\rho \approx 0.89$) was taken at 1136.06 UT (Figure 48c). The brighter node shows a more distinctive difference in structure which has already been pointed out. Within one linear polarization a single detail of size 2.4″, and within another polarization a double one with distance between maxima $\sim 1.4''$, were observed. The degree of linear polarization in this case is quite high and represents significant changes $(+25\%, -11\%, +18\%)$.

The sb flare on 20.07.1982 at 1256.30 UT ($\rho \approx 0.53$) showed that the fine structure of the biggest node I differs significantly for two orthogonal polarizations (Figure 48d). The degree of linear polarization also changes effectively in this case $(+6\%, -7\%, +4\%)$.

The brightest points of the principal knot of the sn flare on 19.07.1982 at 0920 UT ($\rho \approx 0.85$) did not show any difference (Figure 48e). At the same time, the brightness of two separate, adjacent, very fine and extended emissions were significantly different, one being brighter in linear polarization and the other in orthogonal polarization.

By these means, some flare knots represent a crucial structural difference while analyzing images for orthogonal linear polarizations. The most typical case is when a single detail is observed in one linear polarization while a double one takes place in orthogonal polarization. For some flare nodes within several arc seconds the degree of linear polarization changes not only its magnitude, but also the sign within the range of +25% to −11%.

It was pointed out[186] that these results could be doubtful for several reasons. First, the spatial and spectral resolutions of the spectrograms for two linear polarizations are slightly different, while a double emission structure has been observed in both one linear polarization band and in the orthogonal one. The difference in spatial resolution within two spectral bands is evident by looking at the spectrograms.[186] Second, the image at the entrance slit usually flickers, and therefore both bands could not be taken at the same instant because of the limited working time of the photocamera shutter. A thorough analysis of all these sources of errors lead to the conclusion that the observed differences in the structure and brightness of emission knots for orthogonal polarization could not be explained only by instrumental effects.

8.5. *Orientation of the Polarization Plane*

An estimate of the orientation of the polarization plane for work conducted at the Crimean Astrophysical Observatory[185–188] is complicated, because all the observations were carried out for only one position of the analyzer with the axes of the polarizing prism fixed at 45° to the entrance slit position.

For the flare on 26.07.1981 the direction of the entrance slit was directed approximately 29° to the solar disk center. In this case the polarization plane of one beam formed by the analyzer was oriented at 16°, and the other at 74°, to the direction to the solar disk center. Therefore, in this case the orientation of the axes of the analyzer was favorable for the detection and measurement of impact linear polarization. Observation showed that the direction of oscillations of the electric vector in the beam was closer to radial in those cases when the detected linear polarization of emission exceeded the instrumental polarization. When the observed polarization was smaller than the instrumental polarization, the direction of polarization appeared to be preferentially tangential.

The polarization axes of the prism for the flare on 14.06.1982 were oriented at angles of 42° and 48° to the direction to the solar disk center. In this case the geometry was unfavorable for an estimate of the orientation of the polarization plane. On a spectral band with radial direction of the electric vector, a single detail was observed for two flares dated 19.07.1982 of 2n and sn importance, while a double detail was detected for the tangential direction.[186, 187] For some flare nodes, not only the magnitude but also the sign of linear polarization varied within a range of several arc seconds.

8.6. Concluding Remarks

In conclusion, it should be noted that the observational data of the Crimean group enabled the degree of linear polarization to be determined reliably for about 30% of the flares studied. This proportion is, indeed, effectively underestimated because polarization was definitely observed in many other cases.[185–188] However, due to the fixed position of the analyzer axes these results could be regarded only as qualitative, since due to this instrumental restriction, it was impossible to detect and measure the probable polarization in some cases.

Linear polarization of flares was observed either for the whole emission contour or only in the central part of the H_α line. Asymmetry of polarization was found within some emission line contours. The degree of polarization as determined from cuttings made along the dispersion did not exceed 10%, being most often 3–5%.

The specific result of spectropolarimetric observations of the Crimean group is that, when one flare knot emitted polarized optical radiation, the adjacent one did not show any polarization, although it was photographed simultaneously with the first. Often, the degree of polarization was higher for the less bright flare node. Frequency asymmetries of the degree of polarization were observed within the emission line contour in some cases, but typically for these observations the linear polarization was detected in the line core.

We noted above that the degree of polarization as determined from cuttings made along the dispersion did not exceed 10%, being most often 3–5%. For transverse cuttings, the degree of polarization was much higher, reaching 25%. The complicated structure of the ionized medium of the upper chromosphere as an emissive location of a solar flare was detected in the course of spectropolarimetric observations of the Crimean group. The variation in the degree of polarization could be within the range of $+25\%$ to -18% for two closely spaced locations with a $2''$ separation. For some complex flare knots a structural difference takes place even for observations in orthogonal linear polarizations. For example, a single detail was observed in one linear polarization while a double one arises in orthogonal polarization and, for some closely spaced flare knots, within a range of several arc seconds. The degree of linear polarization in this case could change not only the magnitude, but also the sign.

The most important result of the Crimean group is that the orientation of the polarization plane for all studied flares preferably coincides with the direction to the solar disk center, and that spectropolarimetric observations in the H_β hydrogen line confirm, in general, the data for the H_α line.

9

Observations of Polarization Effects at the Irkutsk Institute of Solar and Terrestrial Physics (Russia)

9.1. Introduction

Initial spectropolarimetric observations at the Irkutsk Institute of Solar and Terrestrial Physics were influenced by the fact that, in the 1970s, a large amount of research was undertaken on turbulent electric fields in plasmas under laboratory conditions.[227, 228] Since most theories and models of solar flares rely on the concept of the turbulent state of a flare plasma, the Irkutsk group made some initial attempts to study plasma turbulence with the upper chromospheric layers of a solar flare by different spectroscopic techniques, in particular by the Berange–Mozer method, based on satellites of the forbidden HeI line[228, 229] and estimates of linear polarization.[194]

A criterion for the validity of the turbulent Stark effect, stipulated by the Holzmark fields, is provided by the special shape of linearly polarized profiles of hydrogen lines as the most typical example and caused by the specific positions of τ and σ components, located mainly within the line wings. The local magnetic field also provides polarization of the emission spectral lines, but in the case of the Zeeman effect, polarization profiles of the hydrogen lines have another typical shape. Hence the main problem in wing spectropolarimetric studies is to distinguish between these two sources of polarization and to provide rather strong observational evidence for the existence of the first or second effect.

Observations in Irkutsk did not make it possible to detect the turbulent Stark effect. The reason for this seems to be that the effect of turbulent electric fields is relatively weak compared with thermal effects, and actually it is impossible to detect this effect in the wings of hydrogen lines under an optically thick layer condition, which is the case for the chromospheric medium. At the same time, in the course of these spectropolarimetric observations and initial measurements, the presence of linearly polarized emission of hydrogen lines during solar flares was clearly detected only at the center of the emission line profile. Such behavior of the polarization profiles indicates directly that linear polarization of emission is produced by collisional excitation of the emitting gas. An additional argument in favor of the impact polarization mechanism is that the equivalent widths of the Stark and Zeeman polarization profiles within the orthogonally polarized spectral bands, recorded in the course of photographic observations, must be equal and the difference must be strictly equal to zero. Contrary to this, the spectral line polarization caused by impacts with accelerated particle beams leads to a difference in the amount of energy emitted within two orthogonal polarizations, with the largest difference corresponding to the center that was effectively observed.

These initial motivations, the initial observations and comparative analysis of different possible mechanisms of emission line polarization, lead the Irkutsk group of researchers to concentrate on obtaining more observational spectropolarimetric data, advocating evidence for the collision excitation of the upper chromospheric part of solar flares by accelerated particles and later using this extensive observational material for the needs of spectropolarimetric sensing.[(197)] This chapter presents observational results of the Irkutsk group and spectropolarimetric measurements of the degree of polarization of atomic hydrogen emission lines.

9.2. Spectropolarimetric Parameters of H_α *and* H_β *Lines for the Flare on 15.09.1981*

A strong solar flare of importance 1B (18°N, 22°E) was observed on 15.09.1981 by the automated solar telescope of Sayan Observatory, affiliated to the Institute of Solar and Terrestrial Physics. A Wollaston prism mounted behind the entrance slit of the spectrograph was used to insure reliable spectropolarimetric observations and measurements. An achromatic rotating $\lambda/2$ plate as the principal part of the polarization analyzer was placed in front of the entrance slit of the spectrograph. For the position of the plate corresponding to 0°, oscillations of the electric vector of the studied beam were parallel to the entrance slit and the grooves of the grating of the spectrograph. This position was represented by the first spectral band on the photoplate. For the second spectral band the oscillations of the electric vector were parallel to the dispersion of the spectrograph. Using data within the two spectral bands, corresponding to zero angle, the Stokes parameter Q could be inferred. By rotating the $\lambda/2$ plate through 22.5°, the spectroscopic data incorporated in both spectral bands were used to obtain the Stokes parameter U.

The main importance of this work was that the recording the system yielded simultaneous spectropolarimetric data for the two H_α and H_β hydrogen lines. The simultaneous recording was organized by rotating the grating in such a way that, in the cassette plane, the H_α line was in the third order of the grating, while the H_β line was in the fourth. As a result, each spectrogram taken at some moment included four spectral bands: H_α and H_β lines in two orthogonal linear polarizations.

Detailed observations were made for the decay phase of the flare. Twelve spectrograms were processed within the interval of time 0044–0102 UT. Figure 49 illustrates the profiles for the H_α and H_β lines for these series of observations. The profiles are typical of the emission of flares located above sunspots.

The degree of polarization and orientation of the plane of polarization were determined by means of known formulas, using measured Stokes parameters Q and U within a 30-second interval. Figure 50 shows the time variation in intensity of the H_α and H_β lines and the orientation of the polarization vector.

Linear polarization is more clearly detectable for the H_α than for the H_β line. The degree of polarization for the H_α line as the flare was evolving fluctuated within the range of 2%–5%, but for the red peak of the line it attained 9.5%. The orientation of the polarization plane coincides with the direction to the solar disk center to a higher degree. For the H_β line the maximal degree of polarization was observed to be no more than 3.7%.

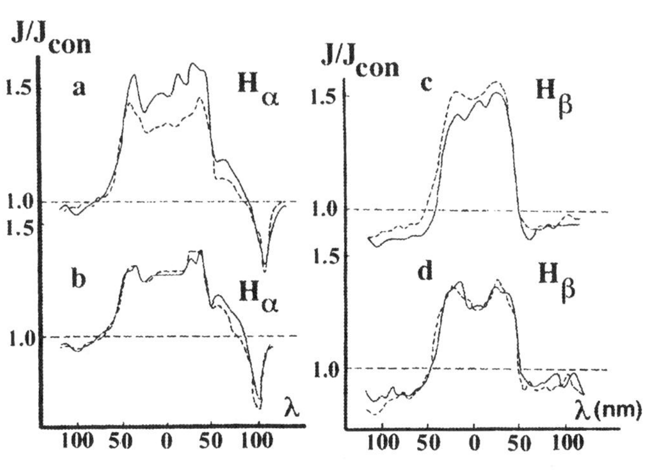

Figure 49. Illustration of the behavior of J/J_{con} for H_α and H_β emission lines at different moments of the decay phase of the solar flare on 15.09.1981.

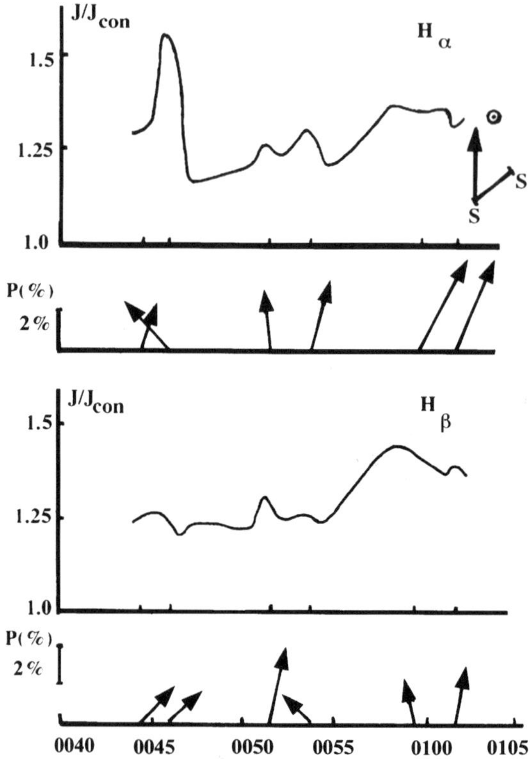

Figure 50. Temporal variations in the intensity (J) of H_α and H_β lines and the vector of linear polarization (P) for the flare on 15.09.1981. The orientation of the spectrograph entrance slit to the direction toward solar disk center is indicated by arrows.

9.3. Spectropolarimetric Parameters of the Active Region 5669 (SGD) on 07.09.1989

This event was observed at the Baikal Solar Vacuum Telescope of the Baikal Astrophysical Observatory affiliated to the Institute of Solar and Terrestrial Physics of the Russian Academy of Sciences.[(50)] The analyzer of the linear polarization consisted of a Wollaston prism place behind the spectrograph entrance slit. In order to equate the intensities of the ordinary and extraordinary beams, a $\lambda/2$ plate was used behind the prism. The other $\lambda/2$ plate placed in front of the slit was used to determine the azimuth of the plane of polarization. The procedure for determining the Stokes Q and U parameters, the degree of polarization P, and the azimuth of the linear polarization is described elsewhere[(96)] and in Section 9.2.

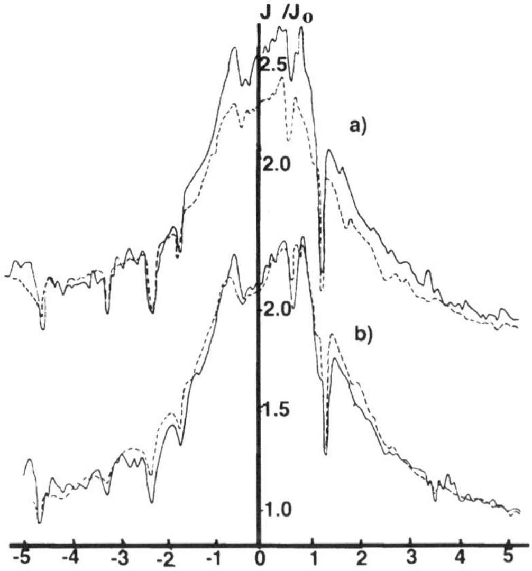

Figure 51. Photometric cutting of the H_α emission line along the dispersion in two orthogonal linear polarizations for the flare on 07.09.1989 at 0245 UT for two nodes (a,b).

For this event the polarization was observed within the active region 5669 appearing on the disk during the period 28 August to 12 September 1989 and, more specifically, on 07.09.1989. A well-defined polarization in the H_α line core was recorded in one of the knots during the period 0245–1346 UT. No polarization was present in the adjacent node on the same spectrogram (Figure 51). The degree of polarization for this flare did not exceed 6% and the polarization was observed only within the central part of the line. The orientation of the plane of polarization for this flare was close to the direction toward the center of the solar disk.

Nine spectrograms for other, weaker flare were obtained during the time interval of 0830–0836 UT. Spectral cuttings of this flare made for different positions of the $\lambda/2$ plate are shown in Figures 51 and 52. The degree of polarization for this flare did not exceed 5% and the plane of polarization was directed toward the solar disk center.

9.4. Detection of the Linear Polarization of the Flare on 16.05.1991

The observations were conducted at the Baikal Solar Vacuum Telescope of the Baikal Astrophysical Observatory affiliated to the Institute of Solar and

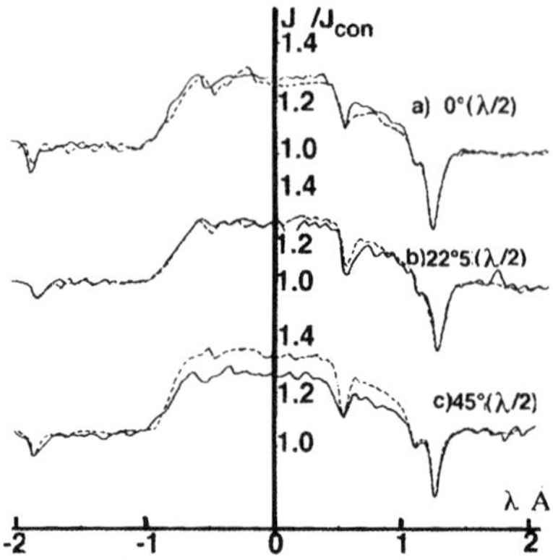

Figure 52. Same as Figure 51, but at 0830–0831.2 UT for different positions of the $\lambda/2$ plate.

Terrestrial Physics of the Russian Academy of Sciences[96] using the same technique as in work of Kazantsev *et al.*[50] In this work the spectrograms were processed with the aid of the CCD matrix.[230] Data processing by the CCD matrix enabled eight cuttings to be undertaken across the spectrum and along the spectrograph entrance slit. Five spectrograms were processed during the first two minutes. Figure 53 displays the original recordings of the H_α line intensities reduced to a continuum in two polarizations for one section ($-3.75''$) and for five spectrograms.

The profiles were recorded two or three times. Averaged and smoothed-out H_α-intensity profiles and the variation in parameter Q for all sections of spectrogram III are presented in Figure 54. Figure 55a,b,c shows a three-dimensional space picture for quantities I, P, and χ, where χ is the azimuth of the plane of polarization measured with respect to the orientation of the entrance slit of the spectrograph. The other two axes represent the wavelength and the number of the cutting. It is evident from this figure that the values of P and χ in the line core are less subjected to the noise. Figure 56 shows the intensity variation of the H_α line core (interval -0.5 Å $< \Delta\lambda < +0.5$ Å) and average behavior of P and χ within this interval. The direction to the solar disk center corresponds in this figure to the axis of ordinates. The length of segments represents the value of the degree of polarization P and the direction of segments indicates the angle χ.

In the course of these observations it was thus possible to obtain a reliable indication of linear polarization only for the central part of the H_α line. The average value of the Stokes parameters Q/I and U/I for the emitting part of the flare was 5–6%. The Stokes parameters changed sign at locations of the flare within which the line profiles resembled those for the mustaches. The average value of the degree of polarization at locations with decreased emission and

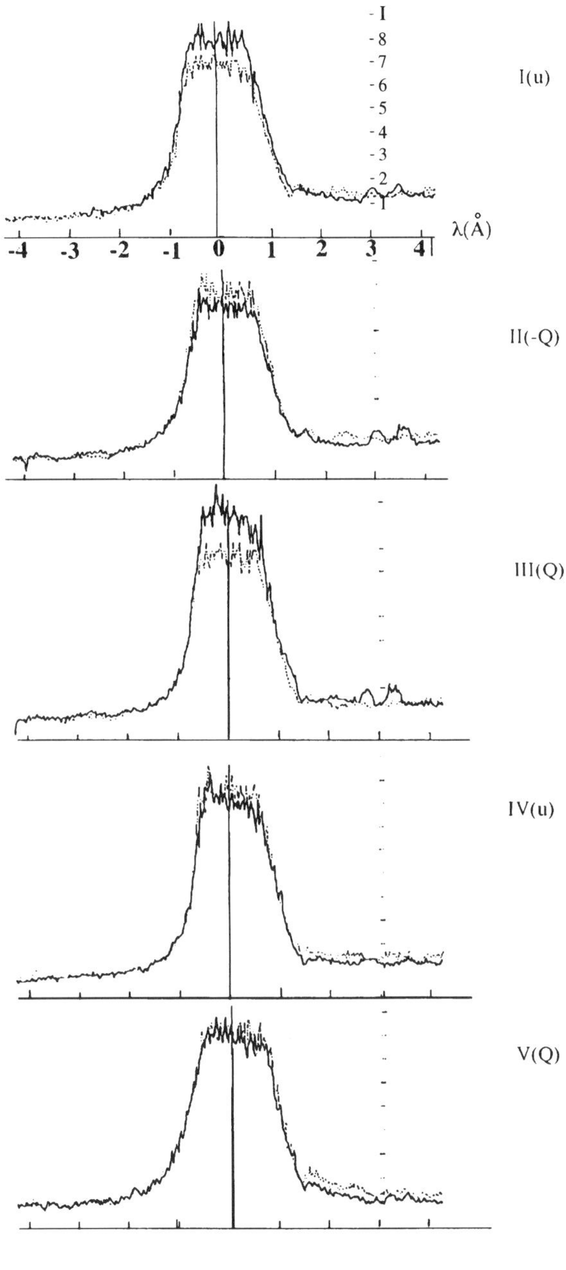

Figure 53. H_α intensity in two orthogonal linear polarizations for cutting $-3.75''$ on five spectrograms.

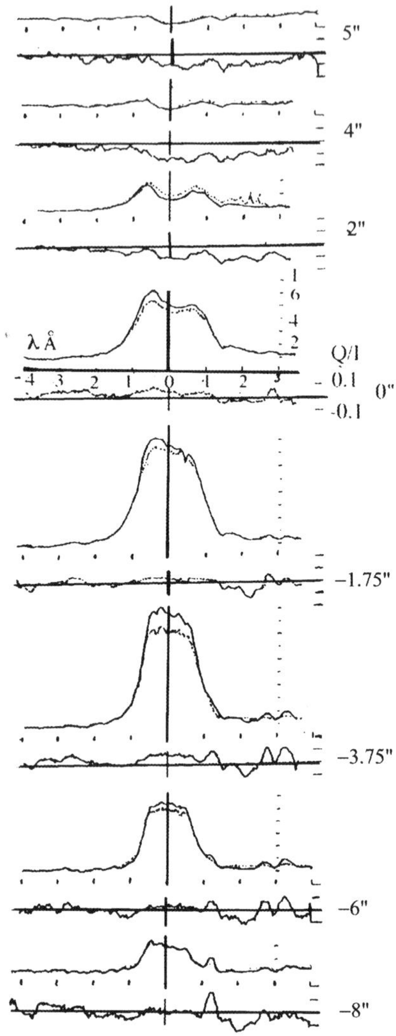

Figure 54. Variation in intensity and Stokes parameter Q with wavelength for all sections of spectrogram III.

within adjacent regions of the nonperturbed atmosphere was found to be slightly higher than 7–10%. The orientation of the polarization plane in most cases coincided with the direction toward the solar disk center, except for these locations at which the line profiles resembled contours, typical of mustaches.

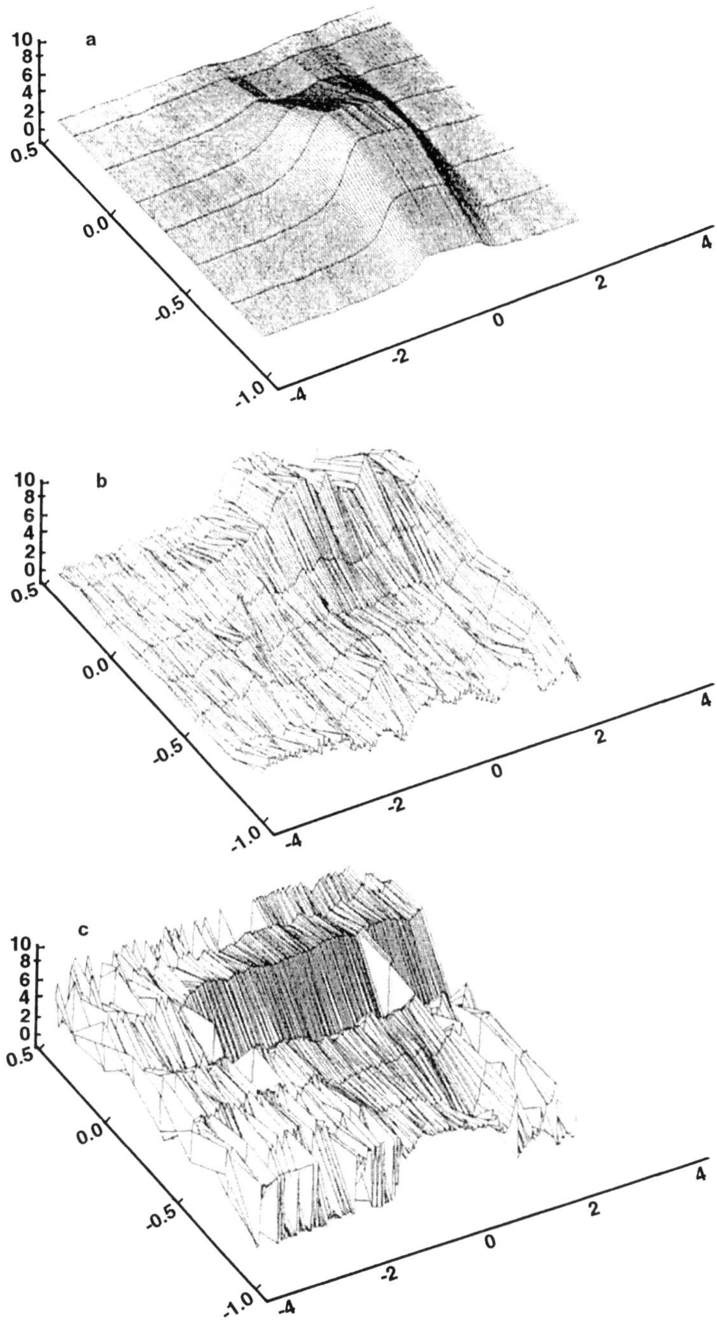

Figure 55. Three-dimensional representation of H_α line intensity I (a), degree of polarization P (b), and azimuth of the polarization plane χ measured from the orientation of the spectrograph entrance slit (c).

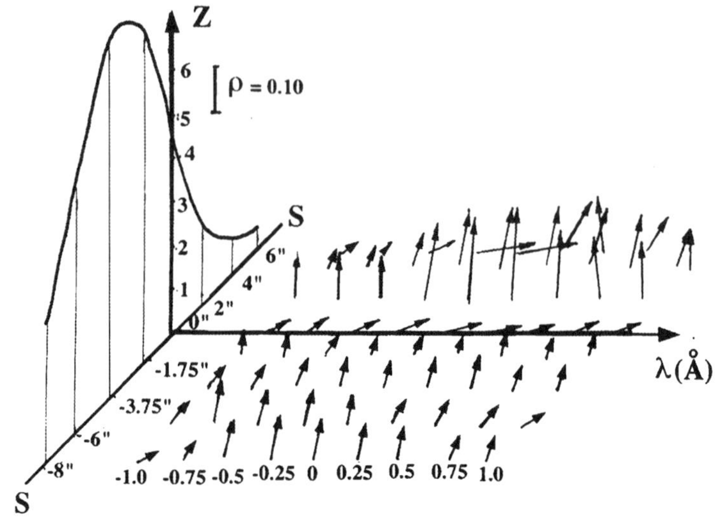

Figure 56. Flare intensity and polarization vector variation along the spectrograph entrance slit.

9.5. Concluding Remarks

All the observational results in this chapter confirm clearly the existence of impact linear polarization of hydrogen lines observed during solar flares. Unfortunately, the instrumental polarization was never fully eliminated. However, there were some effective attempts to minimize instrumental polarization to the level of statistical errors. First, an additional $\lambda/2$ phase plate, placed behind the Wollaston prism, makes it possible to completely eliminate the influence instrumental polarization effects introduced by the diffraction grating. Second, the observed line intensities were normalized to the continuum intensity in each orthogonally polarized band of the spectrum. This operation eliminates the effect of the telescope polarization matrix on Stokes parameter I and reduces considerably the influence of this factor on the other Stokes parameters. However, the maximum factor of polarization distortion of other Stokes parameters, or the instrumental polarization, introduced by the telescope cannot exceed 10%.

The averaged value of the degree of linear polarization observed in flares in the H_α line was about 5% and for some events it can reach 10%. Polarization in the H_β line did not usually exceed 4%. The orientation of the plane of polarization in flares, as a rule, coincided with the direction toward the solar disk center.

When cuttings were made along the emission part of the flare every $2''$ of arc, in the region where H_α profiles had a form typical for flares, the plane of polarization had an orientation directed toward the solar disk center. When the shape of the profiles changed to that resembling H_α profiles in mustaches, the orientation of the plane of polarization approached tangential and, subsequently, the vector of polarization again reverted to a radial orientation. A decrease in emission was usually accompanied by an increase in the degree of polarization.

10

Solar Flare Observations at Meudon Observatory (France)

10.1. Introduction

The first observations of impact polarization in the optically active region of solar flares, presented in this chapter, were conducted by the Meudon group. The significance of these data and general conclusions are very important in connection with the fact that detailed spectropolarimetric information was obtained and analyzed together with X-ray observations. Such a comprehensive approach can provide a clearer and deeper understanding of the evolution of a flare and the determination of the flare phase.

Solar chromospheric flares are usually regarded as secondary phenomena. The free magnetic energy as a primary source is assumed to be released in the corona, producing enhanced local heating, parameter gradients, and particle acceleration. Consequently, energy is transferred from the corona to the chromosphere via heat conduction, particle bombardment, or X-ray irradiation. The observed direction of the polarization plane of emission spectral lines is a very powerful factor in selecting the mechanism of chromosphere heating during a solar flare.

Local physical processes within ionized media in the presence of fields can also produce linearly polarized emissions of spectral lines and, more specifically, a hydrogen H_α atomic line, which is mostly used for spectroscopic studies of the solar atmosphere. Processes include the Zeeman effect, Stark effect, and reabsorption of anisotropic resonance radiation, or so-called optical resonance polarization, which, in connection with the alignment of atomic ensembles under laboratory and astrophysical conditions, have been analyzed and reviewed.[(2)] The influence of the first two mechanisms within the line center is expected to be smaller than the influence of anisotropic collisional excitation. Resonance polarization could compete with impact polarization, but merely in the vicinty of bright H_α patches. It is usually not so for locations outside bright patches, which are usually studied. The influence of the Zeeman effect on the hydrogen line is probable as flares are generally located above active regions, often with complicated magnetic field structures. For ordinary conditions of optical observations of solar flares, the Zeeman effect produces only negligible polarization in the line center ($\leq 0.1\%$) due to Doppler broadening of the H_α line.

Turbulent electric fields originating from plasma turbulence may produce linear polarization via the Stark effect. Under strong turbulence ($E^2/8\pi = 0.1N_ekT$) the turbulent field could reach 170 V/cm for chromospheric flare conditions and the corresponding Stark spacing produced by the dynamical field could be much smaller than the Doppler width. Estimates[(3)] showed that the contribution to the polarization fraction arising from these two effects could not exceed 0.5%.

Another strong argument in favor of impact polarization of the optical emission of solar flares is that the net Stark and Zeeman polarizations within the integrated line profile are zero. On the other hand, impact excitation of an atomic ensemble by a beam of particles may lead to the emission of a line with nonzero net polarization in the integrated profile. Therefore, observations of linear polarization integrated over a line profile provide evidence for the impact nature of polarization, which could then provide information about the velocity distribution function for particles at the origin of impact excitation.

A series of polarimetric observations at Meudon Observatory, determination of the degree of polarization and the azimuth of the polarization plane, was very important in establishing the impact origin of the polarization of the emission spectrum and, furthermore, in using the spectropolarimetric data for quantitative sensing.

10.2. Linear Polarization of Emission of the Flare on 17.05.1980

Four series of observations were made at 0738, 0739, 0740, and 0741 and described elsewhere.[(3)] The maximum brightness in the H_α center occurred at 0737 and the observation period corresponded to the decay phase of a 1B flare.

Observations were performed at the center of the H_α line by a filter with a bandpass of 0.75 Å. For these observations, the existing heliograph was slightly modified and equipped with the Lyot filter of bandpass 0.75 Å. A rotating half-wave plate was inserted in front of the linear polarizer, which was placed at the entrance of the H_α filter. The half-wave plate was rotated through 22.5° per step. Sixteen heliograms of the center of the H_α line were obtained on photographic film every 32 s, and successive sets of 16 images were taken every minute.

The intensity $I(\theta)$ at each point of the filtergrams was a function of the linear polarization of the studied optical radiation described by Stokes parameters (U, Q) and the angular position of the half-wave plate defined by angle θ, satisfying the following expression:

$$I(\theta) = \tfrac{1}{2}(I + U \sin 4\theta + Q \cos 4\theta)$$

Images were processed by a microdensitymeter connected to a computer. A least-squares fit computer code was used in order to adjust the relative positions of the images with an accuracy better than 1 arc sec. The Stokes parameters were subsequently inferred by adding or subtracting appropriate images:

$$S_i = \sum_i \left(I_i - \frac{I_{i-1} + I_{i+1}}{2} \right)$$

where S_i is either U_i or Q_i, depending on the parity of index i, and defines the angular position of the half-wave plate.

The set of 16 images corresponded to a full rotation of the half-wave plate and provided 7 determinations of the Stokes parameters. Linear polarization was observed above H_α kernals and outside these kernels, particularly along H_α ribbons. The polarization observed outside the kernals was generally low, or less than 2%, and subjected to effective temporal variations. At the two brightest H_α locations, observed on each side of the neutral line and outside sunspots, the radiation was polarized above the kernels during the period 0738–0741 with the stable direction of polarization lying in the plane defined by the vertical and the line of sight, while the degree of polarization was between 1 and 3%.

The direction of polarization was observed above the kernals and is in agreement with the supposition of H_α collisional excitation by nonthermal particles with a velocity distribution anisotropy oriented in the vertical direction. The collisional nature of the observed polarization was confirmed by the fact that polarization induced by the Stark effect was very unlikely for the center of the H_α line for which polarization was observed during this event. The Zeeman effect also did not contribute to the observed polarization because linear polarization was not detected above sunspots.

As a result it was safe to claim that the observed polarization during this event was real and its direction invariable for 4 minutes. These observations brought evidence that linear polarization of the spectral emission of a solar flare was linked to the process of energy transfer by suprathermal particles bombarding the chromospheric layers or by heat conduction. Linear polarization during this event was observed not only for the flare nodes but also above the filament of the H_α line.

10.3. Spectropolarimetric Effects of the Flare on 15.07.1980

A flare of importance SB and class C4 was observed on 15.07.1980 from 2248 to 2307 UT.[182, 201] This flare was located at 218″ S, 450″ E and was also observed by the Hard X-Ray Imaging Spectrometer (HXIS) and by the Hard X-Ray Burst Spectrometer (HXRBS) on the SMM. Figure 57 shows the time variation of soft X-ray emission above 3.5 keV observed by HXIS, and the time variation of $33 < E < 55$ keV hard X-ray emission observed by HXRBS. The time variation of UV emission integrated over the polargram's field of view is also shown in Figure 57.

The first brightening of the SI 1437 Å was observed in polargram 12 at 2248. Since the time lag between two consecutive observations of the same point was 148 s, this implies that the UV flare started between 2246.14 and 2248.42. The time profiles of soft X-ray and of UV emission appear to be very similar. However, impulsive brightenings in the chromospheric line could easily have been missed because of the long raster time. Soft X-ray and UV emission rose at the start of the impulsive hard X-ray emission.

Figure 58 shows polargrams 12–14 obtained, respectively, during the rise phase, the maximum phase, and the decay phase of soft X-ray emission. The fine field of view of HXIS covers partially the polargram field of view. The HXIS fine FOV images at $11.5 < E < 16$ keV are also shown in Figure 58. These images were obtained by integrating over the period of observation for each polargram. In the region common to UV polargrams 12 and 13 and the X-ray images, the

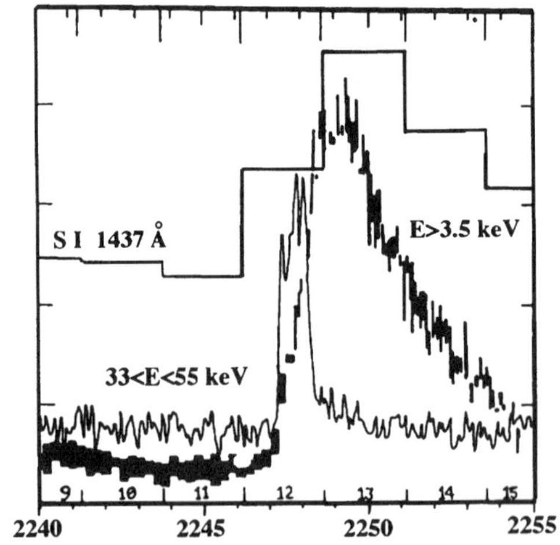

Figure 57. Time variation of soft X-ray ($E \geq 3.3$ keV), hard X-ray $33 \leq E \leq 55$ keV), and SI 1437 Å UV emission observed by HXIS, HXRBS, and UVSP. Starting and termination Moments of observations are shown.

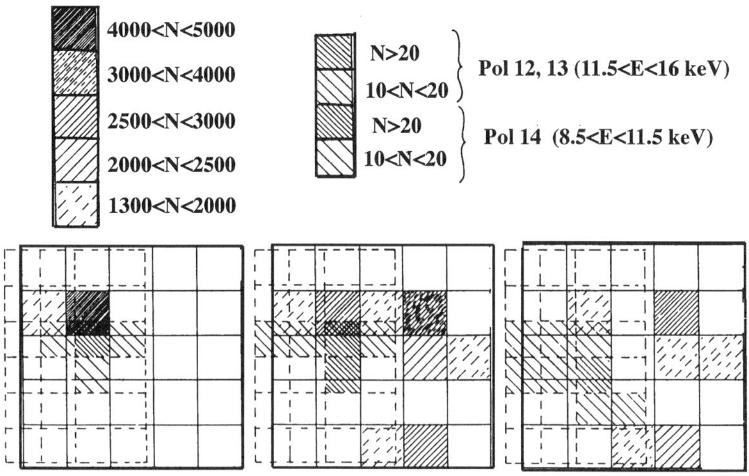

POLARGRAM 12 (11.5<E<16 keV) **POLARGRAM 13 (11.5<E<16 keV)** **POLARGRAM 14 (8.5<E<11.5 keV)**

Figure 58. Polargrams 12 and 13: spatial distribution of X-ray ($11.5 < E < 16\,\mathrm{keV}$) and SI 1437 Å line emissions. Polargram 14: spatial distribution of X-ray ($8 < E < 11.5\,\mathrm{keV}$) and SI 1437 Å line emissions. The X-ray emission is integrated over the rastering time of a full UV polargram, i.e., $t_R \approx 2\,\mathrm{min}\ 30\,\mathrm{s}$.

brightest X-ray point corresponds to the brightest UV point to within $4''$. The difference could result from adjustment incertainties ($< 10''$) or may be the actual difference in the locations of UV and X-ray emissions.

Signal-to-noise ratios for linear polarization observed at each individual pixel of the 19 observed polargrams was generally low, except for a few cases. In order to detect any polarization due to a process taking place over a wide field of view and satisfying symmetry conditions around the solar vertical, the data were integrated over each polargram. For each polargram the $\sum \Sigma S_i$ of the relative intensity variations derived by integrating over 25 pixels, and the resulting polarization amplitude, phase angles, and signal-to-noise ratios were derived. In polargrams 12, 13, 14, and 15 the flaring points were processed separately from the nonflaring points. The distinction was based on the intensity level. All the flaring pixels ($I > 1300$ counts/gate time) were grouped together in each polargram, and all the nonflaring pixels ($I < 1300$ counts/gate time) were also grouped together. Twenty-three values of polarization amplitude P and phase angle ϕ were derived, four of them corresponding to the flare area.

Histograms $N(\phi)$ have been constructed using 23 values of P and ϕ. These histograms, shown in Figure 59, gave the number of times a given phase angle was obtained in the way explained above. The histograms were obtained by taking samples of the phase angle ϕ related to a degree of polarization with signal-to-

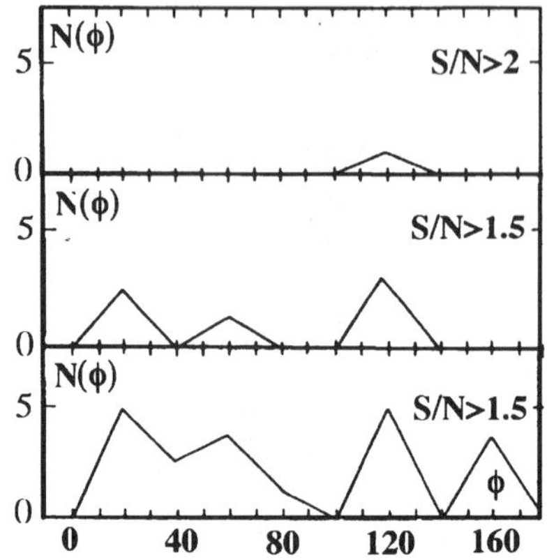

Figure 59. Histograms $N(\phi)$-number of times a given phase angle ϕ is obtained by integration over the nonflaring I < 1300 counts) and flaring (I > 1300 counts) region of 19 polargrams.

noise ratios, exceeding respectively the values 0.5, 1.5, and 2. When all the flaring and nonflaring areas of the 19 polargrams with signal-to-noise ratios as small as 0.5 were considered, there was no indication either of polarization or of a preferential direction of polarization. By increasing the signal-to-noise ratio limit up to 1.5, two peaks appear in the histogram separated by about 90°. The associated phase angles were 120° and 20°, corresponding respectively to the center of the disk direction and to the perpendicular direction. Only the peak at 120° appeared with the signal-to-noise ratio limit was raised to 2. These points are all in the flaring area of polargram 14.

The mean values of intensity, degree of linear polarization P, and phase angle ϕ as the angle between the electric vector and the North–South direction for six flaring pixels (I > 1300 counts) are presented for the SiI 1437 Å line in Figure 60. A vectorial representation of P and ϕ was used. Three of the six flaring points showed linear polarizaation of the SiI 1437 Å emission line with the degree of polarization as high as 25%. The angle between the North–South direction and the electric vector varied from 109° to 117°.

This indicates that the direction of vibration of the electric vector at half of the flaring points was very nearly parallel to the flare-to-disk center direction (this direction made an angle of 116° with the South–North direction). All the signal-to-noise values and probabilities of occurrence discussed below are for photon statistics only and do not include any noise introduced by nonlinear solar intensity variations.

Figure 61 shows the result of Fourier analysis of three sets of data on flaring points: (a) S_i observed at one flaring point for which the signal-to-noise ratio was 2.1 (this point was located in the southwest quarter of polargram 14); (b) $S_s = \frac{1}{6}\sum S_i$, or the mean value of the six S_i values derived at the six flaring

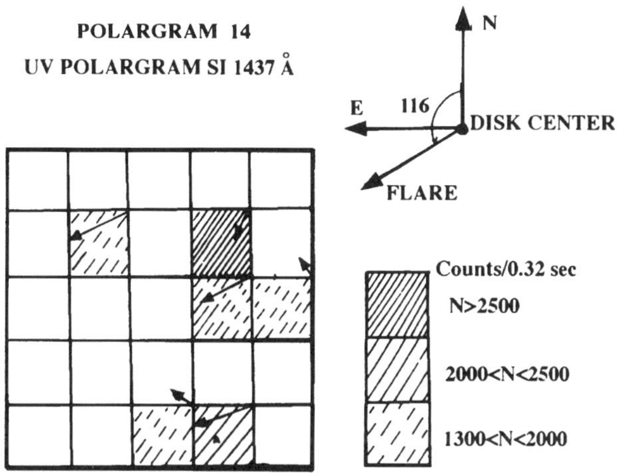

Figure 60. Polargram 14: linear polarization in the SiI 1437 Å emission line is represented by vectors of amplitude proportional to the coefficient of polarization. These vectors are oriented in the direction of the polarized electric vector. The origin of each vector is at the right and upper corner of the point to which it refers. The highest degree of polarization corresponds to the electric vector parallel to the direction from the flare to the center of the solar disk.

points; and (c) $S_s = \frac{1}{3}\sum S_i$, or the mean value of the three S_i values derived with the signal-to-noise ratio more than 1 within three or six flaring points.

The maximum of the lower spectrum was at the 4ω frequency. The peak at this 4ω frequency appeared more clearly by adding data corresponding to different flaring points. The power was enhanced crucially, and up to 46–48% of the power corresponded to the 4ω component. Other Fourier components due to the noise effect are reduced by integration. The degree of polarization, phase angle, and signal-to-noise ratio associated with the three sets of data described above are given in Table 13.

The selection procedure used for data presented in Table 13 and Figure 61 requires further explanation. The degree of polarization (P), phase angle (ϕ), and

TABLE 13
Degree of Polarization, Phase Angle, and Signal-to-Noise Ratio for the Flare Region ($I > 1300$ counts)

Polargram	P(%)	ϕ°	S/N
12	9	121	1.1
13	2	74	0.4
14	12	113	2.1
15	6	20	0.5

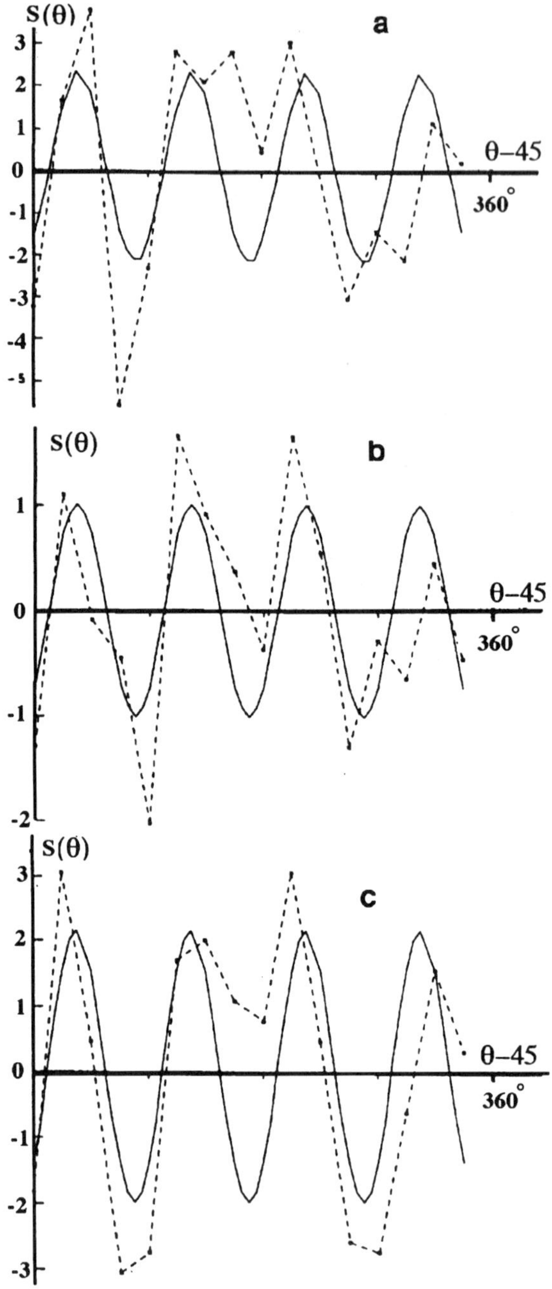

Figure 61. Fourier analysis of (a) intensity modulations $S_I(\theta)$ at point A (see Figure 59): dots are observed values and the full line is the Fourier coefficient at four times the frequency of one waveplate rotation; (b) mean of intensity modulations S_i of the brightest points, $S = \frac{1}{6} S_i(\theta)$; (c) mean of intensity modulations S_i of the three brightest points for which signal exceeds noise.

signal-to-noise ratio (S/N) were obtained by adding all the relative intensity modulation S_i in the nonflaring area ($I < 1300$ counts) and flaring area ($I > 1300$ counts). In the flaring area P, ϕ, and S/N were derived by adding the relative modulation over all flaring points, and also by adding the relative modulation of pixels for which the signal-to-noise ratio was more than 1. It is evident from Figure 60 that the three flaring points with highest signal-to-noise ratio of the polarization amplitude have their phase angles in the same direction. The combined signal-to-noise ratio of the six flaring points was only 2.1 in the polarization amplitude, as shown in Table 13 and Figure 61b, but this value ignores the phase angle information.

When pixels with signal-to-noise ratios of the polarization amplitude higher than 1 were selected, the combined signal-to-noise ratio became 2.9, which occurred in less than 1% of cases, by chance. This happened due to the constructive combination of the phase angles. If an individual pixel has been polarized by chance, their phase angles would have been random and the combined signal-to-noise ratio would likely have decreased below the individual values instead of being increased. The probability of obtaining polarization at a position angle close to that of the sun center direction could be also examined. The probability of the observed polarization being within a 20° band of the sun center direction was 20/180, or 0.11, so that the overall probability of obtaining the observed result was 1.1×10^{-3}. Individual position angles of the three pixels contributing to the result are 109°, 113°, and 117°, compared to the sun center direction of 116°.

These probability estimates were obtained empirically. Theoretical estimates could not be used because of the selection criteria adoped. The mean intensity and linear trend over position angle for each pixel in polargram 14 was studied. Synthetic data created with the same set of mean values and linear trends with no polarization but with photon shot noise have been added. Four hundred such fake data sets were then treated to the same polarization analysis and selection criteria as the real data. In four cases a signal-to-noise ratio of 2.9 or greater was obtained. Therefore, the probability of obtaining a 2.9 signal-to-noise ratio was estimated as 1%.

An additional validity test of these results was obtained by studying the histograms $N(\phi)$. These histograms give the number of times a given angle ϕ is obtained in different data sets. Figure 62 shows the histogram associated with all the observations (14 orbits, 212 polargrams) and the histogram associated with polargram 14. Within statistical accuracy, $N(\phi)$ derived from all observations is independent of ϕ. This demonstrates that there is no systematic instrumental or data processing effect that might select particular polarization directions.

Polargram 14 showed a peak of $N(\phi)$ for $100° < \phi < 120°$. This peak was particularly evident for the data set which included only the brightest points. Selecting only points for which the signal-to-noise ratio was more that 1 lead to a

Figure 63. Filtergrams showing H_α chromospheric brightening during three solar flares observed on 11.07.1982. The filtergram size is about $3' \times 3.3'$ with a pixel of $3'' \times 3''$. The direction of the electric vector in the flaring region and the flare to disk center direction are shown, respectively, by single and double line arrows.

signal-to-noise ratio was obtained by integrating over about 10 minutes. The polarization effect was very critical in this case with respect to the observational interval: one minute beyond the selected integration time, the polarization signal became indetectable.

A degree of linear polarization as high as 2.5% was observed at a few locations in bright H_α patches. Figure 63 shows locations on the filtergrams where polarization was present. The direction of vibration of the polarized electric vector was close to the flare to disk center direction. This result was in agreement with SMM, which showed that the electric vector in the SI line was also directed along the flare to disk center direction. As shown in Figure 63 polarization was observed during a rise in soft X-ray emission and terminates near the moment corresponding to maximum soft X-ray emission. The polarization for this event did not appear to be definitely related to the impulsive hard X-ray emission.

No polarization was observed above sunspots in the absence of flares. Strong Doppler broadening and the observed spatial distribution of the polarization vector averged out the possibility of Zeeman and Stark effects, or polarization induced by reabsorption of anisotropic resonance radiation being an origin of observed polarization. Impact polarization by beams of particles in this case was also the most plausible interpretation.

As a result of these observations it should be noted that the temporal resolution of polarization in this case was very low. Reliable detection with the appropriate signal-to-noise ratio was obtained only by integration of data over a 10-min interval. For all three flares, polarization was observed at the growing phase of soft X-ray emission just before the maximum. At the initial stage of X-ray emission growth and just after the maximum, polarization was not observed. The degree of polarization was at least 2.5% for some bright points for all these flares. The direction of electric vector oscillations was close to the direction to the solar disk center. Analysis of observations lead to the conclusion that the best candidates to explain linear polarization of the H_α line are protons bombarding the solar chromosphere.

10.5. Polarization Effects of the Emission of Two Flares on 17.07.1982

Two chromospheric H_α flares were observed on 17.07.1982 at coordinates 11°N, 38°W and 17°N, 29°W, with starting and terminating moments 1025, 1105 and 1032, 1102, respectively.[(184)] Intense soft X-ray emission was observed, reaching 3×10^{-4} and 8×10^{-5} W·m^{-2} within energy bands of 1–8 Å and 0.5–4 Å. The position of the two flares on the solar disk with respect to the solar disk center is shown in Figure 64.

The H_α line appeared to be linearly polarized during the flares. The time development of the mean H_α line intensity within a $60{,}000 \times 60{,}000\,\text{km}^2$ area, including the flare on the western part of the field of view, is plotted in Figure 65. The time dependence of the relative number of pixels in the selected area with polarization fraction higher than 2% is also presented in this figure. The relative number varies from 10% before the flare to nearly 50% at its maximum. Strong fluctuations occurred around a mean value that appears to be slightly correlated with time.

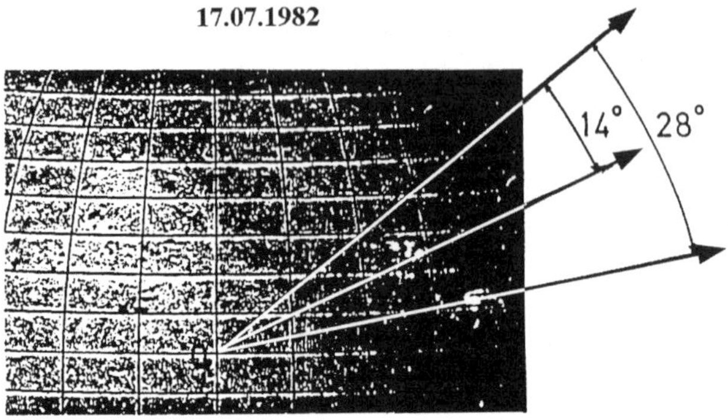

Figure 64. Positions of observed flares on the solar disk. The angles between reference axis *OZ* and the two flare to disk center directions are shown. The limits of the digitized image are plotted as dashed lines.

An image of the spatial distribution of the polarization, including the degree of polarization and the azimuth for the western flare on 17.07.1982 at 1040 near the maximum of H_α emission, is presented in Figure 66. In the brightest H_α patch the direction of polarization is close to the flare to disk center direction. However, polarization was not restricted only to the brightest regions, presumably related to the feet of the magnetic flare loops, but was also present within cooler regions. A great variety of orientations of the polarization direction was seen.

The Stokes parameters have been integrated around the maximum of H_α emission, or over the period from 1038 to 1051. Histograms of the azimuthal

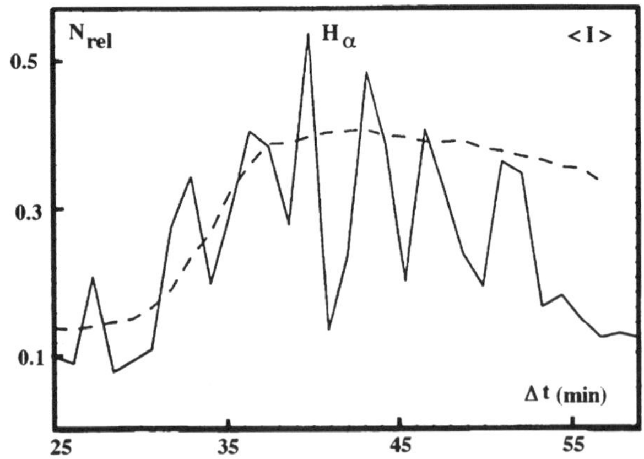

Figure 65. Time dependence of the mean H_α line brightness and the relative number of pixels with polarization fraction higher than 2% on the western 17.07.1982 flare area. Dashed lines refer to the mean H_α intensity and full lines to the relative number of pixels.

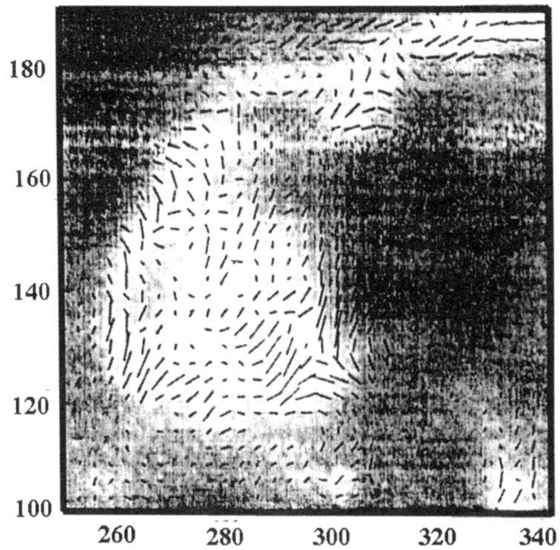

Figure 66. Orientation of the polarized electric vector in the western H_α flare at 1040 for a degree of polarization more than 2%. One pixel of the H_α intensity image is 1 arc sec. The polarization has been integrated over 3 × 3 pixels. On the left an arrow indicates the radial direction for the brightest H_α patch.

distribution of the 14-min time-integrated polarization that exceeds 2% are shown in Figure 67. Three histograms have been prepared for the full field of view and for the two separate flaring regions in this field. All three histograms are clearly peaked. For the regions at the east and west parts of the field of view, for which the azimuth of the flare to disk center direction are, respectively, $14^\circ \pm 2^\circ$ and

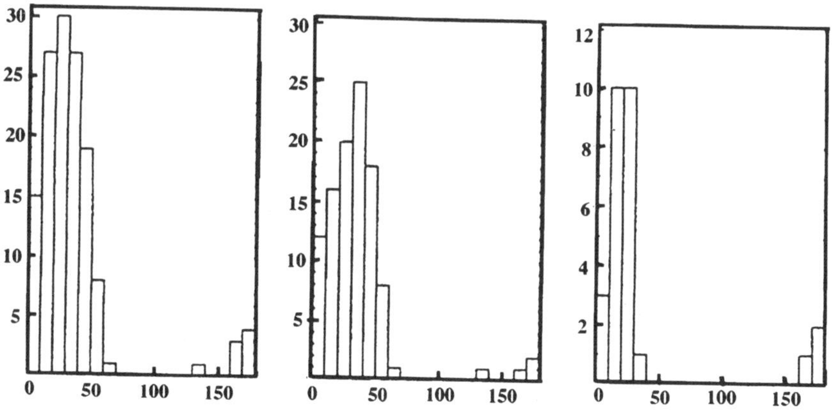

Figure 67. Histograms of the azimuthal distribution of the degree of polarization integrated from 1038 to 1051 that exceeds 2% in the full field of view (left), and in the two flaring regions on the west (center) and east (right). The reference axis for the definition of the azimuth is presented in Figure 65.

$28^\circ \pm 3^\circ$, the maxima of the histograms are at azimuthal angles equal to, respectively, $20^\circ \pm 5^\circ$ and $35^\circ \pm 5^\circ$.

Clearly, the most frequent azimuth of the linear polarization integrated during the maximum of H_α emission differs from the azimuth of the flare to disk center direction by less than 7%. The radial direction of the net polarization can be understood as due to impact polarization produced by particles spiralling around the magnetic field lines within the magnetic loops of the solar flare. In each point of an H_α emitting loop the resulting direction of the polarization vector is the projection of the magnetic field direction on the solar disk. Integration over time does not remove the radial direction of polarization because of the predominant contribution of the vertical component of the field. In comparison with the impact mechanism, polarization due to reabsorption of the anisotropic resonance radiation as a working mechanism leads to a directon of net polarization perpendicular to the radial direction of the solar disk.

Histograms of the azimuthal distribution of the 1-min time-integrated polarization exceeding 2% in the western flare area during some period of time Δt have also been drawn for this flare. The histograms that cover the rise of H_α emissions from 1025 to 1040 and the decay phase in the western flare are shown in Figure 68. They peak at an azimuth angle of $25^\circ \pm 5^\circ$ and should be compared with the azimuth of the flare to disk center direction, equal to $28^\circ \pm 3^\circ$. The peak is more pronounced during the rise of H_α emission. This position of the peak provided the idea that the impact polarization was observed at the feet of flaring loops where the distribution of magnetic lines of force was, presumably, symmetrical around the solar vertical.

In summing up the results for these two flares, we should point out that a degree of polarization of more than 2% was observed within 10% of the

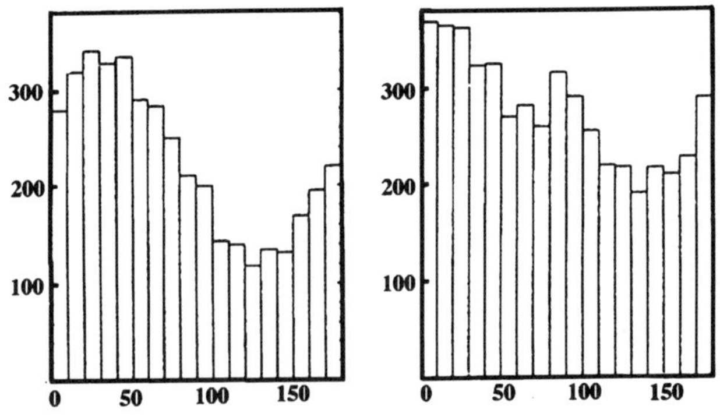

Figure 68. Histograms of the azimuthal distribution of the 1-min time-integrated polarization that exceeds 2% in the western flare area during the rise (1025–1040) and after the maximum (1041–1059) of the H_α emission.

$60{,}000 \times 60{,}000\,\text{km}^2$ region, including those flares on the solar disk. At the moment of maximal intensity of H_α, the number of pixels where polarization was observed already attained 50%. A clear understanding of the orientation of the polarization vector can be obtained only while integrating observations during a period of more than 10 minutes. For the bright part of a flare, the orientation of the polarization vector was radial. The linear polarization observed in cold regions or these flares was characterized by significant variation in the orientation of the polarization vector.

10.6. Concluding Remarks

Observations of several solar flares on 17.05.80, 11.07.82, and 17.07.82 were observed by the Meudon group in the center H_α line with a filter bandpass of 0.74 Å. Polarization of the flare on 17.05.80 remained stable during 4 minutes. Reliable polarization data with an appropriate signal-to-noise ratio were obtained only with time integration over a 10-min interval for three flares on 11.07.82 within the complex active region and for two flares on 17.07.82. The average degree of polarization in these flares was 2–3%. In the majority of cases the direction of electric vector oscillation was close to the direction to the solar disk center.

During observations of two flares on 17.07.82, intense soft X-ray emission was seen reaching 3×10^{-4} and $8 \times 10^{-5}\,\text{W/m}^2$ in the spectral bands 1–8 Å and 0.5–4 Å. For all three flares on 11.07.82 polarization was observed at the growing phase of X-ray emission just before the maximum. AT the initial stage of X-ray emission growth and just after the maximum, polarization was not observed.

A flare of importance 2B and class C4 (15.07.80) was observed in the UV emission in the SI 1437 Å on the SMM. For some knots the degree of polarization attained 25% with a signal-to-noise ratio of 2.9. The degree of polarization averaged over six bright points was 12%. The polarization vector was mainly directed to the center of the solar disk with scatter not exceeding 3°. Linear polarization of this flare was observed at the phase of decaying of soft X-rays and probably at the pulse emission stage of hard X-rays.

A few mechanisms could be invoked to explain the beaming of particles responsible for excitation of the chromospheric layer during a solar flare. This was analyzed by the group at Meudon Observatory. They based themselves on the main result, reproduced in the course of all observations, that the polarization vector was directed to the solar disc center.

1. Local acceleration within discontinuities of potential-like local double layers.[183] Single double layers would have to be strong ($\chi/kT > 1$) in order to accelerate electrons and protons to energies χ of at least 12–20 eV in a medium with a mean thermal energy kT of 1–2 eV. Multiple double layers may be weak

interesting effect.[189] In one linear polarization the mustaches had an ordinary structure and, in an orthogonal polarization, a double structure was observed within the red wing. Photometric processing of mustaches has been conducted in order to analyze the structural features and measure the degree of polarization.

Photometric cuttings were made perpendicular to the direction of dispersion within several spectral bands in the blue and red wings of the H_α line. The close wing region was located at $+1.0$ Å; -1.3 Å, and for the far wing region at $+1.7$ Å; -1.8 Å from the center of the line. The intensity distributions obtained in such a way were compared with the help of the image of a thread fixed across the slit of the spectrograph. Correction for instrumental polarization was made using those locations of the registograms where cutting was within the near-photospheric region.

Polarization spectrograms have been obtained for mustaches on 06.10.1983 at 0717 and 0718.20 UT. Strong differences, especially within the red wing, were detected in the distribution of intensity across the dispersion for the spectral bands in two orthogonal polarizations for the mustaches observed at 0717 UT. The mustaches seem to consist of two emission details separated by 1.2″. One fragment of the mustaches did not show polarization, while for the other the degree of polarization attained 23%. Within the red wing this polarization detail was seen over the whole range of the mustach. Within the blue wing it was observed only for those parts of the wing close to the center of the H_α line but was

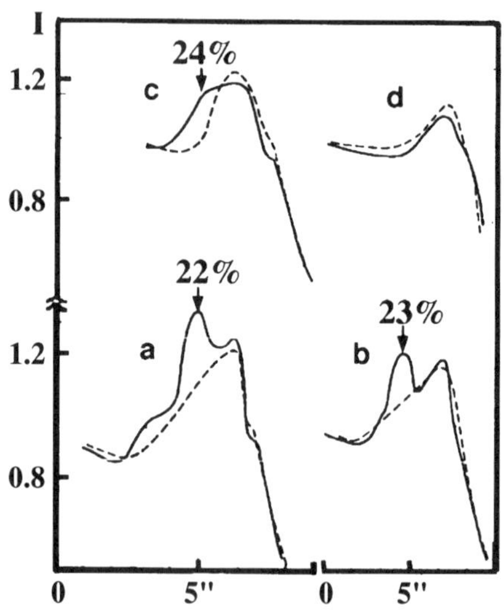

Figure 69. Photometric cuttings of mustaches on 06.10.1983 at 0717 UT perpendicular to the dispersion for the H_α line within the near-red wing (a), far-red wing (b), near-blue wing (c), and far-blue wing (d). Solid and dashes lines indicate orthogonal and linear polarizations while arrows show the points where the degree of polarization was calculated.

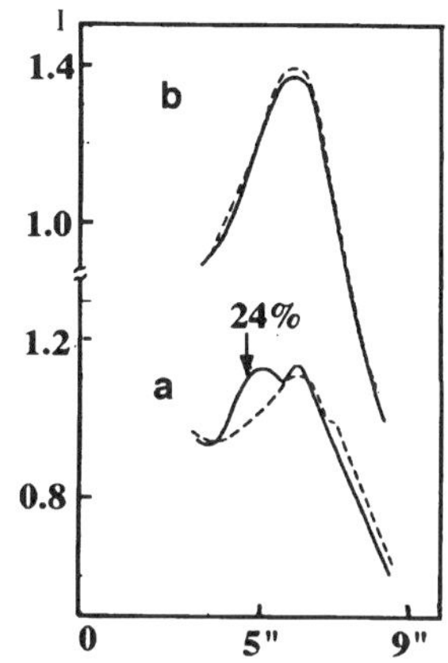

Figure 70. Same as Figure 69 at 0718.20 UT for the (a) far-red wing and (b) far-blue wing.

not seen for remote locations (Figure 69). Similar spectropolarimetric detail within the red wing has been detected for the same mustaches at 0718.20 UT but with lower quality (Figure 70). The degree of polarization was around 24%.

The spectrograms for mustaches on 19.10.1983 were obtained at 0542 UT. Two mustaches were detected in the vicinity of the limb with small spatial separation (1.8″). It follows from Figure 71 that one mustache was not polarized, but the other, weaker and located closer to the limb, showed a significant polarization. The degree of polarization of the latter mustach in the blue wing of H_α line was about 12% and for the red one, 17%.

A spectrogram for the same event was taken later at 0719 UT with the aid of a large solar telescope. Mustaches were observed within the same active zone as for the previous case. Figure 72 shows that for the near-red wing of the H_α line in one polarization split emission of mustaches takes place while a single emission exists in the other polarization. This feature was not detected within both the blue wing and the far-red wing.

Drawings of the edge of the solar disk, the position of mustaches, and the orientation of the analyzer of linear polarization are shown in Figure 73. The axes of the analyzer were oriented at 45° to the direction of the entrance slit.

For the mustaches observed on 06.10.1983, one of the axes of the polarizer was inclined at 50° to the direction to the solar disk center and the other was oriented at 40° to the same direction. The double mustaches at the red wing at 0717 UT and the polarized detail within the same wing 0718.20 UT indicate that the direction of polarization of the electric vector was closer to radial.

For the mustaches on 19.10.1983, taken at 0542 UT and showing polarization of emission, the preferential direction of oscillations of the electric vector was close to tangential. The angle between the polarization direction and

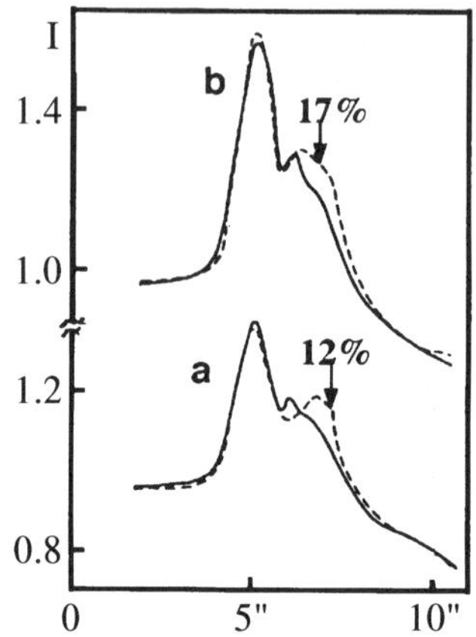

Figure 71. Same as Figure 69 for the flare on 19.10.1983 at 0542 UT.

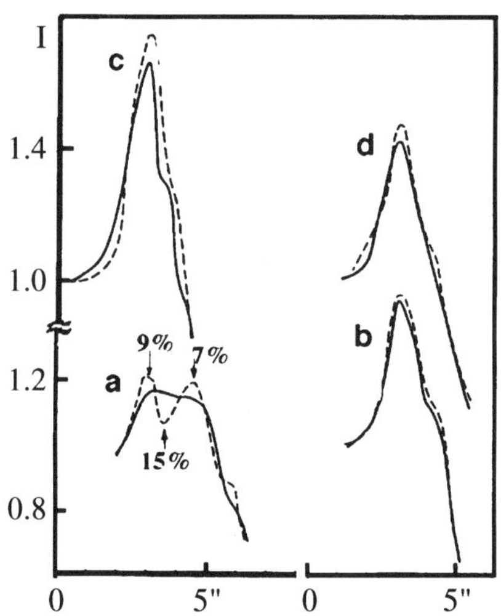

Figure 72. Same as Figure 69 for the flare on 19.10.1983 at 0719 UT.

the radial direction was 78°. For mustaches recorded by a large solar telescope within the same active region after 0137, doubling of the spectrum was observed on that spectral band corresponding to tangential polarization (the angle between the polarization axis and radius was about 65°).

The studied mustaches on 06.10.1983 and 19.10.1983 (0719 UT) were rather spacious (3″ approximately), consisting of at least two fine structure elements which could be at different stages of development. Within the close pair of mustaches on 19.10.1983 (0542 UT) the brighter component was not polarized, but the weaker one was polarized. The same picture was observed by the same astrophysical groups for flares, so they propose that polarization of emission of mustaches or their fine structure elements exists only at the initial stage. For some mustaches close to the solar limb the polarization of H_α emission was predominantly within the red wing and sometimes attained 24%. The preferential direction of oscillations of the electric vector for the mustaches observed on 06.10.1983 was radial, and for the mustaches on 19.10.1983 it was closer to tangential.

11.3. Spectropolarimetric Measurements of Mustaches with an H_α *Filter*

Strong polarization of mustache emission is a short temporal effect which is typical for a certain stage of their development. The temporal variation of polarization within the mustache emission is very difficult to study together with the spectral resolution. Therefore, some observations at the Crimean Astrophysical Observatory were made with a large coronograph crossed with an interference polarization filter.[(191)] One advantage of the interference polarization filter is that, during one exposure, it is possible to obtain a picture of the whole active region with all the mustaches.

The analyzer of linear polarization consisted of an interference polarizer and a rotating half-wave phase plate. The phase plate was fixed instead of the artificial moon on the large coronograph and rotated in its plane with angular velocity 5° per second. The emission of mustaches is concentrated within the wings of the emission lines, so the bandpass of the interference polarizing filter of $\Delta\lambda = 0.5$ Å was shifted from the center of the H_α line at ± 1 Å. An image of the active region with mustaches was fixed at the center of the field of vision and a series of shots for five positions of the $\lambda/2$ plate (0, 22.5, 45, 67.5, 90°) taken. Two exposures were obtained at every position of the $\lambda/2$ plate as a run, and the whole shooting process for a series of five positions took 25 seconds.

A group of four pairs of spots appeared and developed in the northwest part of the disk on 06.09.1984. The studied mustaches were observed within this group of spots. On 07.09.1984, the group was located at a latitude of 11° N at a distance of $\rho = 0.93$ from the center of the solar disk. A system of arch filaments AFS and a large number of mustaches were observed within this group. Spectropolarimetric studies with the help of an interference polarization filter were carried out on 07.09.1984. During the period 0541–0630 UT, 41 series of shots were taken. More than ten mustaches were observed on the filtergrams during some time intervals. Photometric processing was conducted for the very clear and easily detectable mustaches on the subsequent shots.

Measurements were undertaken by a microphotometer connected to a photomultiplier. The slit was chosen of size $16'' \times 1''$ to insure that the size of mustaches is bigger than the slit dimension. The contrast factor for mustaches (ΔI) was determined by

$$\Delta I = \frac{I_{mus}}{I_0} - \frac{(I_n)_{av}}{I_0}$$

where I_{mus} is the measured intensity of a mustache, I_0 is the intensity of a part of the unperturbed chromosphere in the vicinity of the active zone, I_n is the intensity of the background and comprised the averaged intensity of the chromosphere for four measurements made at four opposite sides with respect to the studied mustache.

Detailed measurements ranging between 5 to 24 series were implemented for 12 mustaches having different starting time and lifetime. Around 125 values of the contrast factor ΔI for mustaches as a function of the rotation angle of the phase plate φ were plotted for the whole series of observations. Detailed analysis showed that for some series of one or several mustaches, the contrast depends on the rotation angle of the $\lambda/2$ plate, indicating that the emission of these mustaches is polarized. Figure 74 presents functions $\Delta I(\varphi)$ for four mustaches and for three series of observations (0541, 0544 and 0550 UT).

A very high degree of polarization sometimes appeared for some mustaches two or three times during their observation period. The degree of polarization was not determined because of some difficulties connected to variations in the quality of the image of mustaches and changes in intensity during the shooting of series. Statistical processing of data was undertaken in order to determine weak, but more long-term polarization, not suppressed by the present variations in the brightness and quality of the image of mustaches. By making some

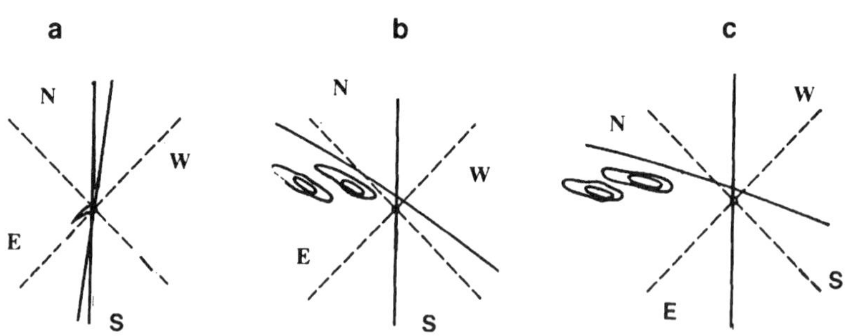

Figure 73. Relative orientation of the axes of the linear analyzer, the entrance slit of the spectrograph, mustaches, and the solar disk edge on: (a) 06.10.1983, (b) 19.10.1983, 0542 UT, and (c) 19.10.1983, 0719 UT. The thin line is the solar disk edge, dashed lines are the axes of the analyzer, and the thick line is the entrance slit position.

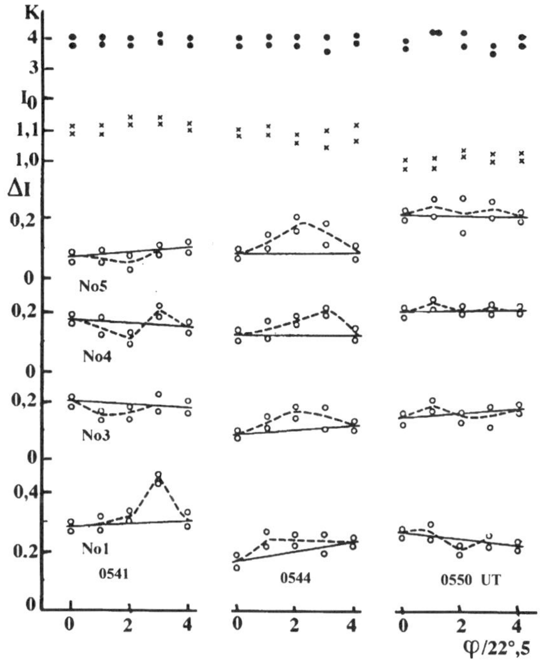

Figure 74. The contrast of four mustaches ΔI (circles) and intensity of the unperturbed chromosphere I_0 (crosses) as a function of the rotation angle of the phase plate φ and the visual estimate of the image quality K (points) for three series of observations.

straightforward calculations with the observed values of $\Delta I(\varphi)$, the mean deviations in the contrast of mustaches $\bar{\Delta}(\Delta I)$ during the series of observations for different positions of the $\lambda/2$ plate were obtained. These results are presented in Figure 75 for six mustaches. For all mustaches the maximum excessive intensity was observed for the second and third positions of the $\lambda/2$ plate, although they are comparable with the least-squares errors of observations indicated in Figure 75. These weak maxima indicate that an emission of mustaches is slightly polarized during the significant part of their lifetime.

By drawing the group of spots, the studied mustaches are given in Figure 76 together with the directions of the axes of the analyzer, corresponding to the positions of the $\lambda/2$ plate for which the shots have been conducted. The arch filaments (AFS), related to mustaches, indicated the formation of the new magnetic field structure and the filaments follow the magnetic lines linking regions with different polarity. Figure 76 shows that the directions of the polarization plane of mustaches (2 and 3) are approximately parallel to the filaments of the system AFS or to the magnetic field lines, and are also close to the direction to the solar disk center.

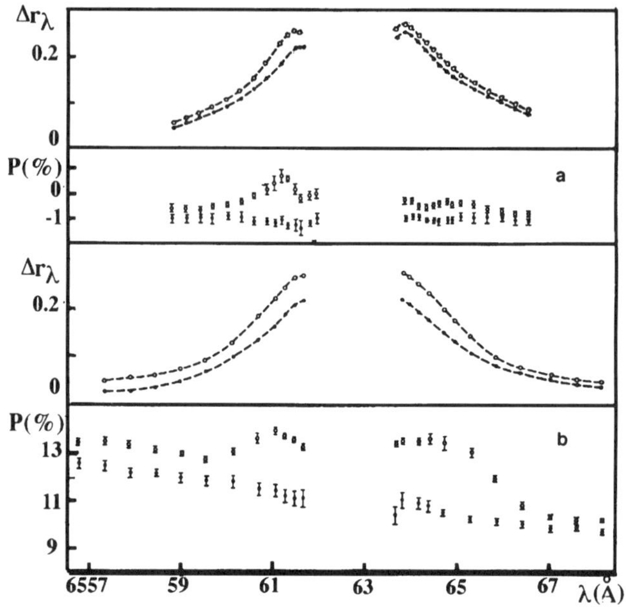

Figure 77. Wavelength dependence of the degree of polarization P for photometric cuttings along a mustache emission of a H_α hydrogen line (circles) and of instrumental polarization (dots) for mustaches on 14.11.1985 at 0919 UT: (a) IV spectral order, (b) II spectral order; Δr_λ is the mustache emission for two orthogonal linear polarizations.

the line intensity. This result indicates that the polarization effect is connected to the mechanism of formation of mustaches.

The emission of sharp clear mustaches possessed the same spectral nature in different polarizations. In some examples, for which emissions with two polarizations had different structures, photometric cuttings perpendicular to the dispersion were made. Some examples of these cuttings are presented in Figures 78 and 79. Figure 78 shows that in cases when the mustaches have rather big dimensions, they consist of different elements with different states of polarization. These elements were reproducible for two spectrograms obtained at an interval of 0120. The results of polarization computation are given in Table 14, where ψ is the azimuth of the oscillation plane of the electric vector, measured from the direction to the solar disk center for the E–N–W–S sense of rotation. The directions of the photometric cuttings as well as the brightness of the mustaches are also presented in Table 14. Figure 80 shows the distribution of the azimuths of the polarization plane of mustaches with respect to the direction toward the solar disk center.

In conclusion, it was detected that the mustache emission was polarized for all the contours and the degree of polarization sometimes exceeded 20%. The

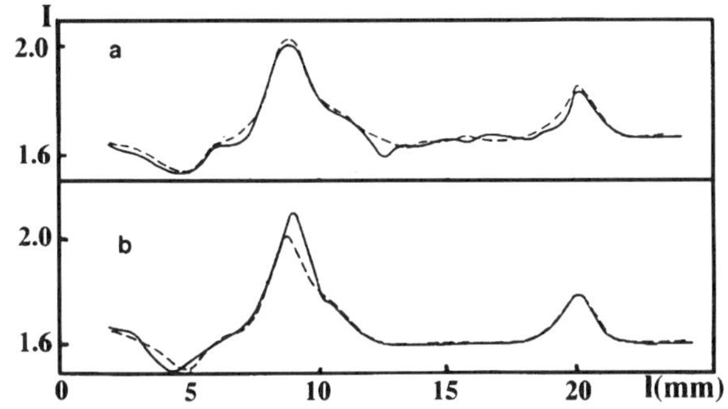

Figure 78. Photometric cuttings of spectrograms within the red wing for the mustache on 14.11.1985 at 0910 perpendicular to the dispersion for two orthogonal polarizations: (a) IV spectral order, (b) III spectral order.

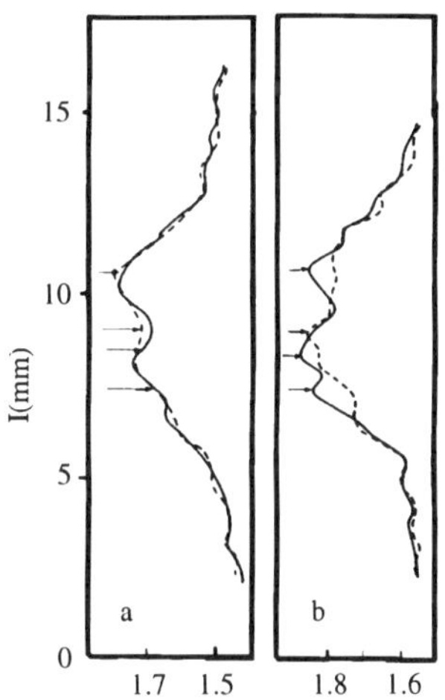

Figure 79. Same as Figure 78, at 0731.31 UT for the blue wing.

direction of polarization varies for different mustaches, even observed within the same active region, and it changed considerably even for close mustaches. Some large mustaches consisted of separate structural elements exhibiting different polarization.

11.5. Observations of Mustaches at Baikal Astrophysical Observatory

Polarization spectroscopic studies have been performed at Sayany Astrophysical Observatory using H_α and H_β hydrogen lines with different techniques

TABLE 14
Degree and Azimuth of Polarization Calculated from Observations[a]

P (%)	$\psi°$	Cutting	Brightness
22	116	Across dispersion	Many mustaches
12	96	—	—
5	−10	—	—
17	116	—	—
2	75	—	—
4	115	—	—
7	93	—	—
3	−16	—	—
4	33	—	Bright
24	120	—	Weak
11	112	—	Bright
11	−2	—	Weak
2	72	—	Bright
4	88	—	—
16	92	Along dispersion	—
17	29	—	Weak
10	67	—	Bright
24	57	—	Weak
8	78	Across dispersion	Many mustaches
10	82	—	—
8	124	—	—
8	128	—	Weak
10	90	—	Bright
6	−35	—	Weak
18	69	Along dispersion	Bright
16	79	—	Weak
9	69	Across dispersion	Bright
12	86	—	Weak
12	58	—	Bright
13	−13	—	Weak
16	72	Along dispersion	Bright

[a] ψ is the azimuth of oscillations of the electric vector, measured from the direction to the solar disk center.

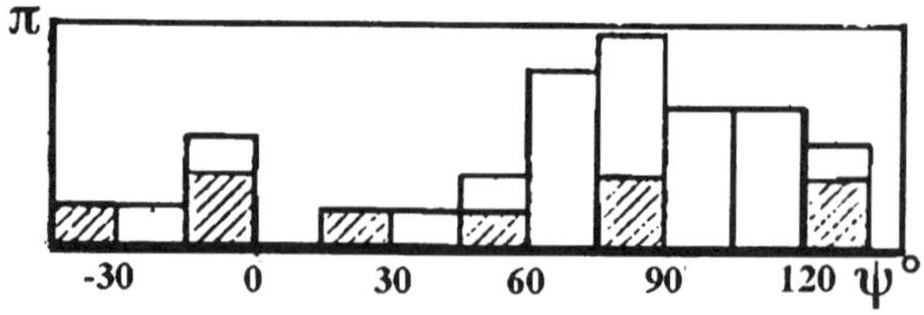

Figure 80. Distribution of observed azimuths of the polarization plane of studied mustaches with respect to the direction toward the solar disk center. Dashed regions refer to weak mustaches.

of observation.[192] Four spectra were recorded simultaneously for two lines in two orthogonal polarizations. One Stokes parameter (Q or U) was obtained at some particular moment. In order to obtain the other Stokes parameters, the phase $\lambda/2$ plate was rotated at an angle of 22°.5.

Mustaches were observed within a large group of spots on 13.09.1981 located at 40°W, 9°N. The spectrograms were processed by means of a self-recording microphotometer and a computer. The slit length of the spectrophotometer was taken equal to $\frac{1}{3}$ of the emission object size. Records of the quiet area of the adjacent chromosphere were obtained with the same slit. The studied wavelength range around the center of the H_α line was ±3 Å and for H_β, ±2.5 Å.

In order to assess instrumental errors, H_α and H_β line profiles in the quiet region were measured. The profiles of the undisturbed photosphere were recorded with the same slit width as for mustaches, and also using a very long microphotometer slit in a region of the spectrogram where no mustaches were recorded. The H_α line profiles within the two spectral bands agreed reasonably, but the H_β line showed a difference in intensities at the line center of about 5%. This difference seems to originate from a slight defocusing of the two bands within this spectral region.

Seven spectrograms for the period 0125–0144 UT were obtained, the first four of which being taken during the first three minutes. The clear difference for H_α and H_β profiles in different orthogonal polarizations indicated the existence of linear polarization of emission. The last three spectrograms do not show any indication of linear polarization of emission: the line contours were exactly the same for two linear orthogonal polarizations.

Figure 81 represents the $(I_1 - I_2)/(I_1 + I_2)$ profiles for the first four spectrograms. The results for the degree of polarization and azimuth of the electric field obtained from these spectrograms are presented in Table 15. Since a

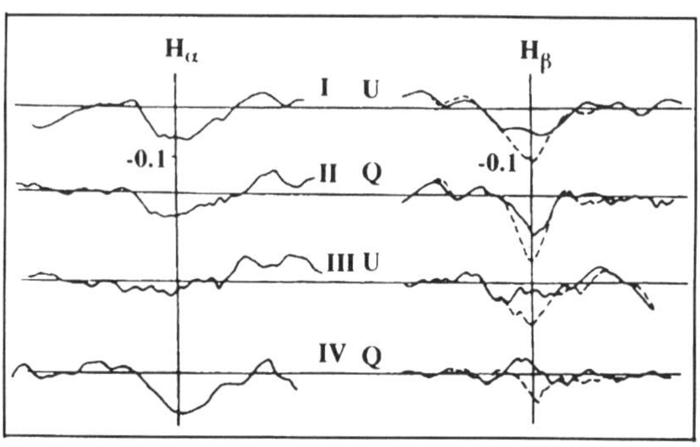

Figure 81. Profiles of $\frac{I_1 - I_2}{I_1 + I_2}$ for spectrograms I–IV for H_α and H_β lines. The dashed profiles for the H_β line were obtained by subtraction of the quiet region contour.

TABLE 15
Results for Time Variation of the Degree and Azimuth of Polarization for Mustaches

UT	$\sqrt{\frac{Q^2+U^2}{J^2}}\,(H_\alpha)$	$\chi^\circ(H_\alpha)$	$\sqrt{\frac{Q^2+U^2}{J^2}}\,(H_\beta)$	$\chi^\circ(H_\beta)$
01.25.00–0125.30	0.078	27	0.090	15
01.25.30–0126.30	0.052	13	0.076	7
01.26.30–0128.00	0.081	8	0.024	72

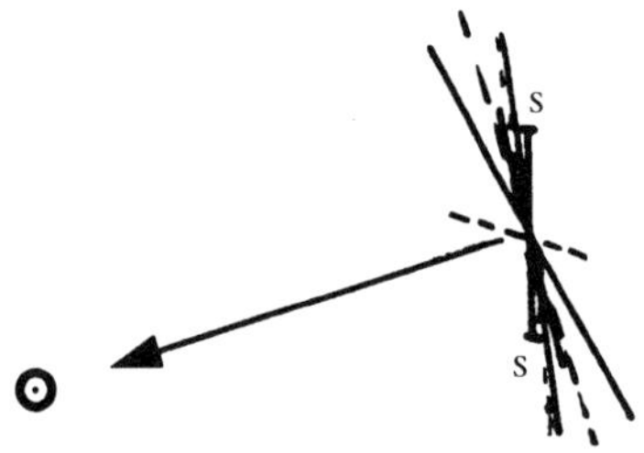

Figure 82. Orientation of the preferential polarization directions for H_α (solid lines) and H_β (dashed line) with indication of the entrance slit position (SS) and direction toward the solar disk center (arrow).

single spectrogram gives either Q/I or U/I, the first column of the table indicates two moments corresponding to successive spectrograms. The geometry of the observations is given in Figure 82.

These observations showed that the degree of linear polarization of mustaches for the H_α line was 5–8%. The degree of polarization in the H_β line was of the same order of magnitude as the instrumental polarization and for these observations it fluctuated significantly. The orientation of the degree of polarization in the mustaches for both lines was tangential with respect to the direction toward the solar disk center. The polarization was detected only during the initial stage of the mustache lifetime.

11.6. Observations of Mustaches at Sacramento Peak Observatory

Independent observations of the H_α line polarization of mustaches or Ellerman bombs were made at Sacramento Peak Observatory on 02.08.1990 in NOAA region 6172 at 25°S 43°W ($\rho = 0.8$),[200]. Mustaches were observed in locations near the spots, where each 50 arc sec square field had one or more very bright mustache. Video images were recorded and summed at each location, and as a result, the averaged polarization maps at each location were obtained.

During a 75-min time interval, 32 mustaches were observed. One mustache showed an 8% degree of polarization, one 5%, two 4%, nineteen $2 \pm 1\%$, and nine were less than 2% linearly polarized. The polarization was as a rule uniform, but in one case a 2% degree of polarization was observed at the center of the line and 5% around the line center. Different polarization was detected for close mustaches within the same location where the degree of polarization was varied in the range ±2%.

11.7. Concluding Remarks

The degree of polarization in mustaches varies over a wide range. A number of mustaches were shown[(189–191)] not to exhibit any polarization. On the other hand, the same group of observers pointed out that, in some cases, the degree of polarization in mustaches could exceed 20%. Such a high degree of polarization seems quite dubious, because for the spectra presented in these works there existed some evident differences in quality for two orthogonal linear polarizations. Such a high degree of polarization, or even a reverse of the sign of polarization, could be related to the instrumental effect of the defocusing of one spectrum with respect to the other. On the other hand, a high degree of polarization was obtained using cuttings along the spectrum, where the influence of this instrumental factor was less significant. This justified an inclination in favor of the reliability of such a high degree of polarization.

These observations indicated a difference in the degree of polarization within closely located pairs or groups of mustaches. The degree of polarization, as a rule, was higher for weaker mustaches. The polarization of emission was observed mainly at the initial stage of formation of mustaches, during the first 1–3 minutes of their lifetime.

A degree of polarization of 7–8% was observed for the centers of H_α and H_β lines,[(192)] while many observations[(189–191)] were associated with the wings of the H_α line. Despite the fact that in some cases polarization was detected in the red wing, it is possible to conclude that mustache emission was polarized for the whole spectral contour. The polarization plane of mustache emission is mainly oriented orthogonally with respect to the direction to the solar disk center and related to the filaments of the AFS system or the orientation of the local magnetic field.

Extensive observations of mustaches were undertaken[(200)] and showed that 28 of 32 mustaches emitted polarized light. Reported[(189–191)] anomalies in the degree of polarization were not confirmed, but reliable measurements of a rather high degree of polarization, 4–8%, have been presented. Observations of elements of mustaches with different polarization were reported[(189–191)] and have also been confirmed.

12

Methodology of Impact Spectropolarimetric Sensing

12.1. Introduction

The aim of this chapter is to construct a methodology and practical ways of carrying out spectropolarimetric sensing of an ionized gas entity. As a basic example we consider the chromospheric location of a solar flare and employ theoretical methods of impact polarization of atomic ensembles, reliable data of spectropolarimetric observations, measurements of solar flare emissions. The practical implementation of impact spectropolarimetric sensing of an ionized media, in general, consists of the following principal problems: Measurement of spectropolarimetric characteristics, determination of the kinetic model of particle motion resulting in the creation of collisional ordering of the ensemble of excited atomic particles, theoretical calculation of spectropolarimetric parameters by a relevant kinetic model with appropriate atomic parameters and cross sections of collisional alignment, and comparison of theoretical and observational data. All these four stages will be reviewed in this chapter, oriented specifically to the problem of remote sensing of energy input to chromospheric locations of a solar flare. Nevertheless, all the methodological approaches and methods of computation can be easily applied to many other natural or artificially created ionized media.[127,133]

The next section is devoted to methodological problems of spectropolarimetric observations in the solar atmosphere using ground based polarization optics. As polarization is a tensorial characteristic, the conditions of observation form a very important factor in the evaluation of spectropolarimetric observables and their subsequent use for sensing purposes. Contrary to intensity spectroscopy, the geometrical arrangement for observation of polarization effects in natural ionized entities is twice as important, because observational spectropolarimetric data must be compared with results of atomic physics computations, which are usually undertaken within a collisional frame of reference.

Another general methodological problem which is very important in a majority of cases, such as spectropolarimetric sensing of solar atmosphere emission or diagnostics of many laboratory plasmas, is the presense of a local magnetic magnetic field. Often, the magnetic field is a factor indispensable for the creation and containing of many ionized media, not only in natural but in a number of laboratory situations. For example, a special configuration of the magnetic field is used to sustain a high temperature plasma generated in many laboratory and technica! installations.[67,68,74]

Despite the conventional concept that the magnetic field destroys polarization via the Hanlé effect,[165] for arbitrary orientation of the local magnetic field strength with respect to the symmetry axis of the excitation process, the polarization of atomic ensembles is not completely destroyed. A magnetized plasma medium can emit polarized radiation stipulated not only by the Zeeman effect, but also due to residual polarization resulting from anisotropic excitation. In the latter case, polarization integrated over the whole line profile will exist. The local magnetic field plays an important role in the chromospheric location of a solar flare, so magnetic field effects are specially considered in this chapter.

A later section is devoted to the implementation of impact spectro-polarimetic sensing on the basis of polarized radiation in H_α and H_β hydrogen lines, which are of the utmost significance for solar and terrestrial physics. Most spectropolarimetric observations of solar flare emissions have been realized on these lines. For this purpose the impact excitation of these lines is analyzed in the framework of Born and Glauber approximations. By numerical integration of the equations governing the impact parameter method, and supposing that the linear polarization of H_α and H_β lines is stipulated by the impact ordering of excited hydrogen atoms in spates with $n = 3$ and 4 respectively, a beam of keV protons is introduced. The idea of generating of beams of such protons during solar flares was put forward in work,[51] in which the relative velocity distribution function was calculated. This velocity distribution function is used in this chapter in the course of actual computation of spectropolarimetric effects of the solar chromosphere. Precipitating energetic protons also play a significant role in the excitation of optical emission in the outer ionosphere of the Earth.

A further section deals with the analysis of impact spectropolarimetric effects in the L_α line due to slow collisions when a drift of hydrogen atoms is present. These results are very important for the development of the theory of impact polarization. The role of a metastable level in the process of impact alignment of an ensemble will be specially studied in this case as a factor of direct astrophysical significance.

12.2. Spectropolarimetric Sensing Using Ground-Based Observation

In most cases measurement of the degree of polarization of the solar atmosphere or any other astrophysical object is undertaken by an observer positioned on the ground. However, most theoretical calculations are carried out in a collisional frame of reference connected to a beam of particles inducing the ordering of angular momenta of an excited atomic ensemble. The principal aim of this section is to present relations between the degrees of polarization calculated within two of these frames of reference.

In order to find a relation between characteristics of the linearly polarized radiation of the solar chromosphere measured by a ground-based observer, and quantities characterizing the degree of ordering of angular momenta in an ensemble of excited particles, we first examine Figures 83 and 84. Figure 83 represents schematically positions of a ground-based observer (point A), a chromospheric location (point S), emitting radiation of hydrogen lines, and the projection of this location on the plane of the solar disk (point S′). The axis $O\tilde{Z}$ is connected to the symmetry axis of the collisional process, and the axis OZ' is the projection of the $O\tilde{Z}$ axis on the solar disk plane; the OX' axis is directed toward the observer while axes $O\tilde{Y}$ and OY' are similar. Let us introduce also the

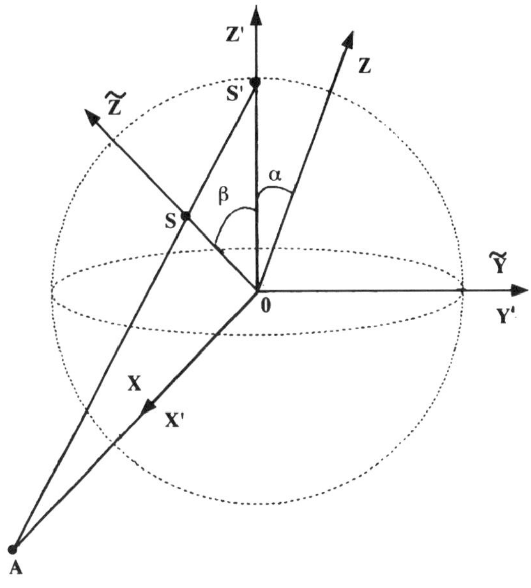

Figure 83. Systems of reference for ground-based observation of spectropolarimetric effects in a solar atmosphere.

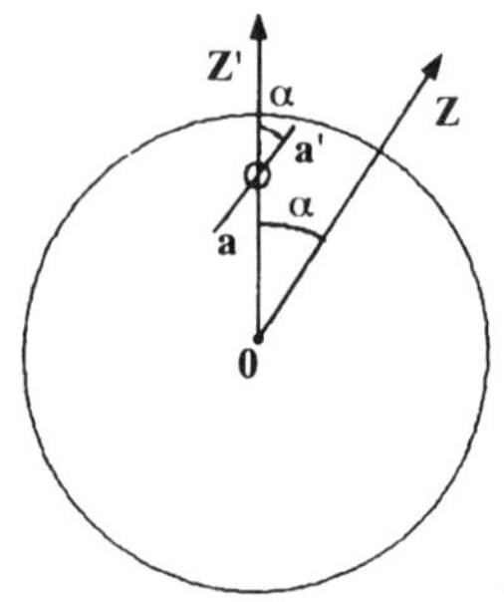

Figure 84. System of reference connected to the solar disk.

observation system of reference *XYZ* in which all observations are effectively implemented, with axis OZ connected to the direction of the spectrograph entrance slit (Figure 84) and inclined at angle α to the OZ' axis, while the OX and OX' axes are parallel. The system of coordinates $(\tilde{X}\tilde{Y}\tilde{Z})$ could be obtained from the system *(XYZ)* by two rotations: around the OY axis at angle β, transforming system $(\tilde{X}\tilde{Y}\tilde{Z})$ into $(X'Y'Z')$; and around the OX' axis at angle α, transforming system $(X'Y'Z')$ into *(XYZ)*.

The components of the alignment tensor in systems $(\tilde{X}, \tilde{Y}, \tilde{Z})$ and *(X, Y, Z)* are related by means of a matrix which is a product of two rotation matrices, namely, $D^{(2)}(0, \beta, 0)$ describing rotation at angle β around the OY axis, and matrix $D^{(2)}(\frac{1}{2}\pi, \alpha, -\frac{1}{2}\pi)$ describing the corresponding rotation at angle α around the OX axis. Taking into consideration that only the zero component of the alignment tensor is non zero in the reference frame $(\tilde{X}, \tilde{Y}, \tilde{Z})$, one can write

$$\rho_q^2 = \tilde{\rho}_0^2 \sum_{q'} D_{0,q'}^{(2)}(0, \beta, 0) D_{q',q}^{(2)}(\tfrac{1}{2}\pi, \alpha, -\tfrac{1}{2}\pi) \tag{12.2.1a}$$

The expression for the elements of rotation matrix (2.4.10) enable the last equation to be written in the form

$$\rho_q^2 = \tilde{\rho}_0^2 \sum_{q'} d_{0,q'}^{(2)}(\beta) d_{q',q}^{(2)}(\alpha) \exp\left[i\frac{\pi}{2}(q - q')\right] \tag{12.2.1b}$$

Quantities $\tilde{\rho}_0^2$ and $d_{qq'}^{(2)}$ are real, so the only terms with an even value of $q - q'$ on the right hand side of the last expression can be obtained by computing Re ρ_q^2, while terms with odd values of $q - q'$ are obtained by computing Im ρ_q^2. Due to axial symmetry of the collisional process in the $(\tilde{X}, \tilde{Y}, \tilde{Z})$ frame of reference, only the two polarization moments ρ_0^0 and ρ_0^2 are not zero. Therefore, Stokes parameter $\tilde{U}$ is equal to zero in this frame of reference, and the polarization characteristics are determined only by Stokes parameter $\tilde{Q}$. The expressions of Section 3.6 for the transition $j_0 - j$ yield

$$I\tilde{Q} = I_{\tilde{Z}} - I_{\tilde{Y}} \sim -K\frac{3}{\sqrt{6}}\tilde{\rho}_0^2(j) \tag{12.2.2}$$

where

$$K_p = \begin{Bmatrix} 1 & 1 & p \\ j & j & j_0 \end{Bmatrix}$$

The intensity I for this radiative transition may be expressed as

$$I = I_{\tilde{Z}} + I_{\tilde{Y}} \sim \frac{2}{\sqrt{3}}K_0\tilde{\rho}_0^0(j) - K\frac{\tilde{\rho}_0^2(j)}{\sqrt{6}} \tag{12.2.3}$$

Hence Stokes parameter $\tilde{Q}$ in the collisional frame of reference, using Eq. (12.2.1a), yields

$$\tilde{Q} = -\frac{3K_2\tilde{\rho}_0^2(j)}{2\sqrt{2}K_0\tilde{\rho}_0^0(j) - K_2\tilde{\rho}_0^2(j)} \tag{12.2.4}$$

The nonzero polarization moments in the (X, Y, Z) reference frame will be $\rho_0^0 = \tilde{\rho}_0^0$ and ρ_q^2, $q = 0, 1, 2$. Expressions for Stokes parameters within this frame of reference will take the form

$$IQ = I_Z - I_Y \sim -K_2\left[\frac{3}{\sqrt{6}}\rho_0^2(j) - \text{Re }\rho_2^2(j)\right] \tag{12.2.5}$$

and

$$IU = I_{ZY^+} - I_{ZY^-} \sim -2K_2 \text{ Im } \rho_1^2 \tag{12.2.6}$$

where I_{ZY^+} and I_{ZY^-} indicate the intensities of radiation linearly polarized along the diagonals of the first and second quarters of the ZY plane.

Relation (12.2.1) and expression (2.4.12) for matrix elements $d_{qq'}^{(2)}$ give

$$IQ \sim -\frac{3}{\sqrt{6}}K_2\tilde{\rho}_0^2(j)\cos^2\beta\cos 2\alpha, \qquad IU \sim \sqrt{6}K_2\tilde{\rho}_0^2(j)\cos^2\beta\sin 2\alpha \tag{12.2.7}$$

where

$$I \sim \frac{2}{\sqrt{3}} K_0 \tilde{\rho}_0^0(j) - \frac{1}{\sqrt{6}} K_2 \tilde{\rho}_0^2(j) + \frac{3}{\sqrt{6}} K_2 \tilde{\rho}_0^2(j) \sin^2 \beta \tag{12.2.8}$$

The polarization moment ρ_0^0 is invariant under rotations, therefore

$$\tilde{\rho}_0^0 = \rho_0^0.$$

Hence Eq. (12.2.4) allows us to obtain

$$Q = \frac{I_Z - I_Y}{I_Z + I_Y} = \frac{\tilde{Q} \cos^2 \beta \cos 2\alpha}{1 - \tilde{Q} \sin^2 \beta} \quad \text{and} \quad U = \frac{I_{ZY^+} - I_{ZY^-}}{I_{ZY^+} + I_{ZY^-}} = \frac{\tilde{Q} \cos^2 \beta \sin 2\alpha}{1 - \tilde{Q} \sin^2 \beta} \tag{12.2.9}$$

Finally, the degree of polarization of radiation detected from the Earth, is given by

$$P = \sqrt{Q^2 + U^2} = \frac{|\tilde{Q}| \cos^2 \beta}{1 - \tilde{Q} \sin^2 \beta} \tag{12.2.10}$$

This expression is used for impact spectropolarimetric sensing for ground-based observations.

Surfaces $Q(\alpha, \beta)$ and $U(\alpha, \beta)$ are presented in Figures 85 and 86 for a fixed value $\tilde{Q} = 0.2$. Parameter Q varies between two values, $-\tilde{Q}$ corresponding to ($\beta = 0$, $\alpha = \pi/2$) and $\tilde{Q}$ corresponding to ($\alpha = 0$, $\beta = 0$). At intermediary values of angular variables α and β, the signs of Q and $\tilde{Q}$ are the same if $0 < \alpha < \pi/4$ and opposite if $\pi/4 < \alpha < \pi/2$. The signs of parameters U and $\tilde{Q}$ are always the same. The parameter $U = 0$ for $\alpha = 0, \pi/2$. For $\alpha = \pi/4, \beta = 0$ this parameter attains its maximal value equal to $\tilde{Q}$.

Figure 87 represents the dependence of the degree of polarization as a function of two variables β and $\tilde{Q}$. For fixed $\tilde{Q}$ the degree of polarization decreases from the value $\tilde{Q}(\beta = 0)$ to zero ($\beta = \pi/2$) once angle β is increasing. For Stokes parameter $\tilde{Q} = 1$ (100% polarization) the degree of polarization (12.2.10) is also equal to 1, while parameters Q and U are equal to $\cos 2\alpha$ and $\sin 2\alpha$, respectively, being less than unity in absolute value.

Two questions are of importance once experimental or observational material is analyzed with the help of formulas (12.2.9) and (12.2.10). The first is connected to uncertainties of quantities Q, U, P computed with the help of these formulas, and determines the errors of the method. The second question is the

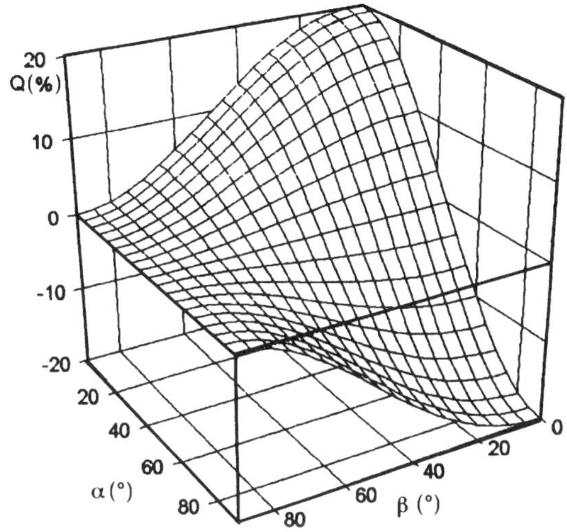

Figure 85. Two-dimensional angular dependency of Stokes parameter Q in the frame of reference of a ground-based observer for $\tilde{Q} = 0.2$.

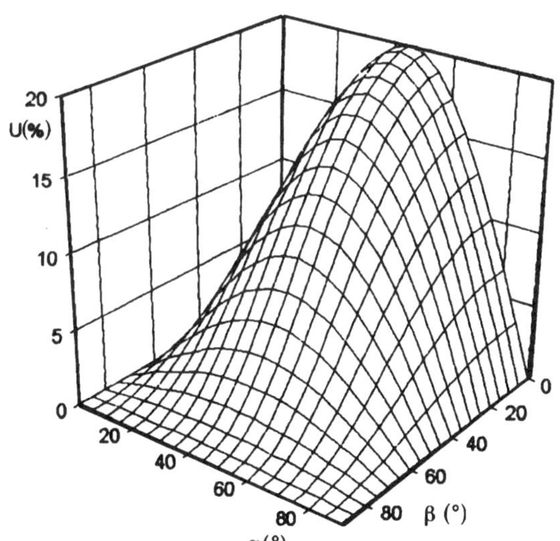

Figure 86. Two-dimensional angular dependency of Stokes parameter U in the frame of reference of a ground-based observer for $\tilde{Q} = 0.2$.

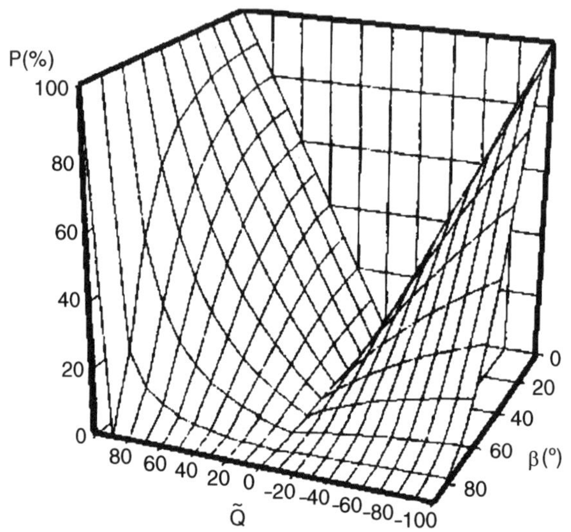

Figure 87. Two-dimensional dependency of the degree of polarization P, measured by a ground based observer.

connection of possible angles α and β with the characteristics of a spectropolarimeter recording the emission.

Concerning the first question, we can assume equal errors of intensity measurements in ordinary and extraordinary beams, which could be approximately of the same value. The relative errors in quantities Q, U, and P are almost the same as for the intensity measurements

$$\frac{\Delta U}{U} = \frac{\Delta I}{I} \quad \text{and} \quad \frac{\Delta Q}{Q} = \frac{\Delta I}{I} \tag{12.2.11}$$

It is evident that, for fixed β formulas (12.2.9) and (12.2.10) enable us to write

$$\Delta Q = -2\Delta\alpha P \sin 2\alpha \quad \text{and} \quad \Delta U = 2\Delta\alpha P \cos 2\alpha \tag{12.2.12}$$

For a fixed value of angle β, it is more effective to measure the Stokes parameter Q when angle α is small, while the most favorable condition for the measurement of the other parameter U is $\alpha \sim \pi/4$. The degree of polarization P for a fixed $\tilde{Q}$ takes the form

$$\frac{\Delta P}{\Delta \beta} = -\frac{\tilde{Q}(1-\tilde{Q})\sin 2\beta}{(1-\tilde{Q}\sin^2\beta)^2} \tag{12.2.13}$$

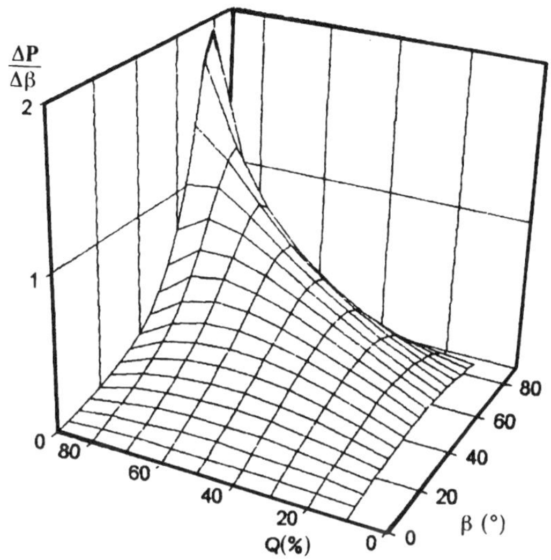

Figure 88. Two-dimensional dependency of function $\Delta P/\Delta\beta$.

Figure 88 shows a surface corresponding to the function on the right hand side of expression (12.2.13). This function for large values of β grows with $\tilde{Q}$. Therefore, large errors could arise, when the polarization characteristics are measured for objects located near the equator.

Concerning the second question, it should be noted that parameters Q and U can be recorded, if the radiation anisotropy ΔI exceeds the sensitivity of a detector δI, namely,

$$\Delta I \geq \delta I \tag{12.2.14}$$

If both sides of this inequality are divided by the sum $(I_Z + I_Y)$, the left hand side would become the quantity

$$\frac{\Delta I}{I_Z + I_Y}$$

proportional to Stokes parameter Q or U, and the right hand side would become the ratio

$$\frac{\delta I}{I_Z + I_Y}.$$

For estimates, the expression for the sum of intensities can be presented with the aid of formulas (12.2.2) and (12.2.8) as follows:

$$I_Z + I_Y = (I_Z + I_Y)_0(1 - \tilde{Q}\sin^2\beta) \tag{12.2.15}$$

where

$$(I_Z + I_Y)_0 = (I_Z + I_Y)_{\beta=0}$$

is the sum of intensities referred to a collision system of coordinates.

The latter relation yields

$$\frac{\Delta I}{(I_Z + I_Y)} \geq \frac{\delta I}{(I_Z + I_Y)_0(1 - \tilde{Q}\sin^2\beta)} \tag{12.2.16}$$

With system (12.2.9), it is possible to write

$$\tilde{Q}\cos^2\beta\cos 2\alpha \geq \frac{\delta I}{(I_Z + I_Y)_0} \quad \text{and} \quad \tilde{Q}\cos^2\beta\sin 2\alpha \geq \frac{\delta I}{(I_Z + I_Y)_0} \tag{12.2.17}$$

which can be unified in the following equation

$$\tilde{Q}\cos^2\beta\cos(\tfrac{1}{2}\pi - 2\alpha) \geq \frac{\delta I}{\sqrt{2}(I_Z + I_Y)_0} \tag{12.2.18}$$

The right hand side of expressions (12.2.17) and (12.2.18) does not depend on angles, so these relations impose certain, not very strict limitations on possible values of angles α and β. It follows from relation (12.2.18) that small angles α and β close to $\pi/8$ are optimal.

12.3. *Role of the Solar Magnetic Field*

A basic concept that was exploited in the development of impact spectropolarimetric sensing is that ordering of angular momenta of an ensemble of excited atomic particles, and the spectroscopic consequencies such as polarization and anisotropy of the optical emission, are caused by the symmetry of internal impact processes. The macroscopic anisotropy of collisional iteraction inside an ionized medium is related to ordering in the space of relative velocities of colliding particles. In other words, impact polarization phenomena refer to the transfer of macroscopic symmetry of collisional interaction to microscopic properties of atomic ensembles, resulting in a particular symmetry of the spatial distribution of angular momenta of particles.

There exist some other physical processes which could also introduce or affect the symmetry of an excited atomic ensemble and, as a result create or influence spectropolarimetric effects. One well-known process of this kind is the interaction of an atomic system with the external magnetic field. The role of the magnetic field cannot be neglected, especially when spectropolarimetric effects are studied in a solar atmosphere, where the magnetic field plays an important role in the formation and maintenance of an ionized gas medium. Therefore, the magnetic field strength is an important local parameter, the influence of which should be specially studied and modeled.[87–90]

Consider the role of the magnetic field in spectroscopic effects of a polarized atomic system. In such a study we examine the variation in the density operator of an ensemble of atoms at a state with electronic moment **J** due to the effect of a constant magnetic field of strength **H**. Such an evolution is described by the Liouville equation

$$i\hbar \frac{\partial}{\partial t}\hat{\rho} = [\hat{V}, \hat{\rho}] \tag{12.3.1}$$

where the interaction operator of atoms with the magnetic field is represented by

$$V = (\boldsymbol{\mu}, \mathbf{H}) \tag{12.3.2}$$

and the magnetic moment

$$\boldsymbol{\mu} = \mu_0 g_J \mathbf{J}$$

is expressed in terms of the Bohr magneton μ_0 and Landé factor g_J.

The matrix element of the interaction operator in the *(kq)* representation is obtained using the momenta coupling technique

$$V_{qq'}^{(k)} = \sqrt{\frac{2k(2k+1)(2k+2)}{12}} \sum_{q''} (-1)^{k+q''} H_{q''} \begin{bmatrix} k & k & 1 \\ q & -q' & q'' \end{bmatrix} \tag{12.3.3}$$

where H_q is the circular component of the magnetic field.

Suppose the atomic ensemble is excited by interaction with an external source, for example, resonance radiation or collisions. Then, within the reference frame with the *OZ* axis in the excitation direction, at least two polarization moments of zero rank (population) and the zero component of the second-order moment (alignment along the excitation direction) will be nonzero. If a constant magnetic field, directed orthogonally to the excitation direction (along *OX*), is applied to such an ensemble, the equation describing the stationary dependence of five components of the alignment tensor on the magnetic field strength, with account for radiative damping and external excitation, takes the form

$$-\gamma_0 \rho_q^2(J) + i \sum_{q'} V_{qq'}^{(2)} \rho_{q'}^2(J) = N_q^2 \delta_{q0} \tag{12.3.4}$$

where γ_0 is the radiative damping constant, N_q^2 is the excitation efficiency of the qth component of the alignment tensor by the external source, and $V_{qq'}^{(2)}$ is a matrix determined by Eq. (12.4.3) and accounting for the effect of the magnetic field on the ordering of angular momenta of an atomic ensemble.

In terms of the Larmour precession frequency

$$\omega_J = \frac{\mu_0 g_J H}{\hbar}$$

the matrix $V_{qq'}^{(2)}$ will take the form

$$V_{qq'}^{(2)} = \begin{pmatrix} 0 & \omega_J & 0 & 0 & 0 \\ \omega_J & 0 & \sqrt{\frac{3}{2}}\omega_J & 0 & 0 \\ 0 & \sqrt{\frac{3}{2}}\omega_J & 0 & \sqrt{\frac{3}{2}}\omega_J & 0 \\ 0 & 0 & \sqrt{\frac{3}{2}}\omega_J & 0 & \omega_J \\ 0 & 0 & 0 & \omega_J & 0 \end{pmatrix} \tag{12.3.5}$$

The magnetic field strength dependence of the alignment of an atomic ensemble for small fields is determined by the dependence of the $\rho_0^2(J)$ component which, with the help of Eq. (12.3.3), assumes the form

$$\rho_0^2(J) \sim \frac{1}{1 + 4\left(\frac{\omega}{\gamma_0}\right)^2} \tag{12.3.6}$$

This means that the magnitude of the alignment signal is decreasing with magnetic field strength. This phenomenon of depolarization by a small magnetic field is known as the Hanlé effect.[165] The magnetic field in this case plays a destructive role, destroying the ordering of angular momenta created by the external excitation process.

In order to illustrate different aspects of the influence of a magnetic field on the polarization characteristics of the emission of a solar atmosphere, we shall first consider the variation in the electronic polarization moments of an excited ensemble of hydrogen atoms in the presence of a nonhomogeneous magnetic field. In this case the nonhomogenous magnetic field with linear gradient of the magnetic field strength may possess a symmetry of alignment and could therefore play a constructive role in creating a new type of ordering of the momenta.[28,141] In a second stage we shall consider the dependence of the net polarization of the H_α hydrogen emission line for the joint influence of two factors on an atomic ensemble: the impact alignment specified by its axis, and the direction of the uniform magnetic field when this polarization is observed at a certain direction within the observer's frame of reference.

Let us start with expression (12.3.1) which, within the (l, m) representation, takes the form

$$i\hbar\frac{\partial}{\partial t}\rho^{ll_1}_{mm_1} = \sum_{l'm'}[\langle lm|V(\mathbf{R})|l'm'\rangle\rho^{l'l_1}_{m'm_1} - \langle l'm'|V(\mathbf{R})|l_1m_1\rangle\rho^{ll'}_{mm'}] \tag{12.3.7}$$

where $\mathbf{R} = \mathbf{R}_0 + \mathbf{r}$; here $\mathbf{R}_0$ and $\mathbf{r}$ are vectors that characterize the coordinates of an excited atom and consequently its electron.

If every circular component of the magnetic field in Eq. (12.3.2) is expanded in terms of a Taylor powers series in $\mathbf{r} = \mathbf{r}(r, \theta, \varphi)$, we have

$$H_q(\mathbf{r}, \mathbf{R}_0) = \sum_{LM} r^L c^{(q)}_{LM}(\mathbf{R}_0)Y_{LM}(\theta, \varphi) \tag{12.3.8}$$

where the spherical function Y_{lm} depends on the angular variables θ, ϕ specifying the position of an electron.

On substituting the last relation into Eq. (12.3.7) and using momenta coupling algebra, after some cumbersome transformations we obtain

$$\begin{aligned} i\frac{\partial}{\partial t}\rho^k_Q(ll_1) = \omega \sum_{l'k_1Q_1p\alpha} & (-1)^{l+l_1+k_1+\alpha}\sqrt{(2k+1)(2p+1)} \\ & \times\left[\begin{Bmatrix} l' & l & p \\ k & k_1 & l_1 \end{Bmatrix}\begin{bmatrix} p & k & k_1 \\ -\alpha & Q & Q_1 \end{bmatrix} W^p_\alpha(ll')\rho^{k_1}_{Q_1}(l'l_1)\right. \\ & \left. -\begin{Bmatrix} l' & l_1 & p \\ k & k_1 & l \end{Bmatrix}\begin{bmatrix} p & k & k_1 \\ \alpha & -Q & -Q_1 \end{bmatrix} W^p_\alpha(l'l_1)\rho^{k_1}_{Q_1}(ll')\right] \end{aligned} \tag{12.3.9}$$

where ω is the Larmour frequency. Coefficients $W_{\alpha}^{p}(ll_1)$ depend on the radial part of the matrix element of operator r^L (r_{l,l_1}^L), computed with functions $|lm\rangle$ and $|l_1 m_1\rangle$ (r_{l,l_1}^L), and are expressed as

$$W_{\alpha}^{p}(ll_1) = -\frac{i}{2}(-1)^{l+l_1+p}\sum_{qLM}(-1)^q r_{l,l_1}^L \sqrt{\frac{2l(2l+1)(2l+2)(2l_1+1)}{4}} \times \begin{bmatrix} l_1 & L & l \\ 0 & 0 & 0 \end{bmatrix}\begin{bmatrix} 1 & L & p \\ -q & M & \alpha \end{bmatrix}\begin{Bmatrix} 1 & L & p \\ l_1 & l & l \end{Bmatrix} c_{LM}^{(q)}(\mathbf{R}) \qquad (12.3.10)$$

The use of formula (12.3.9) is connected to the solution of a system of higher order. To simplify this problem we utilize the secular approximation, omitting nondiagonal polarization moments with respect to the orbital moment. Denoting

$$\rho_q^k(ll) = \rho_q^k(l)$$

and taking into account the process of natural damping of an excited state characterized by constant γ_o, formula (12.3.9) can be transformed to

$$\frac{\partial}{\partial t}\rho_Q^k(l) = -\rho_Q^k(l) - i\omega_l \sum_{k_1 Q_1} H_{Q,Q_1}^{k,k_1}(l)\rho_{Q_1}^{k_1}(l) + \frac{N_Q^k(l)}{\gamma_o(l)} \qquad (12.3.11)$$

where $\omega_l = \mu_0 H/\hbar\gamma_l$ is the dimensionless Larmour frequency and the term $N_Q^k(l)$ describes the efficiency of impact excitation of the state with principal quantum number $n = 3$. The latter term can be expressed in terms of the cross section of impact polarization $\sigma_q^k(ll')$

$$N_Q^k(l) = n_0 v \sigma_Q^k(l, l')\delta_{Q0}\delta_{l,l'}$$

while the matrix element of the operator of the interaction of the system with a magnetic field is reduced by a simplification of relation (12.3.9), to the form

$$H_{QQ_1}^{kk_1}(l) = 2\sum_{p\alpha}(-1)^{k+\alpha}\sqrt{(2p+1)(2k+1)}\begin{bmatrix} p & k & k_1 \\ -\alpha & Q & Q_1 \end{bmatrix}\begin{Bmatrix} l & l & p \\ k & k_1 & l \end{Bmatrix} W_{\alpha}^{p}(ll) \qquad (12.3.12)$$

where $p + k + k_1$ taken an odd value.

Equation (12.3.12) has been studied in detail elsewhere.[28,29]

Let us analyze some features of relaxation of angular momenta ordering in a local magnetic field, similar to a physical situation that could arise in the solar atmosphere and in the outer ionosphere of the Earth. A local magnetic field is an inalienable feature of a diluted medium in which polarization effects created by particle collisions take place. We shall use the basic Eqs. (12.3.9) and (12.3.12) to model the effect of the magnetic field. In the simpler case of a homogeneous field where only

$$c_{00}^{(q)}(\mathbf{R}_0) \neq 0 \qquad (q = 0, \pm 1)$$

relation (12.3.12) is reduced to (12.3.3).

In the general case

$$c_{L,M}^{(q)}(\mathbf{R}_0) \neq 0 \qquad (L \geq 0, M = 0, \pm 1, \pm 2, \ldots, \pm L)$$

the polarization moments of different ranks are mixed by Eqs. (12.3.9) and (12.3.11). Hence an inhomogeneous magnetic field can provide ordering of angular momenta, different from the initial one created by the collisional excitation process. Equation (12.3.11) shows more clearly than the more general equation (12.3.9) that an inhomogeneous magnetic field can create a new ordering of an atomic ensemble. In order to illustrate all these features, Eq. (12.3.11) was solved under stationary conditions for an inclined alignment, which is created by collisions in an ensemble of excited atoms ($N_{\pm 1}^2 \neq 0$).

The results of these calculations are presented in Figure 89, in which

$$\langle J_y \rangle = 2 \text{ Im } \rho_1^1, \qquad \langle J_z J_x + J_x J_z \rangle = \sqrt{2} \text{ Re } \rho_1^2$$

are plotted as a function of the ratio ω/γ.

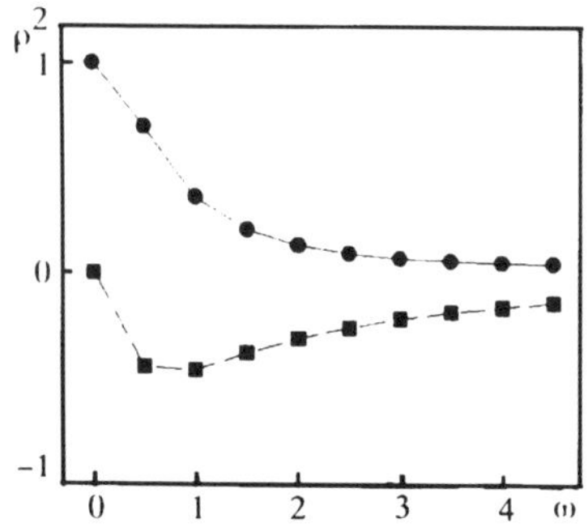

Figure 89. Transfer of alignment tilted at some angle into orientation in an inhomogeneous magnetic field created by a circular current. Dots and squares indicate $\langle J_z J_x + J_x J_z \rangle$ (●) and $\langle J_y \rangle$ (■) as functions of the dimensionless Larmour frequency.

The dependence of the 45° inclined alignment on ω/γ is Lorenzian. Therefore, a magnetic field dependence of this kind of alignment is characterized by the Hanlé effect.[165] Orientation $\langle J_y \rangle$ as dependent on ω/γ is dispersive. At a lower magnetic field the inclined alignment is transformed into orientation, and in this respect a low magnetic field could play a constructive role. At higher magnetic fields the destruction of the existing ordering of momenta predominates over the creation of a new dipole ordering and, as a result, the orientation signal decreases.

Active processes in the solar chromosphere, the role of the local magnetic field in the polarization of atomic ensembles and their spectropolarimetric observables were analyzed optically. The results enabled us to attribute the source of the magnetic field to sun spots, modeling the anisotropy of the magnetic field at the chromospheric level of the solar atmosphere by a magnetic field created by a circular current of radius r_0. The magnetic field strength components at a point, the position of which is characterized by vector $\mathbf{R} = \mathbf{R}(R, \theta, \varphi)$, can be written in the form

$$\begin{aligned} H_x &\sim \frac{\cos\theta}{R}\frac{r_0}{R}\int_0^{2\pi} \cos\varphi' F(\theta, \varphi, \varphi')d\varphi', \\ H_y &\sim \frac{\cos\theta}{R}\frac{r_0}{R}\int_0^{2\pi} \sin\varphi' F(\theta, \varphi, \varphi')d\varphi' \\ H_z &\sim \frac{1}{R}\frac{r_0}{R}\int_0^{2\pi}\left[\frac{r_0}{R} - \sin\theta\cos(\varphi - \varphi')\right]F(\theta, \varphi, \varphi')d\varphi' \end{aligned} \tag{12.3.13}$$

where

$$F(\theta, \varphi, \varphi') = \left[1 - 2\frac{r_0}{R}\sin\theta\cos(\varphi - \varphi') + \left(\frac{r_0}{R}\right)^2\right]^{-3/2} \tag{12.3.14}$$

If Eq. (12.3.14) is expanded in terms of Gekenbauer polynomials,[136] and the series is restricted to the first nonvanishing terms of inverse powers of R, the following approximate expressions for the components of the magnetic field strength vector (12.3.13) may be derived:

$$H_x \sim \frac{3XZ}{2R^5}, \qquad H_y \sim \frac{3YZ}{2R^5}, \qquad H_z \sim \frac{1}{R^3} - \frac{3(Y^2 + X^2)}{2R^5} \tag{12.3.15}$$

It follows from formula (12.3.10) that in the framework of the circular approximation for the expansion of Eq. (12.3.8), only terms with $L = 0,\ 2$ must be taken into account because the term with $L = 1$, which contains spherical functions Y_{1M}, does not have any effect on the value of H^{k,k_1}_{Q,Q_1}.

The latter relations and formula (12.3.8) enable one to express the anisotropic part of coefficients $c_{LM}^{(q)}$ in the form

$$\begin{aligned} c_{20}^{(1)} &\sim -15\frac{f_1(Y_0,Z_0)}{\sqrt{20}}, & c_{22}^{(0)} &\sim 15i\frac{f_3(Y_0,Z_0)}{2\sqrt{15}}, \\ c_{20}^{(2)} &\sim 3i\frac{f_2(Y_0,Z_0)}{\sqrt{10}} & c_{22}^{(1)} &\sim -15\frac{f_1(Y_0,Z_0)}{\sqrt{30}}, \\ c_{2-2}^{(1)} &\sim -05\frac{Y_0^3Z_0}{2\sqrt{30}} & c_{21}^{(1)} &\sim -3i\frac{f_2(Y_0,Z_0)}{2\sqrt{15}} \\ c_{2-1}^{(1)} &\sim -15i\frac{f_3(Y_0,Z_0)}{\sqrt{30}}, & c_{21}^{(0)} &\sim 15\frac{f_1(Y_0,Z_0)}{\sqrt{15}}, \end{aligned} \qquad (12.3.16)$$

where

$$\begin{aligned} f_1(Y,Z) &= (3Y^2-4Z^2)ZY, \\ f_2(Y,Z) &= (3Y^4-24Y^2Z^2+8Z^4), \\ f_3(Y,Z) &= (Y^2-6Z^2)Y^2 \end{aligned}$$

while the other coefficients can be calculated using the symmetry property

$$(c_{L,M}^{(q)})^* = (-1)^{q+M+1}c_{L,-M}^{(-q)} \qquad (12.3.17)$$

Within the framework of the circular approximation in expansion (12.3.8) only terms with $L=0$ and $L=2$ must be taken into account: The term with $L=1$ containing spherical function $Y_{1,M}$ has no effect on the value of matrix element H_{Q,Q_1}^{k,k_1}.

With the previous relations, and by expressing quantity $N_Q^k(l)$ in Eq. (12.3.11) in terms of cross sections of impact excitation of the level of the hydrogen atom with principal quantum number $n=3$,[144,199] system of equations (12.3.11) was solved for a stationary regime. The results of these calculations for a simpler case, when the (X, Y, Z) and $(\tilde{X}, \tilde{Y}, \tilde{Z})$ systems of reference coincide, are illustrated in Figures 90–92. Figure 90 presents Stokes parameters Q, U and V versus the dimensionless Larmour frequency. The dependence of parameter Q is Lorenzian, while the similar dependence for the other two parameters is dispersive. This means that in the absence of a magnetic field, only longitudinal alignment is present. With growth is the magnetic field strength, both inclined alignment appears and is related to polarization moment $\rho_1^{(2)}$, as well as a new type of ordering, transverse orientation, related to polarization moment $\rho_1^{(1)}$. At large values of the magnetic field strength all three Stokes parameters tend to a limiting value. Parameter V, which is present only for

Now we turn to another case, more closely connected to the study of the magnetic field effect on the spectropolarimetric parameters of H_α, emitted by the upper chromosphere. Here we assume that the ordering angular momenta of an ensemble of hydrogen atoms in state $n = 3$ is determined by the interaction of hydrogen with the proton beam propagating along the radius of the solar sphere and specifying the $O\tilde{Z}$ axis of the collision reference frame $(\tilde{X}, \tilde{Y}, \tilde{Z})$ (Figures 83 and 84). The excited atomic ensemble of hydrogen atoms is affected by the magnetic field, originating from a sun spot located at the level of the photosphere. As the vertical dimension of the optically active upper region of the solar chromosphere is very small in comparison to all the other vertical dimensions and the distance to the photosphere, it is possible to use a model of a homogeneous magnetic field within the upper layer of the chromosphere. Due to the axial symmetry of the collisional excitation process, interaction with the proton beam in the collision frame of reference could induce only population and the zero-order component of the alignment tensor.

Polarization moments induced by collisions are affected by the magnetic field, tilted at angle β to the axis of collisional excitation. Only the tangential component of the magnetic field strength can destroy the polarization characteristics of radiation. Therefore, for simplicity, we first assume that the vector **H** belongs to the $(\tilde{Z}, \tilde{Y})$ plane having only one normal component if the emitting chromospheric location is strictly above the sun spot. This situation is described in Figure 93. A second case, with normal and tangential components of the magnetic field strength, corresponds to a situation in which the chromospheric location of a solar flare is positioned above the margin or in the vicinity of a sunspot.

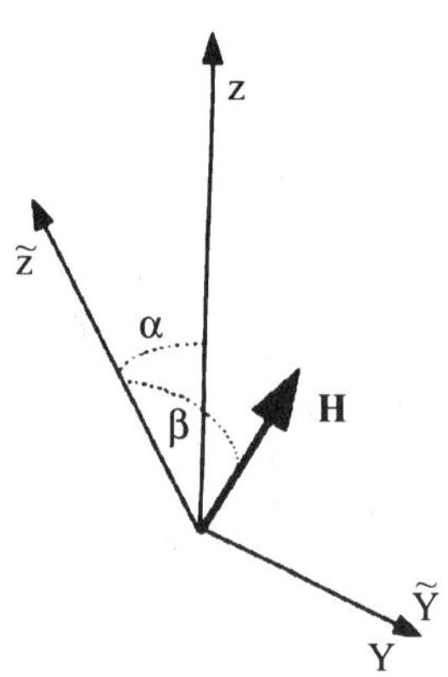

Figure 93. Orientation of the magnetic field strength vector relative to the collisional system of reference $(\tilde{X}, \tilde{Y}, \tilde{Z})$ and a ground-based observer reference system (X, Y, Z) for computation of the effect of the local magnetic field on spectropolarimetric parameters.

The polarization moments of an excited atomic ensemble of hydrogen atoms with account for proton beam excitation, radiation damping and the effect of the magnetic field within a collision frame of reference satisfy the relationship

$$\frac{\partial}{\partial t}\tilde{\rho}_q^k(ll') = -\frac{\gamma_l + \gamma_{l'}}{2}\tilde{\rho}_q^k(ll') + i\omega\sum_{q'}\tilde{V}_{qq'}^{(k)}\tilde{\rho}_q^k(ll') + \tilde{N}_q^k(ll') \tag{12.3.18}$$

Here γ_l is the radiative damping constant for the studied level, ω the Larmour precession frequency, $\tilde{N}_q^k$ the excitation efficiency, and $\tilde{V}_{qq'}^{(k)}$ the matrix element of interaction with the magnetic field, given by expression (12.3.3).

Due to the axial symmetry of proton impact excitation, the efficiency of excitation of state $n = 3$ is expressed in terms of the cross section of impact polarization $\tilde{\sigma}_q^k(ll')$ as follows

$$\tilde{N}_q^k(ll') = n_0 v \tilde{\sigma}_q^k(ll')\delta_{q0} \tag{12.3.19}$$

A ground-based observer measures spectropolarimetric characteristics in a reference frame connected to the solar disk (*X, Y, Z*) system and are expressed in terms of polarization moments $\rho_q^k(ll')$ related to this system. The alignment tensor within these two systems of coordinates is related by the rotation matrix

$$\rho_q^2 = \sum_{q'} D_{qq'}^k(0, \alpha, 0)\tilde{\rho}_{q'}^2(ll') \tag{12.3.20}$$

Only one term, $D_{0q}^{(2)}$, is nonzero on the right hand side of Eq. (12.3.20) in the absense of a magnetic field.

Equation (12.3.18) was solved in the stationary regime for a velocity of beam protons $v = v_0 = e^2/\hbar$ ($E = 24.8$ keV) and different values of the magnetic field strength. Data for the cross sections of impact polarization of the hydrogen state $n = 3$ were obtained using published data.[(144,199)] After solving the system for various magnetic field strengths and radiative damping constants,[(164)] or after determining $\tilde{\rho}_0^0(ll')$ and $\tilde{\rho}_0^2(ll')$ using formulas in Section 3.6 for a number of angles, the polarization moments and Stokes parameters Q and U were found within the (*X, Y, Z*) reference frame with the help of relations (12.3.20) and (3.12.8). Due to the large number of parameters (magnetic field strength H, angles α and β), we shall first examine the influence of a magnetic field on the polarization characteristics of H_α emission within the reference frame $(\tilde{X}, \tilde{Y}, \tilde{Z})$, (where the direction of observation is orthogonal to the plane fixed by the direction of the $O\tilde{Z}$ axis and the vector $\mathbf{H} - \alpha = 0$. The polarization characteristics for a fixed value of the magnetic field strength will then be considered as a function of the angular parameters.

Results are presented in Figures 94–102 . Figures 94–99 give dependencies of Stokes parameters $\tilde{Q}$ and $\tilde{U}$ on magnetic field strength for several values of angle β. The magnetic field strength dependence of Stokes parameter $\tilde{Q}$ is presented by Lorentzian curves (Figure 94). All curves reach saturation at a magnetic field strength of $H \geq 80\,\mathrm{G}$. Similar dependencies of parameter $\tilde{U}$ are presented in Figure 95. At $\beta = 0°$ and $\beta = 90°$ the value of this parameter is zero; it is saturated at $H \geq 60\,\mathrm{G}$. Stokes parameter $\tilde{U}$ changes sign when angle β is close to the "magic" angle ($\beta_0 = 54.73°$). The dependencies of Stokes parameters $\tilde{Q}$ and $\tilde{U}$ on angle β for the same range of magnetic field strengths are given in Figures 96 and 97. Parameter $\tilde{Q}$ takes its minimal value close to the "magic" value of angle β. For angles β exceeding the "magic" value, parameter $\tilde{Q}$ increases with angle, and vice versa for smaller angles. Such behavior is in accordance with Figure 94. The dependence of parameter $\tilde{U}$ is given in Figure 97, which is in accordance with Figure 95. A reversal of the sign of Stokes parameter $\tilde{U}$ can be seen in this figure, as well. These computations of parameters Stokes in a collision reference frame are unified in two three-dimensional dependencies presented in Figures 98 and 99.

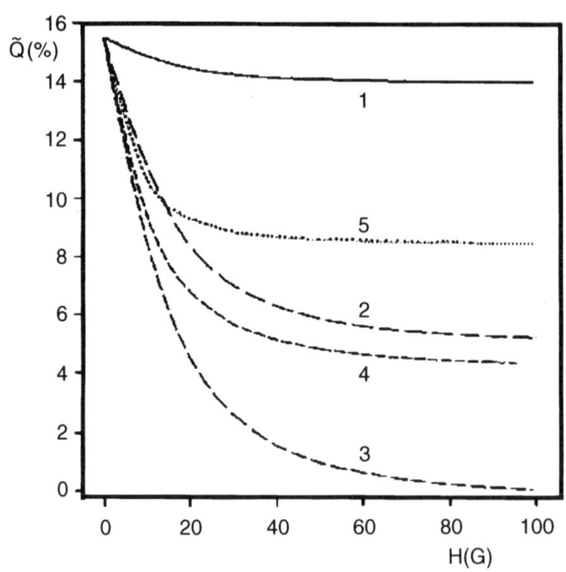

Figure 94. Dependence of Stokes parameter $\tilde{Q}$ on magnetic field strength for different inclination angles of the $O\tilde{Z}$ axis of the collision reference frame with respect to the magnetic field strength vector, β: 10°(1), 30°(2), 50°(3), 70°(4), 90°(5).

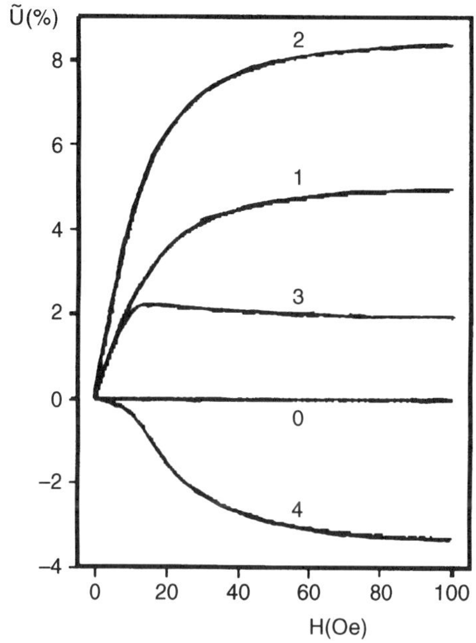

Figure 95. Dependence of Stokes parameter $\tilde{U}$ on magnetic field strength for different inclination angles of the $O\tilde{Z}$ axis of the collision reference frame with respect to the vector of the magnetic field strength, β: 10° (1), 30° (2), 50° (3), 70° (4), 0°, 90° (5).

Figure 100 illustrates the degree of polarization in a collision frame of reference as a function of magnetic field strength H and angle β

$$\tilde{P} = \sqrt{\tilde{Q}^2 + \tilde{U}^2}$$

The degree of polarization attains its maximum at $\beta = 0$, $H = 0$, and its minimum near the "magic" angle.

Spectropolarimetric computations in a collision frame of reference enable one to model the behavior of spectropolarimetric data as recorded by a ground-based observer. Figures 101 and 102 represent Stokes parameters Q and U referred to the coordinate system (X, Y, Z) of a ground-based observer as functions of angles α and β, determined in Figure 93, for a magnetic field strength of $H = 20\,\text{G}$. These parameters can be easily obtained by using the transformation of polarization moments (12.3.20) and the formulas of Section 3.6. As shown in Figure 101, Stokes parameter Q decreases with growth in α for $\beta \leq 70°$. It hardly

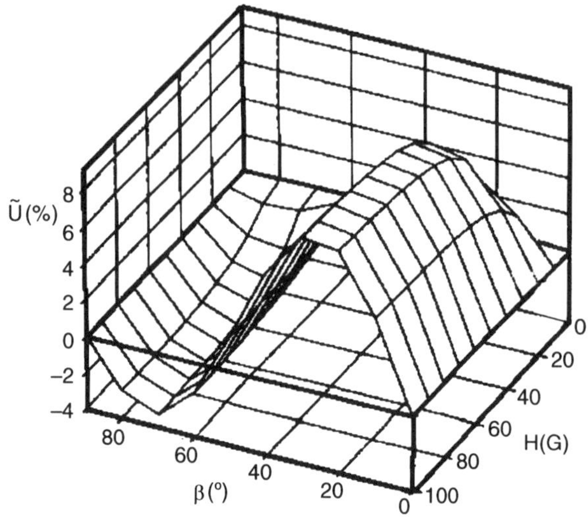

Figure 99. Three dimensional dependence of Stokes parameter $\tilde{U}$ on magnetic field strength H(Gs) and angle β.

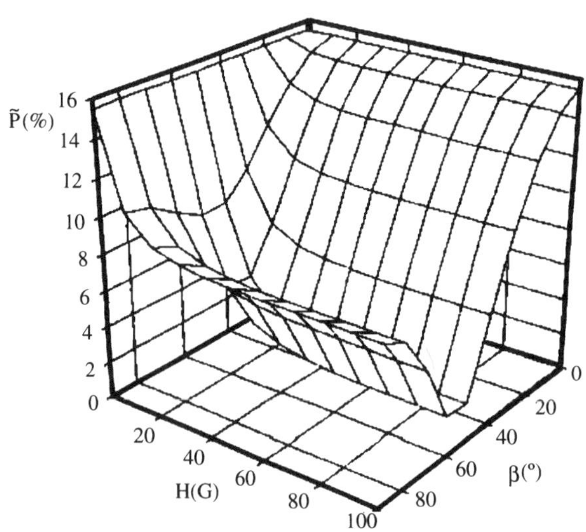

Figure 100. Dependence of degree of polarization $\tilde{P}$ within a collision reference frame on magnetic field strength and inclination angle β.

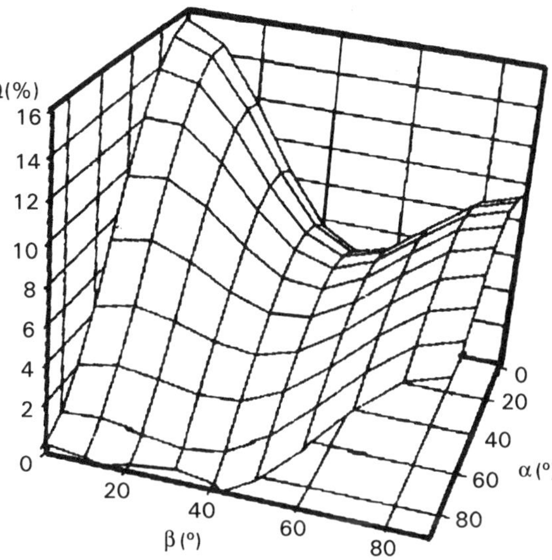

Figure 101. Dependence of recorded Stokes parameter Q on angles α and β, as determined in Figure 93.

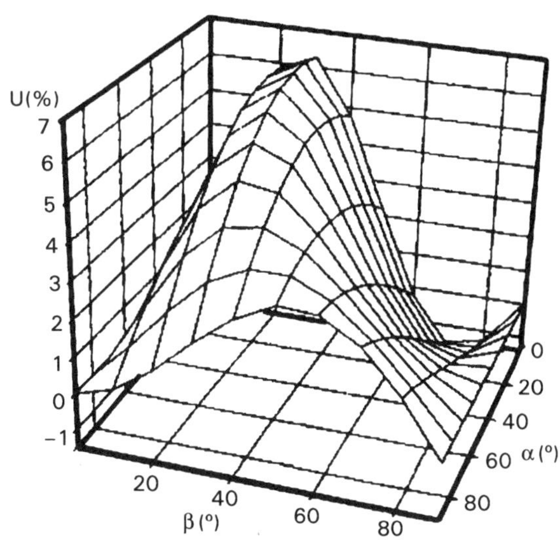

Figure 102. Dependence of recorded Stokes parameter U on angles α and β, as determined in Figure 93.

To calculate this amplitude we expand the exponent in terms of spherical harmonics[33,40,65,139]

$$e^{i(\mathbf{q},\mathbf{r})} = \sum_{LM} i^L j_L(q,r) Y_{LM}(\theta,\phi) Y_{LM}(\vartheta,\varphi) \tag{12.4.2}$$

where $j_l(q,r)$ are spherical Bessel functions while the pairs θ, ϕ and ϑ, φ are angular variables specifying the transmitted impulse during the collision and the vector $\mathbf{r}$.

Substitution of expansion (12.4.2) into Eq. (12.4.1) yields

$$f_{nlm} = \frac{2me^2}{\hbar^2 q^2} i^l \langle nl | j_l(q,r) | 10 \rangle Y_{lm}(\theta\phi) \tag{12.4.3}$$

The cross section of collisional excitation of the energy state n with angular momentum l is expressed as

$$\sigma^{ll'}_{mm'} = f_{nlm} f^*_{nl'm'} \tag{12.4.4}$$

Axial symmetry of the collisional process and the formula for the addition of spherical functions, applied to the cross section $\sigma^k_0(ll'; 0; v)$ given in (3.8.12), yield

$$\sigma^k_0(ll'; 0; v) = 8\pi\left(\frac{v_0}{v}\right)^2 \sqrt{(2l+1)(2l'+1)} \begin{bmatrix} l & l' & k \\ 0 & 0 & 0 \end{bmatrix} \int d\boldsymbol{v} f(\boldsymbol{v}) P_k(\cos\gamma)$$
$$\times \int \frac{dq}{q^3} \langle nl | j_l(qr) | 10 \rangle \langle nl' | j_{l'}(qr) | 10 \rangle P_k(\cos\theta) \tag{12.4.5}$$

where $v_0 = e^2/\hbar$ is the atomic velocity, γ is the inclination angle of the velocity vector to the OZ axis, and $f(\boldsymbol{v})$ is the velocity distribution function.

Let us use the velocity distribution function of type (5.4.12), which corresponds to many astrophysical and laboratory applications. Integration with respect to angular variables specifying different orientations of vector $\boldsymbol{v}$ leads to the following expression for the cross section (12.4.5)

$$\sigma^k_0(ll'; v) = \frac{32\pi^2}{2\kappa+1} f_k(v) \sqrt{(2l+1)(2l'+1)} \left(\frac{v_0}{v}\right)^2$$
$$\times \int \frac{dq}{q^3} \langle nl | j_l(qr) | 10 \rangle \langle nl' | j_{l'}(qr) | 10 \rangle P_k(\cos\theta) \tag{12.4.6}$$

where $f_k(v)$ is the multipole moment of velocity distribution function (3.8.9) (Figure 29).

In order to compute the cross section of the impact ordering of momenta, we must know the matrix elements $\langle nl|j_l(q,r)|10\rangle$. These elements may be calculated in terms of hydrogen wave functions

$$\psi_{nlm}(r) = R_{nl}(r)Y_{lm}(\vartheta\varphi) = N_{nl}r^l L_{n-l-1}^{2l+1}\left(\frac{2r}{na_0}\right)e^{-r/na_0} \tag{12.4.7}$$

Expressions for these elements during transitions from the ground state to states with principal quantum numbers $n = 3$ and $n = 4$ as functions of dimensionless parameter $s = qa_0$ take the form

$$\begin{aligned}
\langle 30|j_0(qr)|10\rangle &= \frac{432\sqrt{3}s^2(27s^2+16)}{(9s^2+16)^4}, \\
\langle 31|j_1(qr)|10\rangle &= \frac{288\sqrt{6}(27s^2+16)s}{(9s^2+16)^4} \\
\langle 32|j_2(qr)|10\rangle &= \frac{6912\sqrt{30}s^2}{5(9s^2+16)^4}, \\
\langle 40|j_0(qr)|10\rangle &= \frac{8192s^2(256s^4+352s^2+125)}{(16s^2+25)^5} \\
\langle 41|j_1(qr)|10\rangle &= \frac{2048\sqrt{15}s(256s^4+352s^2+125)}{(16s^2+25)^5} \\
\langle 42|j_2(qr)|10\rangle &= \frac{65536\sqrt{5}s^2(32s^2+25)}{5(16s^2+25)^5}, \\
\langle 43|j_3(qr)|10\rangle &= \frac{262144s^3}{25(16s^2+25)^5}
\end{aligned} \tag{12.4.8}$$

These matrix elements enable one to calculate parameters integrated over a transmitted impulse and, as a result, obtain the cross sections of collisional creation of an ordering of momenta of a certain type ($k = 0, 2, \ldots$). Under stationary conditions these cross sections determine the polarization moments of an atomic ensemble

$$\rho_0^\kappa(ll') = \frac{2}{\gamma_l + \gamma_{l'}} n_0 v \sigma_0^\kappa(ll'; 0; v) \tag{12.4.9}$$

where γ_l are radiative damping constants for level l.

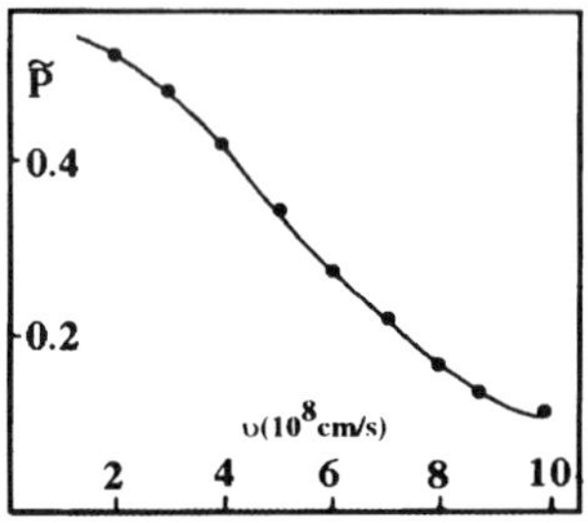

Figure 103. Degree of polarization $\tilde{P}$ as a function of velocity in the collision frame of reference calculated for the Born approximation.

Knowing the values of the polarization moments, it is possible to compute the intensities of polarized radiation and Stokes parameter $\tilde{Q}$ during the transition $n \to n_0$, with the help of formulas (3.6.16)

$$\tilde{Q}_{n\to n_0} = -\frac{3\sum_{ll'} W^{(2)}(ll')\rho_0^2(ll')}{\sum_{ll'}[2\sqrt{2}W^{(0)}(ll)\rho_0^0(ll)\delta_{ll'} - W^{(2)}(ll')\rho_0^2(ll')]} \tag{12.4.10}$$

where coefficients $W^k(ll')$ are determined by relations (3.6.17) and (3.6.18).

The degree of polarization of H_α and H_β lines as functions of the relative velocity of a colliding proton with a hydrogen atom were calculated by this technique. Results for the hydrogen H_α line are presented in Figure 103. We note that, as for large relative velocities, the integral on the right hand part of expression (12.4.6) depends weakly on velocity. The decrease in the degree of polarization with growth in velocity is stipulated mainly by kinetic reasons or a decrease in the multipole moments ratio $f_2(v)/f_0(v)$.

12.5. Linear Polarization of the H_α Hydrogen Line in Glauber Approximation

A criterion for applicability of the Born approximation follows directly from simple considerations concerning the relations between parameters of the collisional problem. If E is the kinetic energy of the projectiles, $\bar{U}$ is the avaraged interaction energy of the colliding pair, R is the effective radius of action of interatomic forces, and $k = mv/\hbar$, then this criterion takes the form

$$\frac{\bar{U}}{E}kR \ll 1 \tag{12.5.1}$$

If $\bar{U} \sim e^2/R$, the latter inequality becomes

$$2\left(\frac{v_0}{v}\right) \ll 1$$

where $v_0 = e^2/\hbar$ is the atomic velocity.

Therefore the Born approximation is valid if $v \geq 10v_0$. For proton energies typical of impact excitation of chromospheric hydrogen during a solar flare ($E \leq 300\,\text{keV}$), $v \leq 8v_0$. This means that for the particular problem of impact spectropolarimetric sensing of solar flares, the Glauber approximation (Section 4.3) could be more fruitful.[39,40,44]

The scattering amplitude corresponding to the impact population of magnetic sublevel m of level l with principal quantum number n could be written in a form (5.3.6)

$$f_{nlm} = \frac{imv}{2\pi\hbar}\int_0^{\infty} \rho'\, d\,\rho' \langle 100|1 - e^{2i\chi(\rho',r)}|nlm\rangle e^{iq\rho} \tag{12.5.2}$$

where the phase $\chi(\rho', r)$ is given by relation (4.3.7)

$$\chi(\rho', r) = -\frac{1}{\hbar v}\int_{-\infty}^{\infty} V(\mathbf{r}, \mathbf{r}')dz' \tag{12.5.3}$$

Here, $V(r', r)$ is the interaction energy of a hydrogen atom with precipitating proton, the position of which is characterized by radius vector r', while $\mathbf{r}$ describes the position of the atomic electron.

Integration in formulas (12.5.2) and (12.5.3) is carried out within the cylindrical system of reference (ρ', z', φ') over the coordinates of the projectile. Making use of a technique developed elsewhere,[46] integral (12.5.2) can be expressed in terms of hypergeometric functions. For the particular case of impact excitation of hydrogen into the state with $n = 3$, the following amplitudes are nonzero

$$\begin{aligned} f_{300} &= -\frac{9\sqrt{3}}{1024}a_0 s^4 \Gamma(1+i\eta)\Gamma(1-i\eta)\left[s\frac{\partial}{\partial s} + s^2\frac{\partial^2}{\partial s^2} + s^3\frac{\partial^3}{\partial s^3}\right] \\ &\quad \times \left[s^{-2-2i\eta}F(1-i\eta, 1-i\eta; 1', -s^2)\right] \\ f_{31\pm1} &= \mp\frac{27a_0}{256}e^{\pm i\varphi}\Gamma(1+i\eta)\Gamma(2-i\eta)s^5\left(2s\frac{\partial}{\partial s} + \frac{1}{4s^2}\frac{\partial^2}{\partial s^2}\right) \\ &\quad \times [2s^{-2i\eta-2}F(2-i\eta, 1-i\eta; 2, -s^2)] \end{aligned}$$

$$f_{320} = \frac{27\sqrt{6}}{256} a_0 s^7 \Gamma(i\eta) \frac{\partial}{\partial s}$$

$$\left[s^{-2i\eta^{-4}} \times \sum_{j=0}^{3} A_j \frac{\Gamma(1+j+i\eta)}{j!} F(1+j+i\eta, -i\eta, 1, -s^2) \right]$$

$$f_{32\pm2} = -\frac{27}{2048} a_0 e^{\pm 2i\varphi} \Gamma(1+i\eta)\Gamma(3-i\eta)(1-i\eta)s^7 \frac{\partial}{\partial s}$$

$$\left[s^{-2i\eta-2} \times 3F(3-i\eta, 2-i\eta; 3, -s^2) - (3-i\eta)F(4-i\eta, 2-i\eta; 3; -s^2) \right] \tag{12.5.4}$$

where the dimensionless parameter s is related to the impulse transmitted during collision by $s = 4/3a_0q$, parameter η is the ratio of velocities: $\eta = v/v_0$, and A_j is an element of the four-dimensional vector $\mathbf{A}(1, -6, 9, -4)$, and $F(\alpha, \beta; \gamma; -s^2)$ is a hypergeometric function.

Knowledge of amplitude $f_{lm}(z)$ enables one to compute the cross section of the collisional ordering of momenta

$$\sigma_0^k(ll'; 0; v) = \int d\boldsymbol{v}\, f(v)|\boldsymbol{v}|P_k(\cos\theta)\tilde{\sigma}_0^\kappa(ll'; 0; v) \tag{12.5.5}$$

where the last factor within the integral on the right hand side takes the form

$$\tilde{\sigma}_0^k(ll'; v) = \frac{16}{9a_0^2k^2} \sum (-1)^{l'-m} \begin{bmatrix} l & l' & k \\ m & -m & 0 \end{bmatrix} \int \frac{ds}{s^3} f_{lm}(s) f_{l'm}^*(s) \tag{12.5.6}$$

Expressions (12.5.5) and (12.5.6) with formula (12.2.9) make it possible to compute the polarization moments induced by collisional excitation of the hydrogen state with $n = 3$, and then the degree of polarization of the H_α hydrogen line with the help of expression (12.2.10).

Results of such calculations[51] for the H_α hydrogen line with the aid of the velocity distribution function of fast protons propagating within the magnetic loop of a solar flare of the form (5.4.12), are presented in Figure 104. By comparing these data with those for the the same for Born approximation, it is seen that the degree of polarization of the H_α hydrogen line resulting from proton impact differs considerably for the Born and Glauber approximations. Therefore, we evidently require a more accurate description of the collisional part of the problem of spectropolartimetric sensing of solar flares in the course of the development of this technique for solar studies.

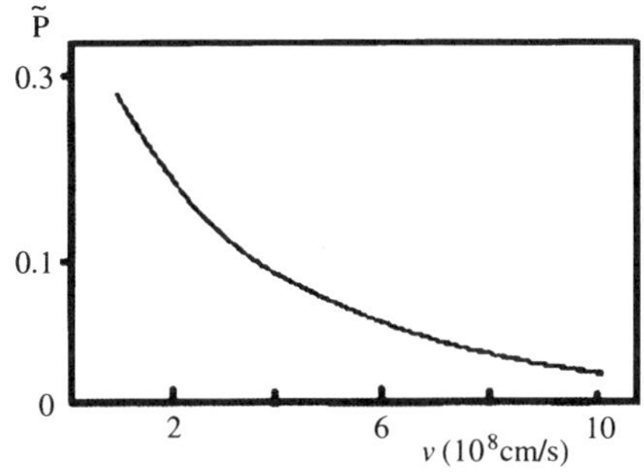

Figure 104. Velocity dependence of the degree of polarization $\tilde{P}$ of the H_α hydrogen line in a collision frame of reference resulting from impact excitation by fast protons propagating within a solar flare loop.

12.6. Linear Polarization of H_α *and* H_β *Hydrogen Lines by the Impact Parameter Method*

For a description of the impact parameter method, we consider two frames of reference: a collisional $(\tilde{X}, \tilde{Y}, \tilde{Z})$ frame with $O\tilde{Z}$ axis antiparallel to the direction of the relative velocity of colliding particles, and a subsequent $(\tilde{\tilde{X}}, \tilde{\tilde{Y}}, \tilde{\tilde{Z}}$ frame with $O\tilde{\tilde{Z}}$ axis directed to the projectile moving in a straight line trajectory with impact parameter b and constant velocity v. (Figure 105). Here and below the parameters related to systems $(\tilde{X}, \tilde{Y}, \tilde{Z}\}$ and $(\tilde{\tilde{X}}, \tilde{\tilde{Y}}, \tilde{\tilde{Z}})$ will be denoted by symbols ~ and $\tilde{\tilde{}}$, respectively. Within the frame of the impact parameter method, the wave function of an electron in the field of two protons (atomic and projectile) can be represented as an expansion in terms of a certain system of functions

$$\tilde{\tilde{\Psi}}(\mathbf{r}, t) = \sum_{nlm} \tilde{\tilde{c}}_{nlm}(t)\tilde{\tilde{\psi}}_{nlm}(\mathbf{r}, t) \exp\left(\frac{-iE_n t}{\hbar}\right) \tag{12.6.1}$$

where $\tilde{\tilde{\psi}}_{nlm}(\mathbf{r}, t)$, $l = 0, 1, 2, \ldots, n-1$; $|m| = 0, 1, 2, \ldots, l$ are a number of wave functions of an hydrogen atom at energy state E_n.

These functions could be expressed as

$$\tilde{\tilde{\psi}}_{nlm}(\mathbf{r}) = \tilde{\tilde{R}}_{nl}(r)\tilde{\tilde{Y}}_{lm}(\theta, \varphi)$$

The radial part of the wave function is represented by

$$\tilde{\tilde{R}}_{nl} = \sqrt{\frac{(2\varepsilon_n)^{2l+3}(n-l-1)!}{2n(n+l)!}} r^l e^{-\varepsilon_n r} L^{2l+1}_{n-l-1}(2\varepsilon_n r) \tag{12.6.2}$$

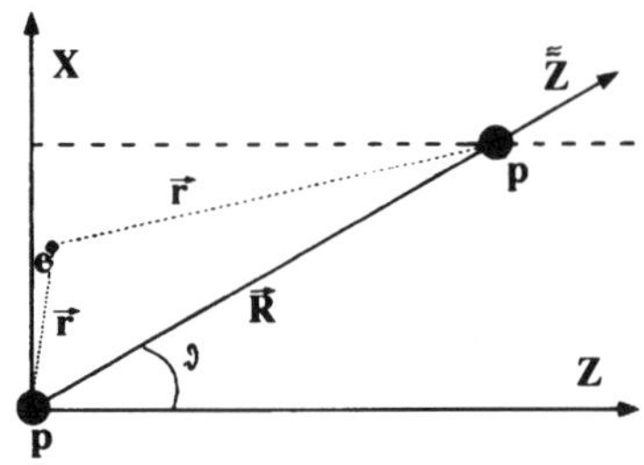

Figure 105. Reference systems for a description of collisions by the impact parameter method.

where $\varepsilon_n = 1/na_0$, $L_n^{\alpha}(x)$ is a Lager polynomial, and $Y_{lm}(\theta\varphi)$ is a spherical function.

The expansion (12.6.1) is restricted to only the wave functions describing two states: the ground ($|100\rangle$) and excited ($|nlm\rangle$) states. Hence, the approximate wave function of an electron in the field of two protons takes the form

$$\tilde{\tilde{\Psi}}(\mathbf{r}, t) = \tilde{\tilde{c}}_{100}(t)\tilde{\tilde{\psi}}_{100}(\mathbf{r}, t)e^{-i(E_1 t/\hbar)} + \sum_{l,m}\tilde{\tilde{c}}_{nlm}(t)\tilde{\tilde{\psi}}_{nlm}(\mathbf{r}, t)e^{-i(E_n t/\hbar)} \tag{12.6.3}$$

This function is a solution of the nonstationary Schrödinger equation

$$i\hbar\frac{\partial\Psi}{\partial t} = H\Psi \tag{12.6.4}$$

The Hamiltonian of this equation consists of two parts. The first part is the Hamiltonian operator of a free hydrogen atom and the second describes interaction with a precipitating proton. This second part, in turn, consists of two contributions, Coulomb repulsion of protons (the R^{-1} law) and interaction of the proton with the atomic electron

$$W(\mathbf{r}, R) = -\frac{e^2}{|\mathbf{r} - \mathbf{R}|} \tag{12.6.5}$$

By substituting (12.6.3) into (12.6.4) and introducing the dimensionless parameter

$$\eta = \frac{v/v_0}{\varepsilon_n b},$$

we obtain in terms of the new variable

$$\vartheta = \arctan\frac{b}{vt}$$

a system of differential equations for the coefficients of the set (12.6.3):

$$\frac{d}{d\vartheta}\tilde{\tilde{c}}_{nlm} = -i\eta\sum_{l'}\langle nlm|W_1(\mathbf{r},R)|nl'm\rangle\tilde{\tilde{c}}_{nl'm}\exp(-i\omega_{nn'}\cot\vartheta) + \frac{\delta_{nn'}\delta_{ll'}}{2}\left[\sqrt{(l-m)(l+m+1)}\tilde{\tilde{c}}_{nlm+1} - \sqrt{(l+m)(l-m+1)}\tilde{\tilde{c}}_{nlm-1}\right] \tag{12.6.6}$$

Here, function $W_1(\mathbf{r}, R)$ is related to the energy or atomic electron–projectile interaction (12.6.5)

$$W_1(\mathbf{r},R) = \frac{\varepsilon_n R^2}{e^2}W(\mathbf{r},R) = \frac{\varepsilon_n R^2}{|\mathbf{r}-\mathbf{R}|} \qquad \text{and} \qquad \omega_{nn'} = \frac{n'^2-n^2}{2nn'^2}\frac{v_0}{v}\varepsilon_n b$$

where, as usual, $v_0 = e^2/\hbar$ is the atomic velocity.

The matrix elements in Eq. (12.6.6) may be calculated by resolving the dimension-dependent part in function $W(\mathbf{r}, R)$ in terms of Legendre polynomials

$$\langle nlm|\frac{1}{|\mathbf{R}-\mathbf{r}|}|n'l'm\rangle = \varepsilon_n\sqrt{\frac{(2l'+1)}{(2l+1)}}\sum_k f^{(k)}_{nl,n'l'}(R)\begin{bmatrix} l' & k & l \\ 0 & 0 & 0 \end{bmatrix}\begin{bmatrix} l' & k & l \\ m & 0 & m \end{bmatrix} \tag{12.6.7}$$

where

$$f^{(k)}_{nln'l'}(R) = \frac{1}{\varepsilon_n}\left[\frac{1}{R^{k+1}}\int_0^R dr\, r^{k+2}\tilde{\tilde{R}}_{nl}(r)\tilde{\tilde{R}}_{n'l'}(r) + R^k\int_R^\infty r^{1-k}\tilde{\tilde{R}}_{nl}(r)\tilde{\tilde{R}}_{n'l'}(r)dr\right] \tag{12.6.8}$$

describes the radial part of interaction (12.6.7).

Analytical expressions of these integrals for transitions $1s \to 3l$ and $3l \to 3l'$ are presented below as functions of parameter $x = \varepsilon_3 R = R/3a_0$

$$
\begin{aligned}
f_{1010}^{0} &= \frac{1-(3x+1)e^{-6x}}{x},\\
f_{1030}^{0} &= -\frac{\sqrt{3}}{16}(8x^2-12x-3)e^{-4x}\\
f_{1031}^{1} &= \frac{3\sqrt{6}}{128x^2}[3+(32x^4-16x^3-24x^2-12x^{-3})e^{-4x}]\\
f_{1032}^{2} &= \frac{\sqrt{30}}{256x^3}[9-(64x^5+96x^4+96x^3+72x^2+36x-9)e^{-4x}]\\
f_{3030}^{(0)} &= \frac{9-(4x^5-4x^4+12x^3+12x^2+15x+9)e^{-2x}}{(9x)}\\
f_{3031}^{(1)} &= -\frac{\sqrt{2}}{3x^2}[9-(2x^6+6x^4+12x^3+18x^2+18x+9)e^{-2x}]\\
f_{3032}^{(2)} &= \frac{\sqrt{10}}{9x^3}[45-(2x^7+x^6+12x^5+30x^4+60x^3+90x^2+90x-45)e^{-2x}]\\
f_{3131}^{(0)} &= \frac{9-(2x^5+2x^4+6x^3+12x^2+15x+9)e^{-2x}}{9x}\\
f_{3131}^{(2)} &= \frac{5}{9x^3}[36-(2x^7+2x^6+9x^5+24x^4+48x^3+72x^2+72x+36)e^{-2x}]\\
f_{3132}^{(1)} &= -\frac{\sqrt{5}}{30x^2}[45-(4x^6+12x^5+30x^4+60x^3+90x^2+90x+45)e^{-2x}]\\
f_{3132}^{(3)} &= -\frac{7\sqrt{5}}{45x^4}[225-(2x^8+6x^7+20x^6+60x^5+150x^4+300x^3+450x^2\\
&\quad +450x+225)e^{-2x}]\\
f_{3232}^{(0)} &= \frac{45-(2x^5+10x^4+30x^3+60x^2+75x-45)e^{-2x}}{45x}\\
f_{3232}^{(2)} &= \frac{1}{5x^5}[126-(2x^7+10x^6+33x^5+84x^4+168x^3\\
&\quad +252x^2-252x-126)e^{-2x}]\\
f_{3232}^{(4)} &= \frac{1}{5x^5}[1575-(2x^9+10x^8+40x^7+140x^6+420x^5\\
&\quad +1050x^4+2100x^3+3150x^2+3150x-1575)e^{-2x}]
\end{aligned}
\tag{12.6.9}
$$

The same expressions for transitions $1s \to 4l$ and $4l \to 4l'$ in terms of parameter $x = \varepsilon_4 R$ will take the following form

$$f^{(0)}_{1s1s} = [1 - (4x + 1)]e^{-8x}$$

$$f^{(0)}_{1s4s} = 32(123x^3 - 550x^2 + 405x + 81)/(9375e^{-5x})$$

$$f^{(1)}_{1s4p} = 32\sqrt{5}[48 - (625x^5 - 2125x^4 + 325x^3 + 600x^2 + 240x + 48)]/(x^2e^{-5x})$$

$$f^{2}_{1s4d} = 32\sqrt{5}[1152 + (15625x^6 - 21875x^5 - 30000x^4 - 24000x^3 - 14400x^2 - 5760x - 1152)]/(1171857x^3e^{-5x})$$

$$f^{(3)}_{1s4f} = 32\sqrt{35}[1152 - (15625x^7 + 25000x^6 + 30000x^5 + 30000x^4 + 24000x^3 + 14400x^2 + 5760x - 1152)]/(5859375x^4e^{-5x})$$

$$f^{(0)}_{4s4s} = [36 - (4x^7 + 20x^6 - 54x^5 + 36x^4 - 54x^3 - 54x^2 - 63x - 36)]/(36xe^{-2x})$$

$$f^{(1)}_{4s4p} = -\sqrt{15}[45 - (2x^8 - 8x^7 + 20x^6 + 30x^4 + 60x^3 + 90x^2 + 90x + 45)]/(30x^2e^{-2x})$$

$$f^{(2)}_{4s4d} = \sqrt{5}[135 - (x^9 - 2x^8 + 6x^7 + 12x^6 + 36x^5 + 90x^4 + 180x^3 + 270x^2 + 270x + 135)]/(9x^3e^{-2x})$$

$$f^{(3)}_{4s4f} = -\sqrt{35}[4725 - (4x^{10} + 4x^9 + 30x^8 + 120x^7 + 420x^6 + 1260x^5 + 3150x^4 + 6300x^3 + 9450x^2 + 9450x + 4725)]/180x^4e^{2x}$$

$$f^{(0)}_{4p4p} = [60 - (4x^7 - 12x^6 + 26x^5 + 20x^4 + 50x^3 + 90x^2 + 105x + 60)]/(60xe^{-2x})$$

$$f^{(2)}_{4p4p} = [225 - (2x^9 - 6x^8 + 16x^7 + 16x^6 + 58x^5 + 150x^4 + 300x^3 + 450x^2 - 450x - 225)]/(6x^3e^{-2x})$$

$$f^{(1)}_{4p4d} = -\sqrt{75}[45 - (x^8 - x^7 + 4x^6 + 12x^5 + 30x^4 + 60x^3 + 90x^2 + 90x + 45)]/75x^2e^{-2x}$$

$$f^{(3)}_{4p4d} = -7\sqrt{75}[3367 - (4x^{10} - 4x^9 + 26x^8 + 88x^7 + 300x^6 + 900x^5 + 2250x^4 + 4500x^3 + 6750x^2 + 6750x - 3375)]/(900x^4e^{-2x})$$

$$f^{(2)}_{4p4d} = \sqrt{525}[315 - (x^9 + 2x^8 + 8x^7 + 28x^6 + 84x^5 + 210x^4 + 420x^3 - 630x^2 - 630x - 315)]/(315x^3e^{-2x})$$

$$f^{(4)}_{4p4f} = \sqrt{525}[33075 - (4x^{11} + 8x^{10} + 46x^9 + 210x^8 + 840x^7 + 2940x^6 + 8820x^5 + 22050x^4 + 44100x^3 + 66150x^2 + 66150x + 33075)]/(700x^5e^{-2x})$$

$$
\begin{aligned}
f_{4d4d}^{(0)} &= [180 - (4x^7 + 4x^6 + 18x^5 + 60x^4 + 150x^3 + 270x^2 + 315x \\
&= 180)]/(180xe^{-2x}) \\
f_{4d4d}^{(2)} &= [567 - (2x^9 + 2x^8 + 12x^7 + 48x^6 + 150x^5 + 378x^4 \\
&\quad + 756x^3 + 1134x^2 + 1134x + 567)]/(18x^3e^{-2x}) \\
f_{4d4d}^{(4)} &= [14175 - (2x^{11} + 2x^{10} + 19x^9 + 90x^8 + 360x^7 + 1260x^6 + 3780x^5 \\
&\quad + 9450x^4 + 18900x^3 + 28350x^2 + 28350x + 14175)]/(10x^5e^{-2x}) \\
f_{4d4f}^{(1)} &= -\sqrt{175}[315 - (2x^8 + 8x^7 + 28x^6 + 84x^5 + 210x^4 \\
&\quad + 420x^3 + 630x^2 + 630x + 315)]/(1050x^2e^{-2x}) \\
f_{4d4f}^{(3)} &= -\sqrt{175}[10125 - (4x^{10} + 16x^9 + 66x^8 + 258x^7 + 900x^6 + 2700x^5 \\
&\quad + 6750x^4 + 13500x^3 + 20250x^2 + 20250x + 10125)]/(900x^4e^{-2x}) \\
f_{4d4f}^{(5)} &= -11\sqrt{175}[297675 - (4x^{12} + 16x^{11} + 84x^{10} + 420x^9 + 1890x^8 + 7560x^7 \\
&\quad + 26460x^6 + 79380x^5 + 198450x^4 + 396900x^3 \\
&\quad + 595350x^2 + 595350x + 297675)]/(6300x^6e^{-2x}) \\
f_{4f4f}^{(0)} &= [1260 - (4x^7 + 28x^6 + 126x^5 + 420x^4 + 1050x^3 + 1890x^2 \\
&\quad + 2205x + 1260)]/(1260xe^{-2x}) \\
f_{4f4f}^{(2)} &= [2835 - (2x^9 + 14x^8 + 66x^7 + 246x^6 + 753x^5 + 1890x^4 \\
&\quad + 3780x^3 + 5670x^2 + 5670x + 2835)]/(126x^3e^{-2x}) \\
f_{4f4f}^{(4)} &= [51975 - (2x^{11} + 14x^{10} + 73x^9 + 330x^8 + 1320x^7 + 4620x^6 + 13860x^5 \\
&\quad + 34650x^4 + 69300x^3 + 103950x^2 + 103950x + 51975)]/(70x^5e^{-2x}) \\
f_{4f4f}^{(6)} &= 13[3274425 - (4x^{13} + 28x^{12} + 168x^{11} + 924x^{10} + 4620x^9 \\
&\quad + 20790x^8 + 83160x^7 + 291060x^6 + 873180x^5 + 2182950x^4 \\
&\quad + 4365900x^3 + 6548850x^2 + 6548850x + 3274425)]/(1260x^7e^{-2x})
\end{aligned}
\tag{12.6.10}
$$

Almost all matrix elements $f_{nln'l'}^{(k)}$ given in analytical form by Eqs. (12.6.9) and (12.6.10) consist of two additive terms, the first of which decreases with increasing internuclear distance as a power law dependence while the second is an exponentially decreasing factor. Matrix element $f_{1s3s}^{(0)}$ is an esxception. Because of the orthogonality of functions, ψ_{100} and ψ_{300} this element decreases exponentially. The power-law-dependent part of matrix elements $f^{(1)}_{1031}$, $f^{(1)}_{3031}$, and $f^{(1)}_{3p3d}$ is proportional to $1/R^2$. The similar part corresponding to zero order of the diagonal matrix elements $f^{(0)}_{nlnl}$ is equal to $1/R$.

Mathematically, Coulomb interaction can be excluded by multiplying all coefficients c_{nlm} of system (12.6.4) by a phase factor not dependent on n, l, and m. By neglecting all exponentially decreasing terms and additive factors decaying faster than $1/R^2$, one obtains a dipole approximation for the description of collisional interaction (see Section 3.8).

Figure 106 represents matrix elements

$$\langle 100|\frac{1}{\varepsilon_3|\mathbf{r}-\mathbf{R}|}|3p0\rangle = \frac{f^1_{1s3p}}{\sqrt{3}}$$

$$\langle 300|\frac{1}{\varepsilon_3|\mathbf{r}-\mathbf{R}|}|310\rangle = \frac{f^1_{3s3p}}{\sqrt{3}}$$

$$\langle 31m|\frac{1}{\varepsilon_3|\mathbf{r}-\mathbf{R}|}|32m\rangle = f^0_{3p3p} + A_m f^2_{3p3p}$$

$$m = 0, 1, \qquad A_0 = 0.4, \qquad A_1 = -0.2$$

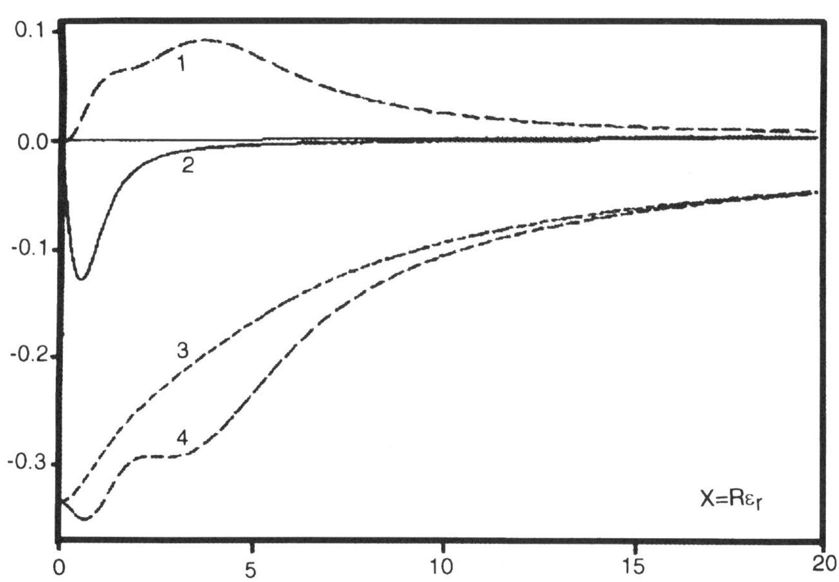

Figure 106. Matrix elements $\langle 100|W|3p0\rangle$, (1), $\langle 300|W|310\rangle$, (2), $\langle 310|W|320\rangle$, (3), $\langle 311|W|321\rangle$ (4).

to a $2p$ state (lifetime $\tau = 1.6 \times 10^{-9}$ s) and subsequent radiation of polarized light have been studied in Chapter 3. The nonstationary case was studied corresponding to the population of a $2s$ state of a hydrogen atom at time $t = 0$, subsequent creation of collisional alignment of a $2p$ state, and, as a result of this process, polarization of L_α emission ($2p \to 1s$). Linear polarization was preferentially along the excitation beam direction and the maximal degree of polarization attained a value of 80%.

Let us analyze a situation when the population of an $n = 2$ hydrogen excited state is implemented by isotropic resonance radiation, or isotropic collisions, providing steady excitation of an atomic ensemble. A steady-state population of $2s$ or $2p$ states without ordering or angular momenta of $2p$ states takes place in this case. When hydrogen atoms are excited in such a way and are affected by collisions with a beam of slow charged particles, collisional alignment of $2p$ states takes place as well as an accompanying process of collisional transfer of the population between $2s$ and $2p$ states.

Such a mechanism of collisional alignment of atomic hydrogen excited preliminarily by a stationary isotropic process may be valid for collisions of beams of slow protons with the hydrogen of interplanetary gas. As the lifetime of a $2p$ state is 10^{-7} times less than that for a $2s$ state, hydrogen atoms forming these nebular clouds and diluted interplanetary gas could be stored in the metastable $2s$ state, resulting in high densities of the metastable hydrogen atoms. Collisional excitation of atomic hydrogen in the state $n = 2$ resulting from the isotropic excitation process and all spectroscopic consequences will be examined in more detail.

If the initial excitation process of the $n = 2$ state is isotropic and these atoms are affected by collisional excitation of a beam of slow protons along the OZ axis, only three polarization moments $\rho_q^\kappa(ll_1)$ will be nonzero: $\rho_0^0(0,0)$, $\rho_0^0(1,1)$, and $\rho_0^2(1,1)$.

These elements determine the populations of $2s$ and $2p$ states

$$n_{2s} = \rho_0^0(0,0), \qquad n_{2p} = \sqrt{3}\rho_0^0(1,1)$$

and the longitudinal alignment of the hydrogen atom in the $2p$ state along the beam direction

$$\rho_0^2(1,1) = \frac{\langle 3L_z^2 - L^2 \rangle}{2}$$

where the angle brackets denote ensemble averaging in the $2p$ state.

The values of these polarization moments with allowance for the stationary isotropic population of states $2s$ ($l = 0$) and $2p$ ($l = 1$) [with efficiencies $N_0^0(0)$ and $N_0^0(1)$], the radiative damping of these states with constants $\gamma_0 = 1/\tau_{2s}$ and

$\gamma_1 = 1/\tau_{2p}$, and collisions with a proton beam can be determined by solving the system of three linear algebraic equations

$$n_0 v \sum_{k'l'} \sigma_0^{kk'}(ll, l'l'; v)\rho_0^{k'}(l'l') - \gamma_l \rho_0^k(ll) = N_0^k(l)\delta_{k0} \tag{12.7.1}$$

where (k, l) and (k', l') take the values (0,0), (2,0), and (2,1), n_0 is the proton number density in the beam, and v is the proton velocity. The effective cross sections for the destruction and mutual transformation of polarization moments $\sigma_0^{kk'}(ll, l'l'; v)$ for hydrogen atoms in a state with $n = 2$ under proton collisions with different velocities are presented in Table 4.

System (12.7.1) was solved for different cases of the isotropic population of an excited hydrogen state with $n = 2$

(1) for the population of a pure $2s$ state [for $N_0^0(0) = 0$, $N_0^0(1) = 0$],
(2) for the population of a pure $2p$ state [$(N_0^0(0) = 0, N_0^0(1) = 1)$],
(3) for the mixed case with the ratio of the efficiencies of populations of $2p$ and $2s$ states of

$$q = \sqrt{3}\frac{N_0^0(1)}{N_0^0(0)}$$

For all these cases the polarization moments $\rho_0^0(0, 0)$, $\rho_0^0(1, 1)$, and $\rho_0^2(1, 1)$, related to populations of levels $2s$ and $2p$ and alignment of the $2p$ state, were found for different velocities v and densities of the proton beam N.

The results of calculations of the steady-state $2s$–$2p$ population ratio are presented in Figures 108 and 109.

In case of the exclusive $2s$ state excitation (Figure 108), collisional population transfer plays a role for a proton beam density of $N \sim 10^{13}\ \text{cm}^{-3}$. For $N \sim 10^{16}\ \text{cm}^{-3}$ the population ratio n_{2p}/n_{2s} is practically equal to the equilibrium value, which is the ratio of the statistical weights of the studied levels (3 : 1). In case of exclusive isotropic excitation of the $2p$ state (Figure 109), collisional population transfer is noticeable already at $N \sim 10^5\ \text{cm}^{-3}$ and for $N \sim 10^8\ \text{cm}^{-3}$ an equilibrium population ratio $n_{2s}/n_{2p} = \frac{1}{3}$ is practically attained. Such a sharp difference in the collisional establishment of the equilibrium population ratio during the excitation of $2s$ and $2p$ states is caused by the large difference in the radiative lifetimes τ_{2s} and τ_{2p} of these levels.

The degree of linear polarization of the Lyman α line for observation along the OY axis in terms of polarization moments is expressed by

$$P = \frac{(I_z - I_x)_y}{(I_z + I_x)_y} = \frac{3\rho_0^2(ll)}{2\sqrt{2}\rho_0^0(ll) - \rho_0^2(ll)} \tag{12.7.2}$$

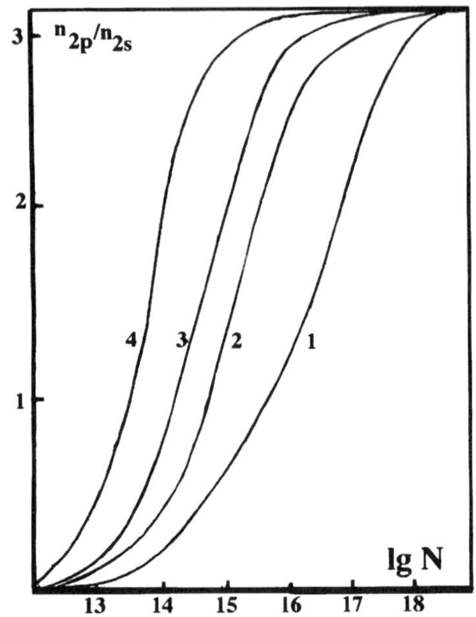

Figure 108. Population ratio n_{2p}/n_{2s} for $2p$ and $2s$ hydrogen states under continuous exclusive excitation of the $2s$ state as a function of the density of a proton beam $N(\text{cm}^{-3})$ for different proton velocities: 1, 200 m/s; 2, 600 m/s; 3, 1000 m/s; 4, 5000 m/s.

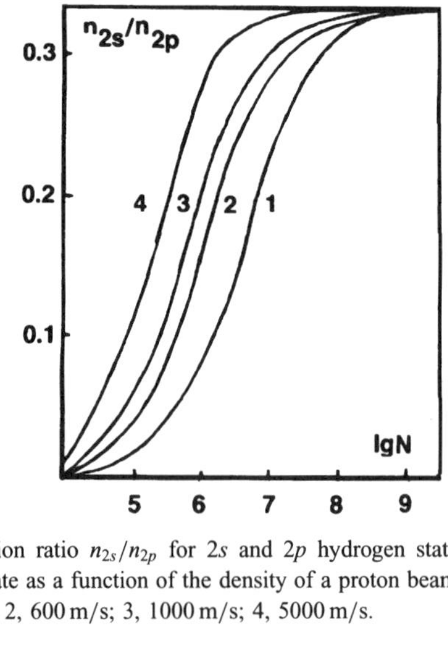

Figure 109. Population ratio n_{2s}/n_{2p} for $2s$ and $2p$ hydrogen states under continuous exclusive excitation of the $2p$ state as a function of the density of a proton beam $N\,(\text{cm}^{-3})$ for different proton velocities: 1, 200 m/s; 2, 600 m/s; 3, 1000 m/s; 4, 5000 m/s.

where I_x, I_y, and I_z are the total intensities of light emitted by hydrogen atoms under transitions $2p$–$1s$ linearly polarized along the corresponding axes.

It should be noted that two-photon light emission at $2s$–$1s$ transition will result in a continuous nonpolarized spectrum to the red side with respect to the Lyman α line[(171)] and will not affect its degree of polarization.

Anisotropic collisional alignment of angular momenta and the resulting linear polarization of emission are very much connected to the degree of the inequilibrium population of collisionally mixed atomic levels. The collisional creation of alignment does not happen under an equilibrium population ratio.[(34)] Therefore, alignment of a $2p$ state is principally possible only when the density of the proton beam is $N \leq 10^{16}\,\mathrm{cm}^{-3}$ under the excitation of a pure $2s$ state, and $N \leq 10^{8}\,\mathrm{cm}^{-3}$ under the isotropic excitation of a pure $2p$ state.

The degree of polarization (12.7.2) is not zero unless the collisional alignment of the $2p$ state, $\rho_0^2(11)$, is of the same order of magnitude as the population $n_{2p} = \sqrt{3}\rho_0^0(ll)$.

In case of an exclusive $2s$ state excitation, quantities $\rho_0^0(ll)$ and $\rho_0^2(ll)$ are nonzero only due to collisional excitation transfer from the $2s$ state and they have the same order of magnitude until the population ratio of $2s$ and $2p$ states does not approach the equilibrium value. Therefore, the dependence of the degree of polarization on the density of the proton beam N (Figure 110) is very weak. It decreases, once the density exceeds a value of $N \simeq 10^{14}\,\mathrm{cm}^{-3}$, and it turns out to be zero for $N \simeq 10^{17}\,\mathrm{cm}^{-3}$.

On the other hand, under isotropic excitation of a $2p$ state, even at a well-expressed population nonequilibrium for $2s$–$2p$ populations, a noticeable degree

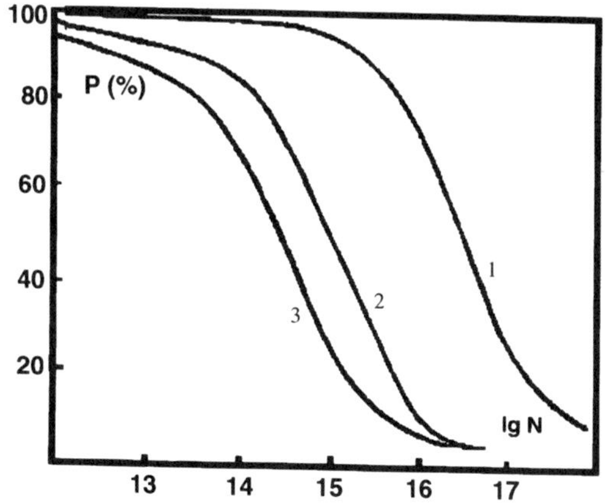

Figure 110. Degree of polarization of radiation of hydrogen atoms under continuous exclusive excitation of a $2s$ state as a function of the density of the proton beam $N\,(\mathrm{cm}^{-3})$ for different proton velocities: 1, 200 m/s; 2, 600 m/s; 3, 1000 m/s.

of polarization is possible only when the efficiency of collisional alignment $\rho_0^2(ll)$ is comparable to the rate of radiative damping, namely,

$$Nv\left[\sigma_0^{20}(ll, 00; v)\rho_0^0(0, 0) + \sigma_0^{20}(ll'll; v)\rho_0^0(ll)\right] \approx \gamma_{2p}\rho_0^2(ll) \tag{12.7.3}$$

If we take into consideration that the orders of magnitudes are $v \simeq 10^5$ cm/s, $\sigma_0^{20} \simeq 10^{-10}\,\text{cm}^2$, and $\gamma_{2p} \simeq 10^9\,\text{s}^{-1}$, then for the density of protons in a beam we have the estimate $N \geq 10^{14}\,\text{cm}^{-3}$. This density does not lie within the possible limits of conditions corresponding to the nonequilibrium populations of $2s$ and $2p$ states. Therefore, for isotropic excitation of a pure $2p$ state, the degree of polarization is extremely low and does not exceed 10^{-9} at each proton beam density.

Figure 111 gives the degree of linear polarization calculated by formula (12.4.2) for isotropic excitation with mixing of $2p$ and $2s$ states, specified by a population efficiency ratio q in the range of 0–12 with a proton beam density of $N = 10^{10}\,\text{cm}^{-3}$. The degree of polarization decreases monotonously with the growth in contribution of the $2p$ state excitation. Such behavior is in accordance with the effect of a small contribution of the population of the $2p$ level to the apparent degree of polarization.

These results show that collisional alignment of an anisotropic proton beam and linear polarization of hydrogen atoms under continuous and isotropic hydrogen excitation into states with $n = 2$ is practically determined by collisional transitions from the metastable $2s$ into the radiative $2p$ state. The degree of polarization in case of continuous exclusive excitation of the $2s$ state may be 80–90% for the proton densities and velocities in the beam under study. This value is the same as for the maximal time-dependent degree of polarization under conditions of pulsed excitation, calculated[94] for the same densities and velocities

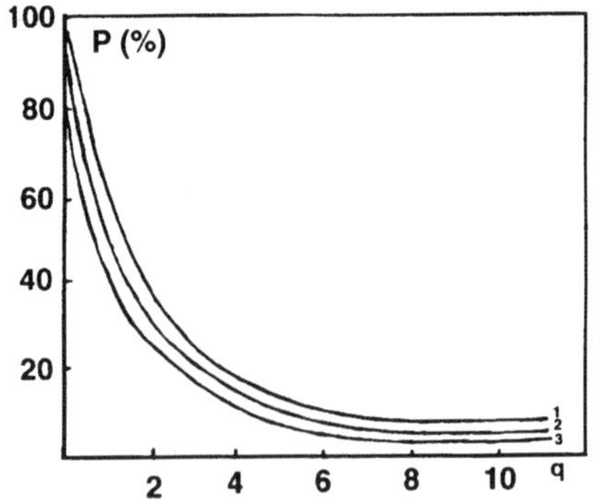

Figure 111. Degree of linear polarization of radiation of hydrogen atoms under continuous excitation of $2s$ and $2p$ states with an excitation efficiency ratio q for the density of the proton beam $N = 10^{10}\,\text{cm}^{-3}$ at different proton velocities: 1, 200 m/s; 2, 600 m/s; 3, 5000 m/s.

of protons in the beam. In case of isotropic excitation of both $2s$ and $2p$ states, the degree of polarization is mainly determined by the fraction of atoms excited in the $2s$ state. This fraction could be quite high due to a similar order of magnitude of effective cross sections for $1s$–$2p$ and $1s$–$2s$ collisional excitation transitions for the hydrogen atom by electrons with above-threshold energies.

12.8. Concluding Remarks

The theory and practical techniques for computing the degree of polarization for H_α and H_β hydrogen lines as a function of proton energy within a collision frame of reference related to a precipitating proton beam have been presented in this chapter. These computations and results for hydrogen lines of prior diagnostic significance have been described in much detail and are indispensable for practical problems of impact spectropolarimetric sensing.

The accuracy of calculations of energy dependencies of impact excitation cross sections of a hydrogen atom, obtained by different approaches and computational techniques, requires verification. To this end it is worthwhile analyzing and comparing these principal atomic characteristics. These quantities are related to zero-order cross sections $\sigma_0^{(0)}(nl, nl)$ defined by

$$\sigma^{(ex)}(n) = \sum_{lm} \rho_{mm}^{ll}(n) = \sum_{l} \sqrt{(2l+1)}\sigma_0^{(0)}(nl, nl) \qquad (12.8.1)$$

The values of $\sigma^{(ex)}(n)$ $(n = 2, 3, 4)$ as dependent on the energy of relative motion are presented in Figure 112. Similar values[115] are presented as well. Data calculated by the impact parameter method and the results of cited work are in very good agreement, in spite of the fact that in some work[115] a more extended basis of functions was used. The results of model calculations, demonstrated in this chapter, took into account only two states: the ground $1s$ and one excited nl states but elsewhere[115] some other excited states were also considered when forming the expansion basis. Figure 113 gives calculated values of $\sigma^{(ex)}(3)$ as dependent on the energy of relative motion in the Glauber approximation. One can see that the Glauber approximation gives results similar to the impact parameter method at energies of relative motion higher than 70 keV.

It is very important to compare the dependence of Stokes parameter $\tilde{\tilde{Q}}$ on the energy of the relative motion subject to the impact parameter, and the Born and Glauber approximations. The latter approximations are not valid for calculations of spectropolarimetric characteristics of H_α radiation because the $\tilde{\tilde{Q}}$ parameter, using the Born approximation, is practically independent of the energy of relative motion, while in the Glauber approximation this value varies in the range 5.5–34% for an energy range of 5.2–330 keV. A supplementary

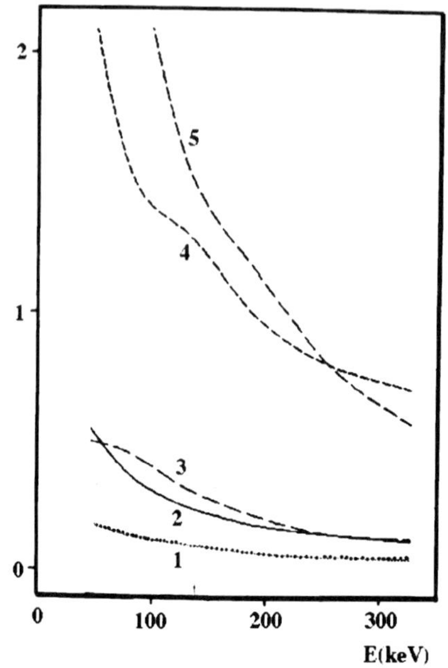

Figure 112. Values of $\sigma^{(ex)}(n)$ for hydrogen lines as dependent on the energy of relative motion: H_β, theory (1); H_α, theory (2); H_α, experiment (3); L_α, experiment (4); L_α, theory (5).

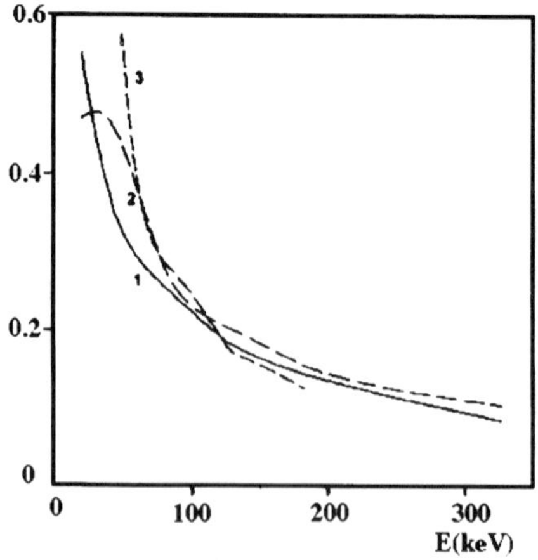

Figure 113. Values of $\sigma^{(ex)}(3)$ as dependent on the energy of relative motion: theory (1), experiment, (2), Glauber (3).

selection rule of the Glauber method, prohibiting transitions between two magnetic sublevels with a change in azimuthal quantum number, probably does not play the last role in this situation.

Figure 114 presents Stokes parameter $\tilde{\tilde{Q}}$, defined by Eq. (12.2.4), computed under the impact parameter approximation for L_α, H_β and H_α radiation together with an experimental curve.[199] The difference between the experimental and theoretical curves for H_α radiation is rather high. The reason for such a discrepancy could be that the theoretical computations, used for the data presented in Figure 114, were undertaken with account only for two states in the sum (12.6.6): the ground and excited states ($n = 2$ for L_α and $n = 3, 4$ for $H_{\alpha,\beta}$ respectively). On the other hand, it is clear that simultaneous impact excitation of levels at all other values of n takes place for a given range of relative motion energies. Therefore, not only will interaction terms in Eq. (12.6.7) comprise matrix elements nondiagonal with respect to the principal quantum number n, but also on the right hand side of the equation, for the evolution of a polarization moment at some level with principal quantum number n, besides the terms accounting for impact excitation there will be some terms describing radiation

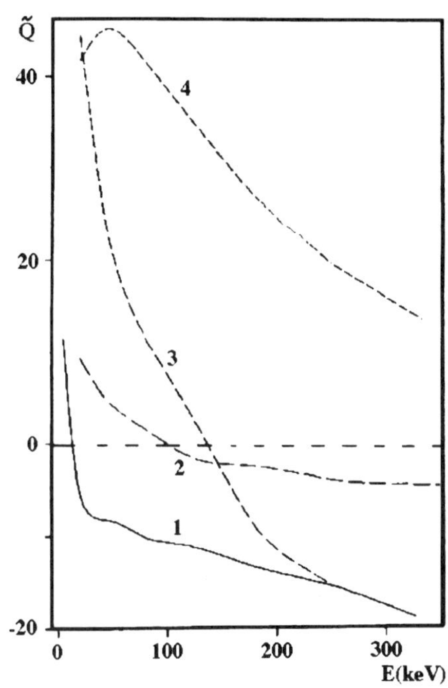

Figure 114. Stokes parameter $\tilde{Q}$ for H_α and H_β hydrogen lines as a function of the energy of protons in the collision frame of reference: H_α, theory (1); H_α, experiment[144,199] (2); L_α, theory (3); H_β, theory (4).

TABLE 16
Computed Degrees of Polarization of an H_β Hydrogen Line Resulting from Proton Impact[a]

v (10^8 cm/s)	E (keV)	$\tilde{P}_{H_\beta}$	P_{H_β}
2	20.7	12	9.45
3	46.6	7.5	6.1
4	82.9	4.2	3.5
5	129	2.5	2.11
6	186	0.1	0.85

[a] $\tilde{P}$ is the degree of polarization in the collision frame of reference; P is the same value in the rame of reference connected to the solar disk, inclined at 61.87°to the first frame.

TABLE 17
Theoretical Data for the Degree of Linear Polarization P of the H_α Hydrogen Line under Proton Impact for Different Angles β (Impact Parameter Technique)

v (10^8 cm/s)	E (keV)	$\tilde{Q}$	60.0°	63.0°	68.5°	49.2°	21.55°	61.8°
1.0	5.2	11.5	3.15	2.61	1.72	5.26	10.1	2.82
2.0	20.8	− 5.50	1.43	1.19	0.78	2.42	4.79	1.28
3.0	46.7	− 8.30	2.21	1.83	1.20	3.72	7.26	1.09
4.0	83.0	− 10.3	2.79	2.31	1.52	4.67	9.04	2.50
5.0	129	− 11.3	3.09	2.56	1.68	5.16	9.93	2.77
6.0	187	− 13.6	3.79	3.14	2.07	6.30	12.0	3.40
7.0	254	− 15.6	4.42	3.67	2.42	7.31	13.7	3.96
8.0	332	− 19.1	5.57	4.64	3.07	9.16	16.9	5.01

decay from upper levels with $n' > n$. This means that spectropolarimetric effects of H_α radiation should be analyzed with account for transitions from a higher-lying state with $n = 4$.

The spectropolarimetric sensing of solar flares is an example of a natural ionized medium and is implemented by comparing observations of polarization parameters in a frame of reference connected to the observer, with computational data in the same frame of reference. Therefore, we conclude this methodological chapter by presenting some computed data in Tables 16 and 17 comprising numerical results of spectropolarimetric parameters for H_α and H_β hydrogen lines in the system of reference connected to the solar disk. These data were used to estimate the energy of protons in practical problems of remote sensing of solar flares.

13

Impact Spectropolarimetric Diagnostics of Nonthermal Phenomena in the Solar Atmosphere

13.1. Introduction

This chapter summarizes all known results on the impact spectropolarimetric sensing of chromospheric locations of a solar flare. These regions of the solar atmosphere are of the utmost importance for the clarification of energy transport modes from the corona, where primary energy release takes place, to the lower parts of the solar atmosphere. The upper layer of the chromosphere is an optically active region, where most emission lines are formed.[(204)] It was shown in the previous chapters that spectropolarimetric data provide direct means for distinguishing the energy transfer mechanism from the corona to the chromosphere at a qualitative level.[(5)] Moreover, by employing all the developed methods, this technique allows one to study diagnostics of the chromosphere plasma using quantitative observational data on the Stokes parameters of emission lines. Observational spectropolarimetric information about solar flares, mustaches, and other emissive objects such as degree of polarization, orientation of the polarization plane, spatial and temporal behavior of the frequency distribution of Stokes parameters along the emission line contour, are reviewed and analyzed in this chapter in more detail for the purpose of the practical implementation of impact spectropolarimetric sensing of different emissive objects in the solar atmosphere.

Spectropolarimetric data presented in this chapter have been inferred by means of two principal methods of observation and registration, utilizing spectrograms and filtergrams. These different techniques of observation were applied for a number of independent events in the solar atmosphere. Unfortunately, neither spectrogram nor filtergram techniques allow one to collect all the indispensible information during solar observations. Filtergrams have a clear advantage in that they allow the state of a solar object to be examined at a particular moment as a whole. On the other hand, however, impact linear polarization, localized at the center of a line, can be determined only with considerable errors caused by the Doppler effect, which shifts the center of the line with respect to the frequency bandpass of the filter. Spectrograms present the whole frequency line contour and so allow one to avoid this uncertainty, but the disadvantage of this method of recording is that only that part of an object projected on the spectrograph slit can be studied.

Theoretical computation of the Stokes parameters of hydrogen lines within a ground-based observational frame of reference is implemented on the basis of some realistic models of proton kinetics, as they represent the most appropriate candidate for energy transport to the upper chromosphere within a magnetic loop, and yield indispensible collisional atomic parameters and cross sections of the impact alignment of a hydrogen atom.[50, 193] Comparison of theoretical and observational data allowed us to evaluate the energy of protons bombarding the solar chromosphere.

Different emissive objects of the active region N5669 SGD on 07.09.1989 were analyzed by means of this technique. The distribution of the orientation of linear polarization along the H_α hydrogen line at different sections across the spectrograph entrance slit allowed the energy of protons to be measured during a flare on 16.05.1991. Simultaneous spectropolarimetric measurements during the solar flare on 15.09.1991 carried out on H_α and H_β hydrogen lines and corresponding independent computations for both lines were used to assess the reliability of this technique. Linear polarization of mustache emission as well as a review of quantitative spectropolarimetric data based on the filtergram technique are also presented in this chapter. We were thus able to make spectropolarimetric estimates of the energy of electrons which, according to astrophysical models of this phenomenon, are responsible for the emission of polarized radiation.

This chapter comprises observational data on the solar chromosphere during solar flares, and processing of these data for the needs of local sensing, using different methods of atomic collisional physics. Chronologically, in the course of developing this technique for solar flare problems, Born, Glauber, and impact parameter methods have been used successively for impact spectropolarimetric sensing of the solar chromosphere. Before starting to describe particular impact spectropolarimetric sensing problems, a few methodological remarks on the designation of reference frames used in this chapter are in order. Earlier, apart from the laboratory reference frame, collisional ($\sim$) and subsequent ($\approx$) systems were employed to describe collisional excitation of atomic ensembles. For diagnostic problems close to a real situation of the excitation of atomic hydrogen by charged particles propagating within the magnetic loop of a solar flare, we shall use the symbol $\approx$ for values related to the subsequent system in which the

$O\tilde{\tilde{Z}}$ axis is directed to the projectile and the $O\tilde{\tilde{Y}}$ axis is parallel to the $O\tilde{Y}$ axis of the collisional reference system.

13.2. Energy Transport into the Chromospheric Region of a Solar Flare

The first spectropolarimetric estimates of energy transport during a solar flare were carried out by spectropolarimetric observations of the H_α emission line by means of a large solar telescope at Baikal Astrophysical Observatory affiliated to the Russian Institute of Solar and Terrestrial Physics.[202, 203] Active regions N 5669 by the SGD classification during the period 28 August–12 September 1989 were examined. Spectrograms of different locations within this group were obtained for the H_α line on 30 and 31 August and 7 September 1989.

Observational data presented in Figure 115 show a drawing of the image of active group No. 5669 in the plane of the entrance slit averaged over the observation time. Positions of the entrance slit are denoted by numbers 1–4 at different moments. The H_α emission line was observed in the third order of the grating with dispersion 0.035 nm/mm and light from other orders was cut out by a KS 10 filter. Sensitive ORWO NP-27 photoplates were used with exposure 0.3–0.5 s.

Spectropolarimetric measurements were undertaken with the help of the analyzer of linear polarization, mounted behind the exit slit of the spectrograph and described in Section 7.4. Every studied location was exposed three times at one photoplate for positions of the $\lambda/4$ phase plate of 0°, 22.5°, and 45°. Vertical

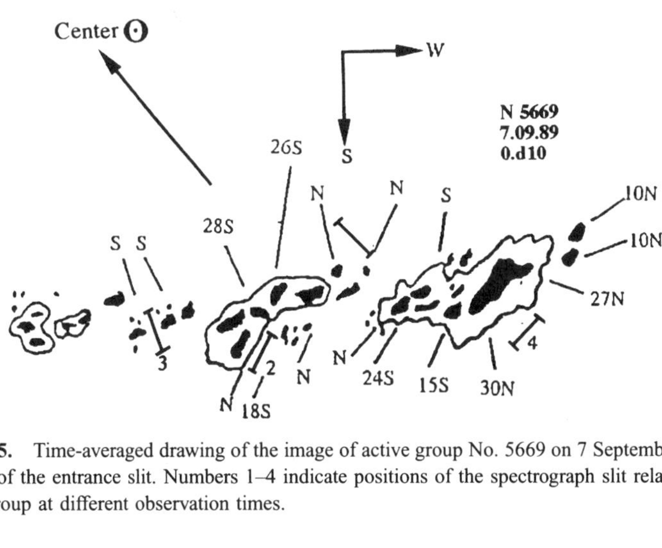

Figure 115. Time-averaged drawing of the image of active group No. 5669 on 7 September 1989 in the plane of the entrance slit. Numbers 1–4 indicate positions of the spectrograph slit relative to the sunspot group at different observation times.

linear polarization was extracted at the first position of the $\lambda/4$ plate, parallel to the grooves of the grating (the first spectral band). The second spectral band corresponded to polarization parallel to the dispersion of the grating. The turn of the $\lambda/4$ plate at 22.5° gave two spectral bands with linear polarizations oriented at 45° to the initial position. When the $\lambda/4$ plate was fixed at 45°, the polarization plane was turned through 90° with respect to the zero orientation. Experimental recordings are given in Figure 116. For the 22.5° position of the $\lambda/4$ plate, there was no polarization. In case (c) there was an inversion of polarization. The minimal time interval of 30 s between exposures and the corresponding slight shift of image did not crucially damage the polarization recordings.

The spectral profiles of two nodes of a relatively strong flare are given in Figure 117a, b. For the first node at the center of the H_α line, the degree of linear polarization was about 4% and the maximal polarization was 6.3%. No polarization was registered in the second node. Figure 117c, d shows the profiles of the H_α line for the same flare after 1 min and 1.5 min. The change in line shape indicates that an image shift took place. In this case the H_α profile is typical of a flare above the shade of a sunspot.

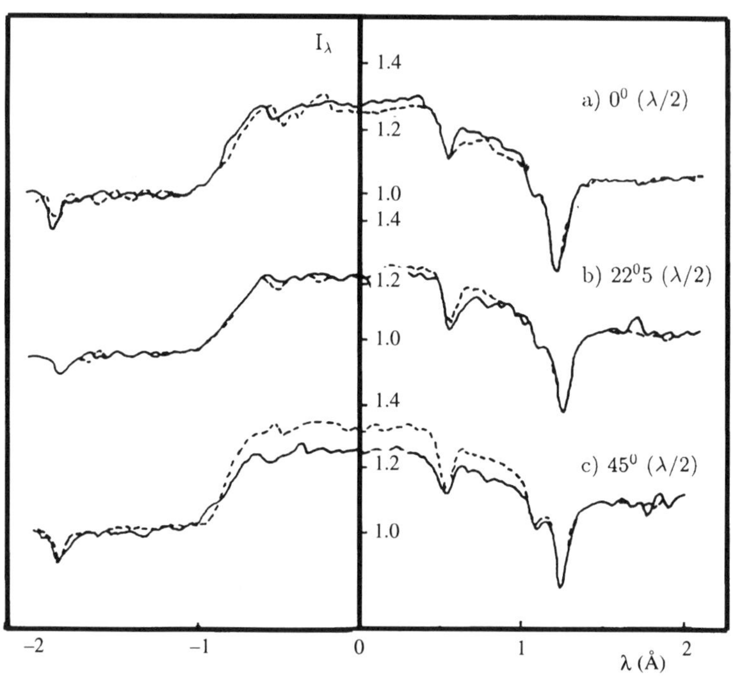

Figure 116. Spectral profiles of the H_α hydrogen line at three positions of the half-wave plate: (a) 0°, (b) 22.5°, (c) 45°.

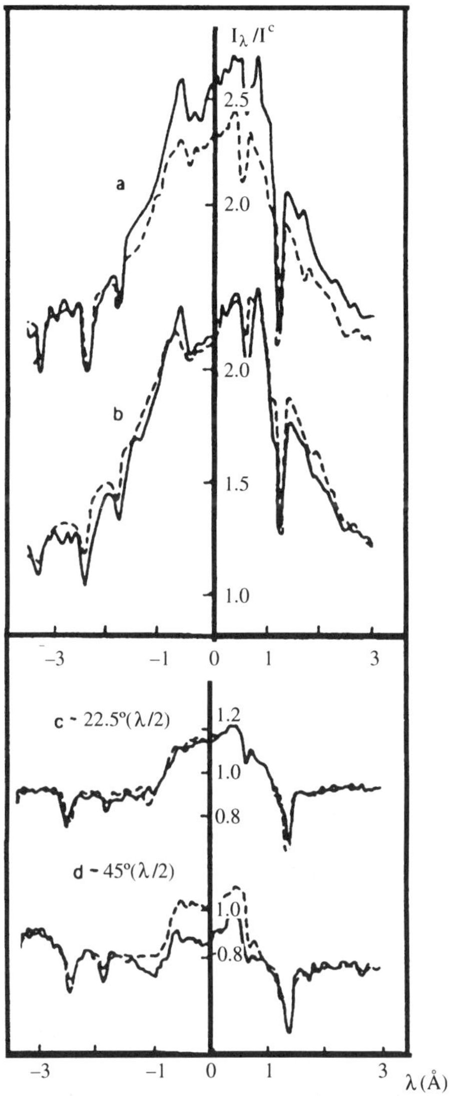

Figure 117. Spectral profiles of the H_α hydrogen line for two power nodes of a flare (a, b); the same contours after 1 min and 1.5 min (c, d).

Figure 118a, b shows H_α profiles for a weak flare above a spot. These contours were observed subsequently with a 40-s exposure and a 22.5° turn of the $\lambda/4$ plate, but in this case there was no image shift in the plane of the entrance slit between the two shots. In the first case the polarization was 3% and in the second there were no polarization at all. The main factor that should be accounted for in

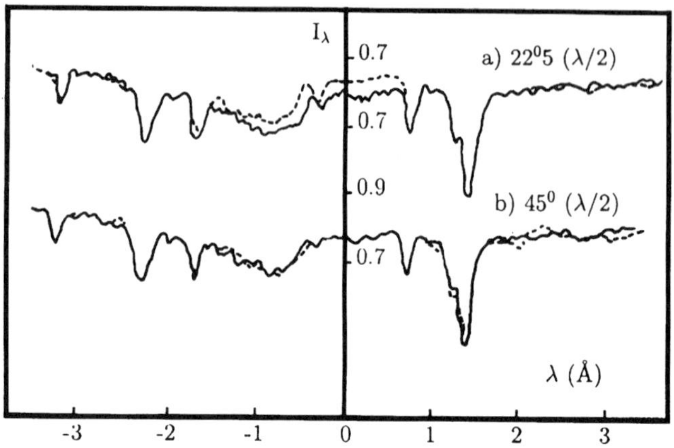

Figure 18. Spectral profiles of the H_α hydrogen line for a weak flare above a sunspot with no image shift at the entrance slit plane: contours (a) and (b) correspond to rotation of the $\lambda/4$ plate and a time interval of 40 s.

spectropolarimetric studies is instrumental polarization. This source of systematic error was reduced to a minimum in the course of these studies by normalizing the emission profiles to the continuum for the same spectral band. These observations and data corrections allowed us to conclude that the central part of the H_α emission line was linearly polarized.

Table 18 presents results of spectropolarimetric observations. Some conclusions as to the structural features of a flare can be inferred by analyzing the spectral profiles for different entrance slit positions. The spectral profiles in Figure 117a, b indicate a flare of second grade. The spectral profiles in Figure 118 indicate that this flare was located above a sunspot. The data presented in the first four rows of Table 18 show that the weak emission in the center of the H_α line cannot be attributed to the flare.

Numbers of the slit positions and the angles between the slit and the direction to the solar center disk are given in the first and second columns of Table 18. Positions of the $\lambda/2$ plate, fixed before the entrance slit of the spectrograph, are indicated in the following columns as well as values of the mean relative intensity $\bar{I}$ of the H_α line core:

$$\bar{I} = \left(\frac{I_1}{I_1^c} + \frac{I_2}{I_2^c}\right)^2$$

where I_i and I_i^c $(i = 1, 2)$ are intensities at the center of the line and in the continuum for the first and second spectral bands, respectively.

TABLE 18
Polarization Characteristics of the H_α Hydrogen Line

Slit number	$\lambda/2$ (°)	$\bar{I}$	Q, U (%)	P (%)	χ (°)
I	0	0.58	$-(4.0 \pm 0.7)$		
	22.5	0.62	$-(2.0 \pm 1.4)$	4.5 ± 1.6	13
$\alpha = -10°$	45	0.67	0		
II	0	2.37	4.2 ± 0.5		
	22.5	1.17	0	5.5 ± 1.7	
$\alpha = 67°$	45	1.75	6.7 ± 1.3		
III	0		0		
	22.5		0	3.5 ± 0.9	
$\alpha = 24°$	45	0.95	$-(3.5 \pm 0.9)$		
IV	0	1.30	0.5 ± 0.4		
	22.5	0.96	1.8 ± 0.6	2.2 ± 0.9	−23
$\alpha = 80°$	45	1.02	2.0 ± 0.7		

The Stokes parameters presented in the fourth column were calculated by the following formulas using data corresponding to different angular positions of the $\lambda/2$ phase plate:

$$\begin{aligned} Q &= \frac{\Delta I}{2\bar{I}}, \qquad \frac{\lambda}{2} \Rightarrow 0° \\ U &= \frac{\Delta I}{2\bar{I}}, \qquad \frac{\lambda}{2} \Rightarrow 22.5° \\ V &= \frac{\Delta I}{\bar{I}}, \qquad \frac{\lambda}{2} \Rightarrow 45° \end{aligned} \tag{13.2.1}$$

where

$$\Delta I = \frac{I_1}{I^c} - \frac{I_2}{I^c} \tag{13.2.2}$$

The next column gives the degree of polarization, calculated by the formula

$$P = \sqrt{Q^2 + U^2} \tag{13.2.3}$$

As the Stokes parameters are measured with a certain time shift, the calculation of the degree of polarization by the last formula is possible if the time interval betwen two shots is rather small ($\Delta t < 0.5$ min). The last column of Table

18 gives the angle between the linear polarization direction and the orientation of the entrance slit:

$$2\chi = \arctan g \frac{U}{Q} \qquad (13.2.4)$$

Due to a flicker of the image one of the Stokes parameters turned out to be zero, therefore these values for the second and third positions of the entrance slit are not included in Table 18.

It has been noted[5,178,179] that the direction of linear polarization varies depending on the mechanism responsible for energy transport into the chromosphere (heat conduction, or beams of fast particles). The direction of the observed linear polarization could be distorted and the degree of linear polarization could be diminished by the local magnetic field present at the centers of activity. Therefore, an estimate of the local magnetic field for an entity under study is of major importance. It is known that the magnetic field strength and the topology of the magnetic lines are essentially different at some level of the chromosphere and photosphere.[204] Practical estimates of the local magnetic field have been conducted by comparing the drawing of the group of spots N 5669 at the entrance slit plane with magnetic field data for the same event from the SGD and solar data. There is also an indirect indication of the local magnetic field strength at the chromospheric level, resulting from the location of the node of the flare with respect to the sunspots.

If a flare is located above a sunspot, the profile of the wings of the H_α line is more abrupt, sometimes almost vertical. On this basis it is possible to infer that for slit number I (Table 18) the location of the flare was not above the sunspot, and it is likely that in these cases the effect of the magnetic field was not major. For slit positions II and III (Table 18), the flare was located above a small sunspot. In these cases the magnetic lines are most probably vertical and the local magnetic field strength, based on a published estimate for the mean vertical field gradient,[204] should not exceed 1000 G. For slit position IV (Table 18) the flare was located above the upper boundary of the semishade of the large sunspot, the leader of the group. In this case the magnetic field strength should not exceed 500 G and the magnetic lines are probably inclined toward the center of the solar disk. In all these cases the magnetic lines are oriented in the plane of the picture toward the center of the disk, making an angle α with the position of the entrance slit, and the magnetic field should change the radial component of polarization.

Thermodynamical parameters of the upper layer of the solar chromosphere, where the core of the line H_α emission is formed, could be estimated making use of F_1 and F_2 semiempirical models of a solar flare.[205] According to this model, the core of H_α line emission is formed within the thin layer of the upper chromosphere and the intermediate region. The thickness of this layer is 22 and

10 km for F_1 and F_2 models, respectively. The local thermodynamic parameters of the optically active layer of a solar flare are presented in Table 19, including the geometrical altitude of the zero point corresponding to zero optical depth (h), mass of the column of unit cross section (m), mass of the column of unit cross section within the intermediary layer (m_0), absolute temperature (T), and concentration of hydrogen atoms (n_H).

We assume that the polarization of the H_α hydrogen line ($n = 3$) is created by collisions with precipitating particles having preferential direction in the velocity space. The ordering of the moments of the excited hydrogen atoms will be described with the help of the polarization moments formalism $\rho_q^{(k)}(l, l')$ [see Eq. (3.4.11)] independently of the nature of the projectiles[1,7]:

$$\rho_q^k(l, l') = \sum_{m,m'} (-1)^{l'-m'} \begin{bmatrix} l & l' & k \\ m & -m' & q \end{bmatrix} \rho_{mm'}^{ll'} \tag{13.2.5}$$

where $\rho_{mm'}^{l,l'}$ are the density matrix elements of the states of the hydrogen atom with orbital quantum numbers l, l' in the representation with magnetic quantum numbers m, m'.

Let us use the same methodology to describe impact polarization, introducing, as in Figure 83, the frame of reference ($\tilde{X}$, $\tilde{Y}$, $\tilde{Z}$) with the $O\tilde{Z}$ axis directed along the symmetry axis of the collisional process and also the system ($\tilde{\tilde{X}}$, $\tilde{\tilde{Y}}$, $\tilde{\tilde{Z}}$) with $O\tilde{\tilde{Z}}$ axis turned through angle ϑ to $O\tilde{Z}$ and oriented along the relative velocity vector between the two colliding particles.

As the collision process in frame of reference ($\tilde{X}$, $\tilde{Y}$, $\tilde{Z}$) is axially symmetric, the nonzero polarization moments in this system will have only zero q-numbers:

$$\tilde{\rho}_0^k(l, l') \neq 0 \qquad \text{where } k = 0, 1, \ldots, 2l$$

TABLE 19
Parameters of Ionized Gas of the H_α Forming Layer of the Upper Chromosphere (F_1 and F_2 Models[205])

h (km)	$m - m_0$ (10^{-6} g·cm^{-2})	T (10^5 K)	n_H (10^{11} cm^{-3})
F_1			
1459	2.012	4.47	0.64
1439	2.575	1.41	2.029
1436	2.735	1.12	2.555
1445 (av. F_1)	2.441	2.33	1.032
F_2			
1120	0.764	7.1	4.197
1113	1.029	4.5	6.619
1110	2.29	2.30	12.93
1114 (av. F_2)	1.361	4.63	7.915

The time variation in the polarization moments of an ensemble of particles within solid angle $d\Omega = 2\pi \sin\vartheta \, d\vartheta$ is expressed by the following equation:

$$\frac{d}{dt}\tilde{\rho}_q^k(l, l') = \frac{\gamma_l + \gamma_{l'}}{2}\tilde{\rho}_q^k(l, l') + n_0 v \tilde{\sigma}_q^k(l, l'|v) \tag{13.2.6}$$

where γ_l is the radiative damping constant; $\tilde{\sigma}_q^k(l, l'|v)$ is the matrix element of the collisional excitation matrix of a hydrogen atom to state $n = 3$ with ordering of momenta in this state.

The last quantity is expressed in terms of the cross section $\tilde{\tilde{\sigma}}_0^k(l, l'|v)$ of momenta ordering of the ensemble of atoms within the solid angle related to the $O\tilde{\tilde{Z}}$ axis.

This value is represented in terms of scattering amplitudes $f_{lm}(\vartheta)$ of the scattered proton at angle ϑ with excitation of the hydrogen atom to the state $|3, l, m\rangle$:

$$\tilde{\tilde{\sigma}}_0^k(l, l'|v) = 2\pi \sum_{mm'} (-1)^{l'-m'} \begin{bmatrix} l & l' & k \\ m & m' & 0 \end{bmatrix} \int_0^\pi f_{lm}(\vartheta) f_{l'm}(\vartheta) \sin\vartheta \, d\vartheta \tag{13.2.7}$$

Let us transform to the system of reference $(\tilde{X}, \tilde{Y}, \tilde{Z})$ related to symmetry of the collisional process, for impact polarization cross sections of the ensemble of hydrogen atoms $\tilde{\sigma}_0^k(l, l'|v)$ in the state with principal quantum number $n = 3$. This yields an expression analogous to Eq. (3.8.14), which under the Born approximation (12.4.6) will take the form

$$\begin{aligned} \tilde{\sigma}_q^k(l, l'|v) = {} & \frac{32\pi^2}{2k+1} \delta_{q0} f_\kappa(v) \sqrt{(2l+1)(2l'+1)} \left(\frac{v_0}{v}\right)^2 \\ & \cdot \int \frac{dq}{q^3} \langle nl | j_l(qr) | 10 \rangle \langle nl' | j_{l'}(qr) | 10 \rangle P_k(\cos\theta) \end{aligned} \tag{13.2.8}$$

where $f_k(v)$ is the multipole moment of the velocity distribution function.

A distribution function of protons (5.4.12) could be used to compute proton kinetics within the solar flare loop:

$$f(\mathbf{v}) = \left[v^4 + \frac{\varphi(v)}{\cos\beta}\right]^{-5/2} \delta(v - v_0) \tag{13.2.9}$$

where $\varphi(v)$ is a function depending weakly on velocity.

The multipole moment of this function, taking into account that $\beta \in [0, \pi/2)$, is

$$f_k(v) = \int_0^{\pi/2} v^3 f(\mathbf{v}) P_k(\cos\beta) \sin\beta \, d\beta \, dv \tag{13.2.10}$$

The ratio f_2/f_0 is given in Figure 119 to illustrate the velocity dependence of these moments. Calculations of the degree of polarization of the H_α hydrogen line, incorporating these numerical data on the kinetic parameters, are presented in Figure 103 for the Born approximation of the collisional problem. A rapid decrease in the degree of polarization with velocity, illustrated by Figure 103, may be explained mainly by a decrease in the multipole moment $f_2(v)$, because the corresponding velocity dependence of the excitation and alignment cross sections in this region of velocities is rather weak.

It is possible to compute the cross section $\tilde{\sigma}_0^k(l, l'|v)$ by calculating the matrix elements on the right hand side of Eq. (13.2.8) by formulas (12.5.8) and the integral over the transmitted impulse with the aid of distribution function (13.2.9).

With the help of Eq. (13.2.6), this cross section enables one to compute, in turn, the polarization moments of excited hydrogen atoms within reference frame $(\tilde{X}, \tilde{Y}, \tilde{Z})$ and the Stokes parameter $\tilde{Q}$ by formulas (12.4.10). Hence we calculate Stokes parameters U and Q and the degree of polarization in the frame of reference connected to the solar disk using relations (12.2.9) and (12.2.10).

Direct calculations using the above technique, showed that the cross section as a function of the energy of a precipitating proton beam (13.2.8) is weakly

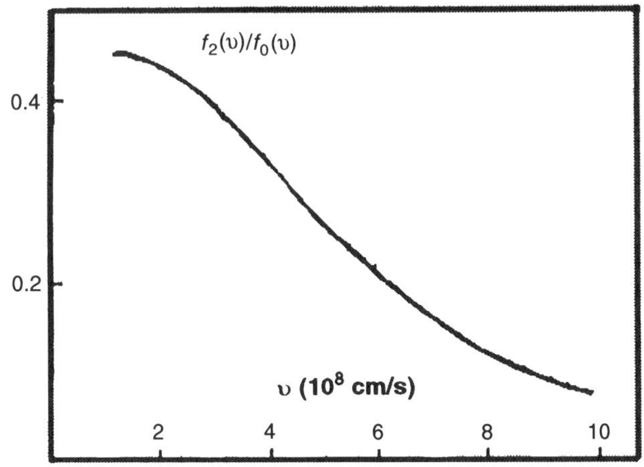

Figure 119. Velocity dependence of the ratio of multipole moments for the velocity distribution function of protons in a solar flare magnetic loop.[51]

dependent on the rank of the impact ordering (parameter k). The dependence of Stokes parameter $\tilde{Q}$ on the relative velocity is mainly caused by the velocity dependence of the multipole moments of velocity distribution function (5.4.12). The resulting dependence of the degree of polarization on the energy of the relative motion for the Born approximation is given in Figure 103. With the help of this dependence and formula (12.2.10) it is possible to obtain the energy dependence of the degree of polarization measured by an observer on the Earth.

Calculations for four angles between the OZ and $O\tilde{Z}$ axes ($\vartheta = 60°$, $63°$, $68.5°$, $49.2°$) corresponding to the observation conditions (entrance slit positions 1–4) are presented in Table 20. Experimental values of the degree of polarization are given in Table 21. With the help of minimum and maximum observed data on the degree of polarization, $P_{max} = P + \Delta P$ and $P_{min} = P - \Delta P$, values of the maximum and minimum proton energies were derived: E_{max} and E_{min}. The mean value of the proton energy was then estimated:

$$E_{\circ} = \frac{E_{min} + E_{max}}{2}$$

The proton energies determined in such a way for the first three positions of the entrance slit (the last row of Table 21) were close.

By analyzing the reliability of the results one should conclude that the Born approximation is not best for computing polarization characteristics of an ensemble of excited particles under conditions of impact polarization within the chromospheric location of a solar flare. Some improvements can be enacted with account for higher orders of this approximation, or using more sophisticated ways to solve the collisional problem. In case of small scattering angles and small transmitted pulses, the second Born correction for the direct and exchange

TABLE 20
Theoretical Degree of Polarization of the H_α Line under the Born Approximation for Conditions of Solar Flare Observations

v (10^8 cm/sec)	E (keV)	P (%) slit 1	P (%) slit 2	P (%) slit 3	P (%) slit 4
1	5.2	9.7	7.8	5.0	17.7
2	20.7	9.4	7.7	4.9	17.2
3	46.6	8.8	7.2	4.5	16.0
4	82.9	7.8	6.4	4.1	14.1
5	129	6.7	5.5	3.5	12.1
6	186	5.6	4.5	2.9	9.9
7	253	4.5	3.7	2.4	7.9
8	331	3.5	2.9	1.9	6.2
9	471	2.9	2.4	1.6	5.1

TABLE 21
Estimate of the Proton Energy Interval Comparing Observations and Computations of the Degree of H_α Hydrogen Line Polarization under the Born Approximation

Slit position	1	2	3	4
Observed P (%)	4.5 ± 1.6	5.5 ± 1.7	3.5 ± 0.9	2.2 ± 0.9
Estimated E (keV)	270 ± 100	140 ± 90	130 ± 70	>700

scattering amplitude is less than that for the first Born correction, at least for the factor $k = P/h$.

As regards an improvement in the reliability of this remote diagnostic, it should be noted that a more adequate description for the studied energy range and small transmitted pulses could be provided by the Glauber approximation. However, this approximation is valid when the transmitted kinetic moment is nearly orthogonal to the initial one. Uncertainties in the results may also be introduced by charge exchange collisions, which could induce alignment of the hydrogen atoms. But for a target atom without an inner shell, the cross section of such a process decreases rapidly at relative velocities of $v > e^2/h^2$. For example, the charge-exchange collision cross section for proton–hydrogen scattering decreases as v^{-12}.

13.3. Determination of Proton Energy for the Flare on 16.05.1991

Spectropolarimetric effects of a solar flare on 16 May 1991 in sunspot group No. 6619 SGD (32°N, 56°W) were observed using the Baikal vacuum telescope of the Russian Institute of Solar and Terrestrial Physics with a Wollaston prism placed behind the spectrograph slit.[202, 203, 230] The flare appeared in the northwest of a large sunspot and was observed from 6 h 18 m to 6 h 20 m. Figure 120 shows a white light sketch of group No. 6619, the position of the spectrograph slit at the instant of the flare recording, and the direction to the solar disk center. Altogether 12 spectrograms were obtained, each of which consisted of two orthogonally polarized spectral bands. The spatial dimension of a single spectrogram was 45 angular seconds.

Spectra were recorded repeatedly at the following positions of the $\lambda/2$ plate: 0°, 22.5°, and 45°. At 0°, the light, polarized along the spectrograph slit and grating grooves, was represented on the upper spectrogram. The light, polarized perpendicular to the slit (Stokes parameter Q), corresponded to the lower spectrogram. For the $\lambda/2$ plate turned through an angle of 45°, the situation was the inverse ($-Q$). For a rotation angle of 22.5°, the light, polarized at 45° to the

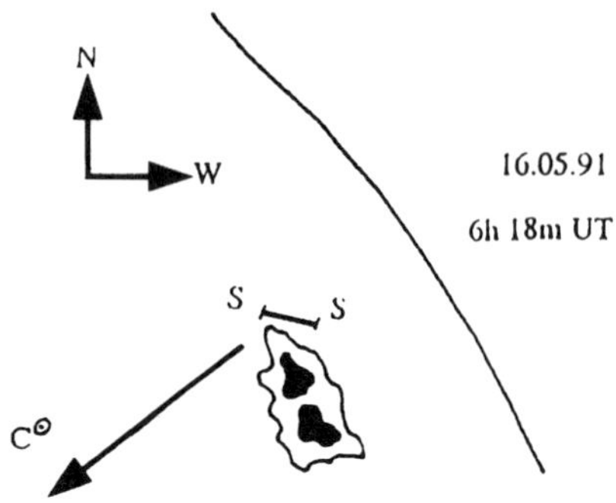

Figure 120. Sketch of group No. 6619. The position of the spectrograph slit (SS) and the direction to the solar disk center are indicated.

slit direction, corresponded to the first spectrogram, and the light polarized at 135° to the slit was represented by the second one (U).

Spectral data obtained by photodetection were processed by means of a CCD matrix. For this purpose, a package of programs has been developed to provide plots of characteristic curves determining the dispersion, converting the photographic densities into intensities, determining the position of the H_α line center, calculating the line profiles in units of the continuous spectrum, and calculating polarization along the dispersion direction.

The selected spectrograms were illustrated by a step wedge, and digitized with a constant gain factor and exposure time. Preliminary studies of the CCD matrix seem to imply that, with a sufficient degree of certainty and within the density range applied, the signal is linear. The characteristic curve was first plotted using exposures with an embedded stepwise attenuator. Then the dispersion, averaged over several spectrogram sections and common to all exposures, was determined from the reference lines at the working spectrograms.

In the first stage, two spectrogram sections along the dispersion (H_α line profiles in two orthogonally polarized spectral bands) were processed in parallel. It was fundamentally important that these sections refer exactly to the same area on the Sun disk. Only the first sections in each spectrogram were set visually; the others were fixed at the same intervals in pixels between two sections. The procedure for setting sections was as follows. A working exposure was positioned on the monitor screen and a horizontal marker visually marked the first section in the most typical part of the H_α.

In this case, two sections that refer to the same position on the solar disk appear on the screen, one above the other, in two orthogonally polarized spectrograms. Then a marker marks the centers of the reference lines in both sections and, on the basis of data obtained at the previous processing stage, the center of the H_α line is established. The marker also marked the beginning and end of each section. The beginning is chosen at the maximum possible distance from the H_α line center, and the continuum value is assigned to the first five points. Then, photographic densities are converted into intensities that were normalized to the continuum value, and this procedure repeated for subsequent sections of the given and other exposures.

The positions of subsequent sections were chosen at intervals of a fixed number of pixels, so that the pixel contour worked uniformly for two spectrograms. Thus, the identity of the section positions in two spectrograms with different polarizations depended only on the accuracy of setting the central section. To avoid ambiguity in setting the central section in each exposure, some exposures were recorded two or three times.

At the output, each file contained a data array of four columns, referring to one section of a single exposure:

(1) the values of $\pm\lambda$ with a zero in the H_α center,

(2) $I_1 = J_1/J_{cont}$ for the H_α line in the first spectrogram,

(3) $I_2 = J_2/J_{cont}$ for the second spectrogram,

(4) Q/I or $U/I = I_1 - I_2/I_1 + I_2$.

Instrumental polarization is an important factor affecting the accuracy of spectropolarimetric measurements. In the course of these measurements, instrumental polarization was estimated and excluded in the following way. In addition to the first one, a second $\lambda/2$ phase plate was inserted after the Wollaston prism, fixed in front of the spectrograph entrance slit, in order that the oscillation of the electric vector in the ordinary and extraordinary beams should be at 45° to the grooves of the grating. The second $\lambda/2$ plate, in the absense of the first one, was adjusted by means of visual equalization of the quiet sun emission intensities in two spectral bands.

Another way to eliminate possible instrumental polarization was to normalize all measured intensities to that of the continuous spectrum. This procedure reduces to a minimum the influence of instrumental polarization. Figure 121 shows profiles of narrow FeI and CrI lines in two spectral bands for the quiet region. Differences in the line profiles of Figure 121 can be explained by ordinary errors in the photographic technique.

Comparison of these two lines in two orthogonal polarizations showed a real absense of instrumental polarization, achieved by mounting a $\lambda/2$ phase plate. This operation did not change the intensity of the continuous spectrum within two spectral bands. A considerable error could arise due to difficulties in fixing the first cutting. The scatter of intensities as a result of repeated fixing of the central cutting for different shots was between 1 and 8%.

Five spectrograms taken during an interval of five minutes were processed. The recorded Stokes parameters are distributed among these spectrograms as follows: I (U), II ($-Q$), III (Q), IV (U), and V (Q). Eight cuttings across the spectrograph slit were made for every spectrogram. Central cutting was

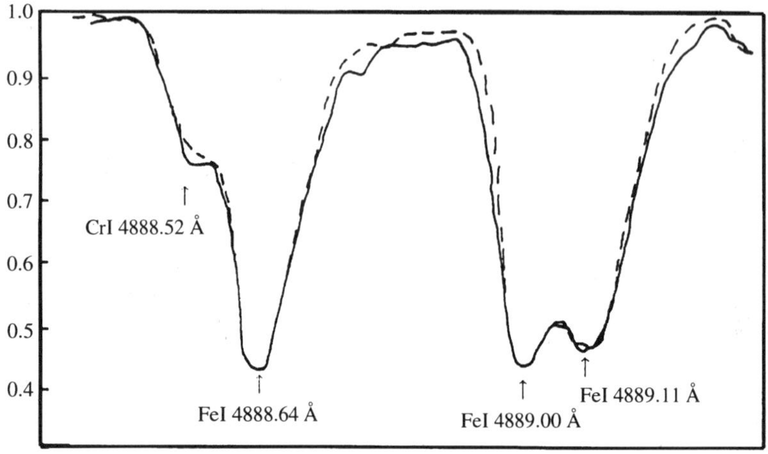

Figure 121. Profiles of FeI and CrI spectral lines in two orthogonal linear polarizations for the quiet solar region.

designated as "0"; the designations of the other cuttings are expressed in terms of solar arc seconds. They were counted on both sides from the central cutting, such as cutting $+4''$ or cutting $-1.75''$. Figure 122 illustrates profiles of the H_α line in two orthogonally polarized spectral bands for the cutting $-3.75''$ in all five spectrograms. The solid line represents the profile of the H_α line in the first spectral band; the dashed line represents the same in the second one. Two peaks at the left wing of the profile are determined only by defects on the protecting glass of the matrix. In this connection, further processing was limited by the long wave boundary of 2.6 Å.

The intensity ratio within the two spectral bands in Figure 122, II ($-Q$) on one side, III (Q) and V (Q) on the other side, during the simultaneous exposure of all five spectrograms, should be strictly opposite. Nevertheless, as they are all taken at different moments, the effect of trembling, mainly the natural temporal changes in intensity and polarization of the emission line, should also influence the results. Figure 114 shows that the intensity ratio within two spectral bands III was opposite to that in II, IV, V, which means that preferential linear polarization was along the slit direction rather than the direction along the dispersion.

The profiles of Figure 122 represent an original recording. The recordings of all the cuttings were made 2 or 3 times, so the H_α profiles for every cutting were averaged and smoothed with the help of the formula

$$I_{k,j} = \tfrac{1}{3}[I_{k,j} + \tfrac{2}{3}(I_{k,j-1} + I_{k,j+1}) + \tfrac{1}{3}(I_{k,j-2} + I_{k,j+2})] \qquad (13.3.1)$$

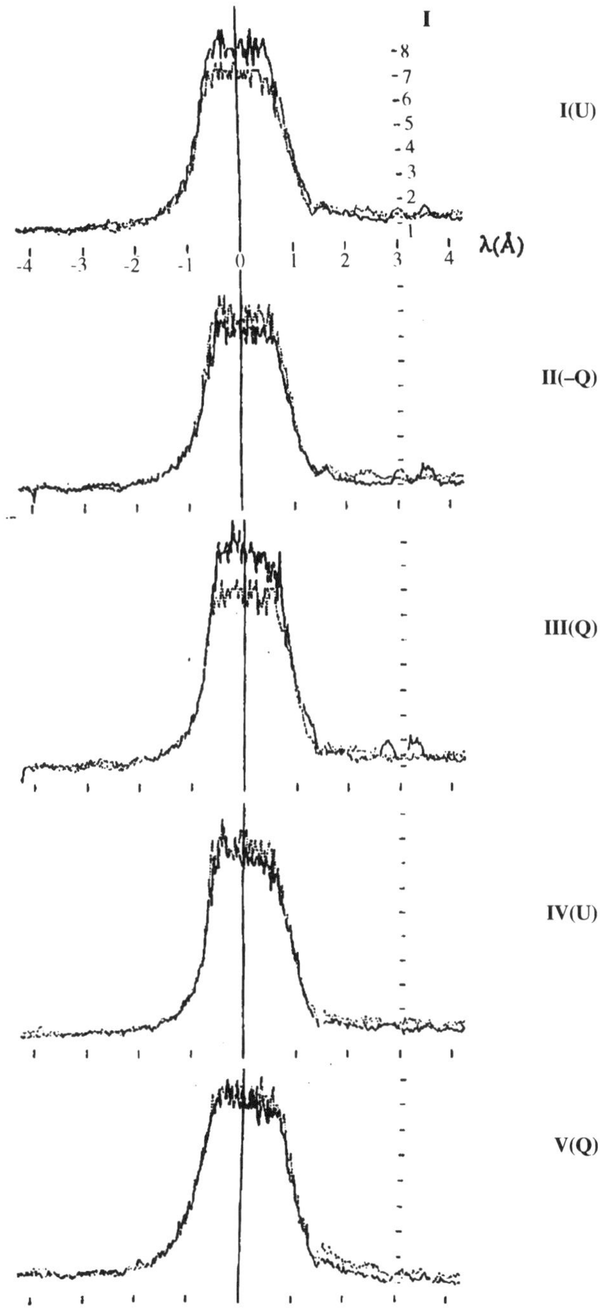

Figure 122. H_α profiles in two orthogonal linear polarizations for the cutting $-3.75''$.

where $k = 1, 2$ indicates the number of the spectral band and j is the index of the current intensity coordinate.

Figure 123 represents averaged I_1 and I_2 values for spectrogram III (Q) for all the cuttings across the spectrograph slit, and the value of parameter Q/I for the same cuttings. The horizontal line corresponds to the zero value of parameter Q/I. Some kind of false increase in the Stokes parameter is seen within the red wing of all the contours, and is connected to defects on the protecting glass of the matrix. The most systematic behavior of the Stokes parameters was observed in the center of the H_α line.

Figure 123a shows the variation in the mean intensity of the H_α line within the interval $-1\,Å < \Delta\lambda < 1\,Å$ with respect to the line center for all five moments of observation. Locations of the parts of the flare along the spectrograph slit, expressed in seconds of the arc on the Sun, are given on the horizontal axis. Figure 123b shows quantities Q/I, U/I, and $-Q/I$ within the same interval. The vertical lines indicate the uncertainties and crosses indicate mean values of the Stokes parameters within the interval of wavelengths $-0.5\,Å < \Delta\lambda < +0.5\,Å$. In order to facilitate comparison of close values of parameter Q, Figure 123b, III shows data from Figure 123b, II with the opposite sign.

Comparison of Stokes parameters from different spectrograms gives the following. Spectrograms I and IV (parameter U) agree fairly well within the quiet region ($0''$–$5''$ cuttings) and not so well for the emission part of the flare. For parameter Q the picture is the opposite: good agreement within the region $-8''$–$0''$ and very strong scatter for the quiet region.

In order to obtain final results, spectrograms IV and V were excluded because their quality is lower than that for I–III. The mean emission intensity of the flare along the entrance slit was determined using spectrograms I–III. For computing the degree of polarization and the azimuth of the polarization plane according to formulas (13.2.2)–(13.2.4), spectrograms II and III (parameter Q) and I (parameter U) were used.

Figures 124–126 represent variation in parameters I, P, and χ along the dispersion for all the cuttings. A vertical line corresponds to the center of the H_α line and the distance between the vertical marks is 1 Å.

Despite a repeated averaging and smoothing procedure, the degree of polarization and mainly the azimuth of polarization are markedly indented, especially within the line wings. When the intensity is decreasing, the uncertainty of determination of the polarization parameters is increasing. Moreover, the nature of the variation in parameters P and χ within the line wings shows quite clearly that it is nothing but noise. Three-dimensional images of the values of I, P, and χ depending on the wavelength and number of the cutting are presented in Figure 127. Data on P and χ in the central part of the H_α line are less noisy, which is in accordance with Figures 125 and 126.

Figure 128 shows data on P and χ in the central part of the H_α line; the photometric cutting of the intensity along the spectrograph slit was made for data

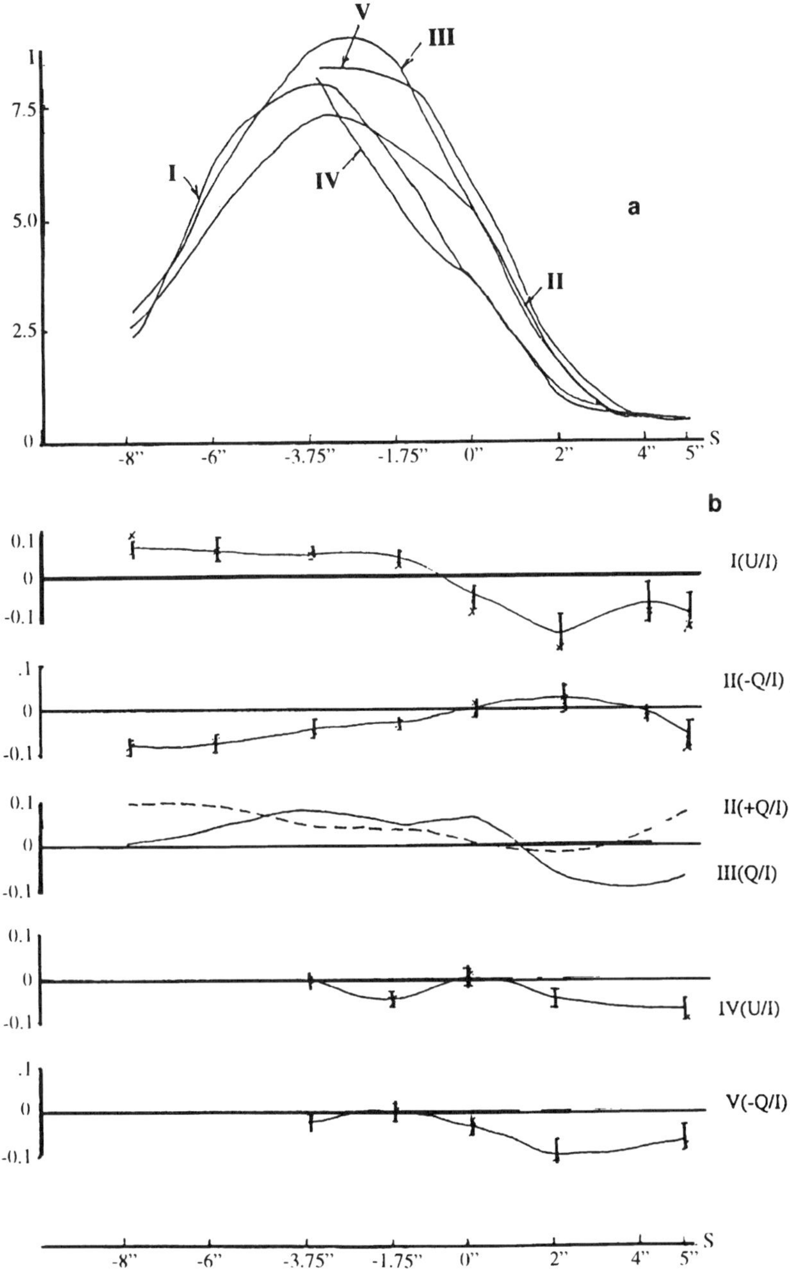

Figure 123. (a) Variation in the mean intensity of the H_α line, and (b) mean values of Q/I, $-Q/I$, and U/I within the interval $-1\,Å < \Delta\lambda < +1\,Å$ along the slit of the spectrograph. Values of parameters Q/I, $-Q/I$, and U/I within the interval $-0.5\,Å < \Delta\lambda < +0.5\,Å$ are represented by crosses.

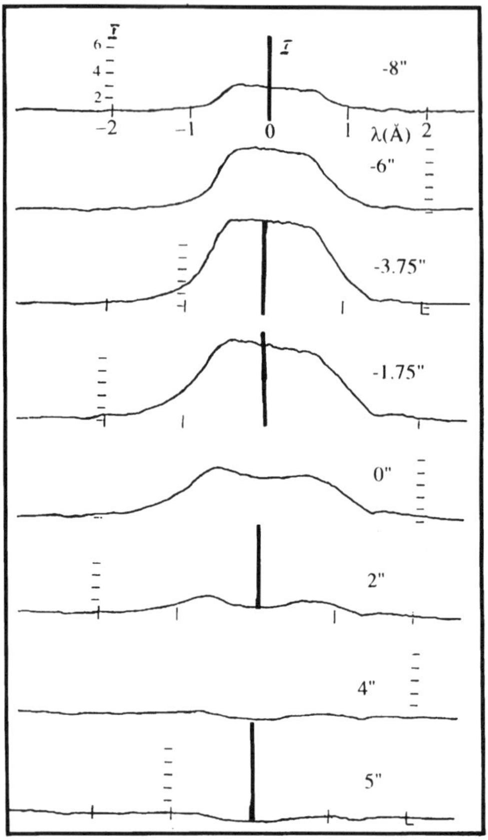

Figure 124. H_α line intensity variation along the dispersion for every cutting.

on the H_α line at $\Delta\lambda = 0$ Å. The length of the small lines corresponds to the magnitude of P, and the direction corresponds to the value of χ. Vector **C** corresponds to the solar center disk direction.

The variation in polarization at the center of the H_α line is rather regular. For cuttings with high emission ($-8''$ to $-1.75''$) the polarization vector is close to the solar disk center direction and the polarization degree decreases from 8% ($-8''$ cutting) to 4% ($-1.75''$ cutting). For the $0''$ cutting, within which the intensity profile of H_α emission is similar to that for mustaches, the degree of polarization again increases and the polarization plane becomes oriented perpendicular to the solar disk center. Then, in the $2''$ cutting in which the intensity profile is again similar to that for mustaches, and in the $4''$ and $5''$ cuttings, corresponding to a practically nonperturbed photosphere, the degree of polarization increases considerably up to 20%. Orientation of the polarization plane for $2''$ and $4''$

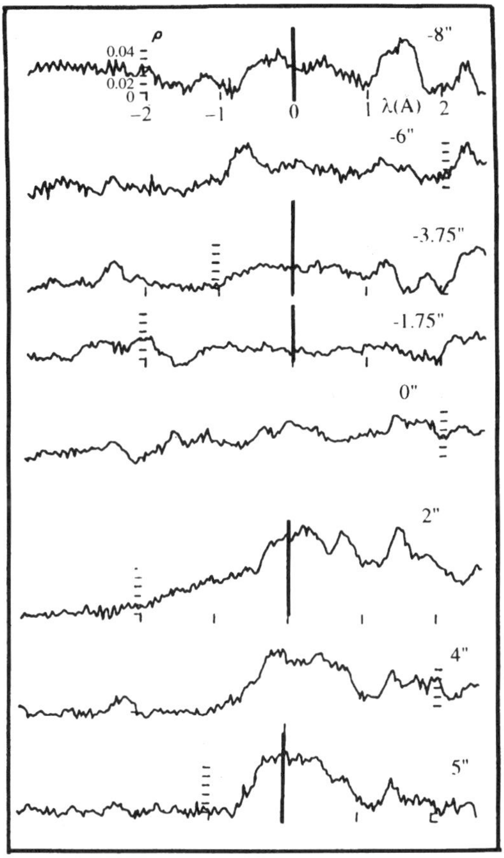

Figure 125. Variation in the degree of polarization of the H_α line along the dispersion for every cutting.

cuttings is close to the direction to the solar disk center. The mean least-squares error of ΔP and $\Delta\chi$ is 0.007 and 4°, respectively, for cuttings from $-6''$ up to $+2''$. This error is much higher for the $8''$ cutting and especially for the $4''$ and $5''$ cuttings.

Table 22 presents a distribution of files over spectrograms and within a single spectrogram. Each vertical column corresponds to a certain spectrogram obtained at a definite position of the $\lambda/2$ phase plate. The upper cell shows the position of the $\lambda/2$ plate with corresponding Stokes parameter Q/I, $-Q/I$, or U/I. Vertical columns of Table 22 correspond to the slit direction transverse to which the cuttings along the dispersion were made. This axis was calibrated with a scale marked in angular seconds on the solar disk, and the angles were

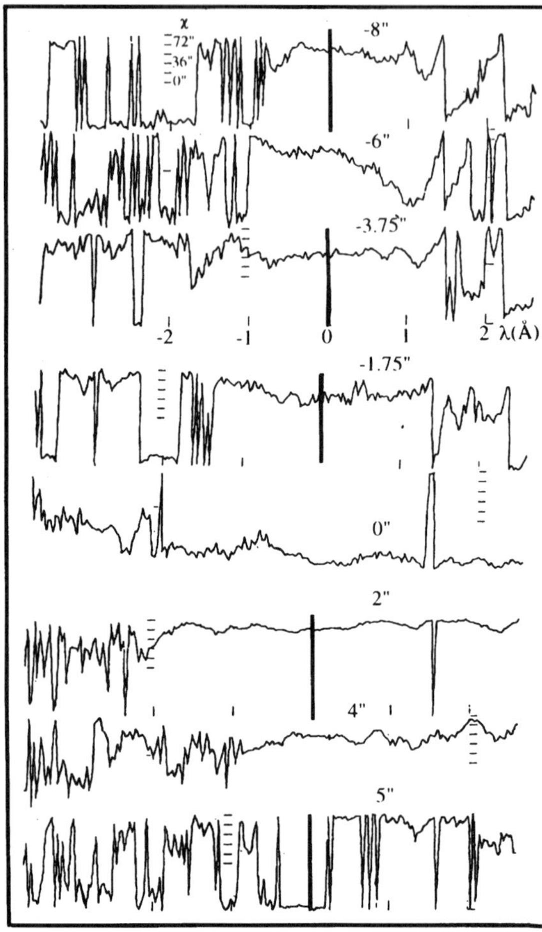

Figure 126. Variation in the azimuth of the linear polarization of the H_α line along the dispersion for every cutting.

measured from zero, corresponding to the first section. The cells denoted as $-1.75''$ and $-3.75''$ include sections obtained at angular distances of $-1.5''$ and $2''$ from the first section and at $-3.5''$ and $4.0''$ from the second section, respectively. It was reasonable to average these sections, because the spatial resolution of the spectrogram obtained is no higher than $1''$. The averaged and smoothed profiles of the H_α line formed new files.

The spatial and temporal distribution of new files (determined by the $\lambda/2$ plate position) is presented in Table 23, where the mean values of I_c, P_c, and χ_c, corrected within the range ± 0.5 Å from the H_α line center, are given. Taking into

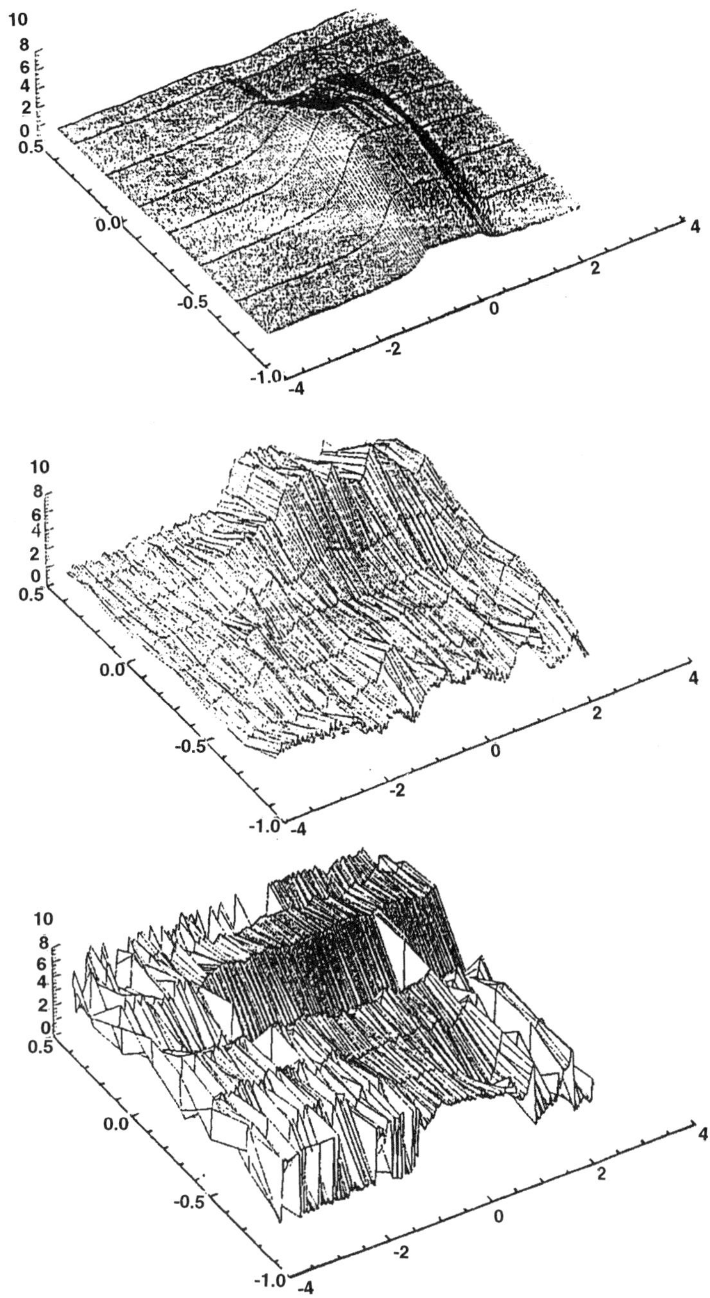

Figure 127. Three-dimensional images showing, respectively, the intensity I (at the top), the degree of polarization (in the middle), and the azimuth angle χ (at the bottom) as functions of wavelength and cutting position along the spectrograph entrance slit.

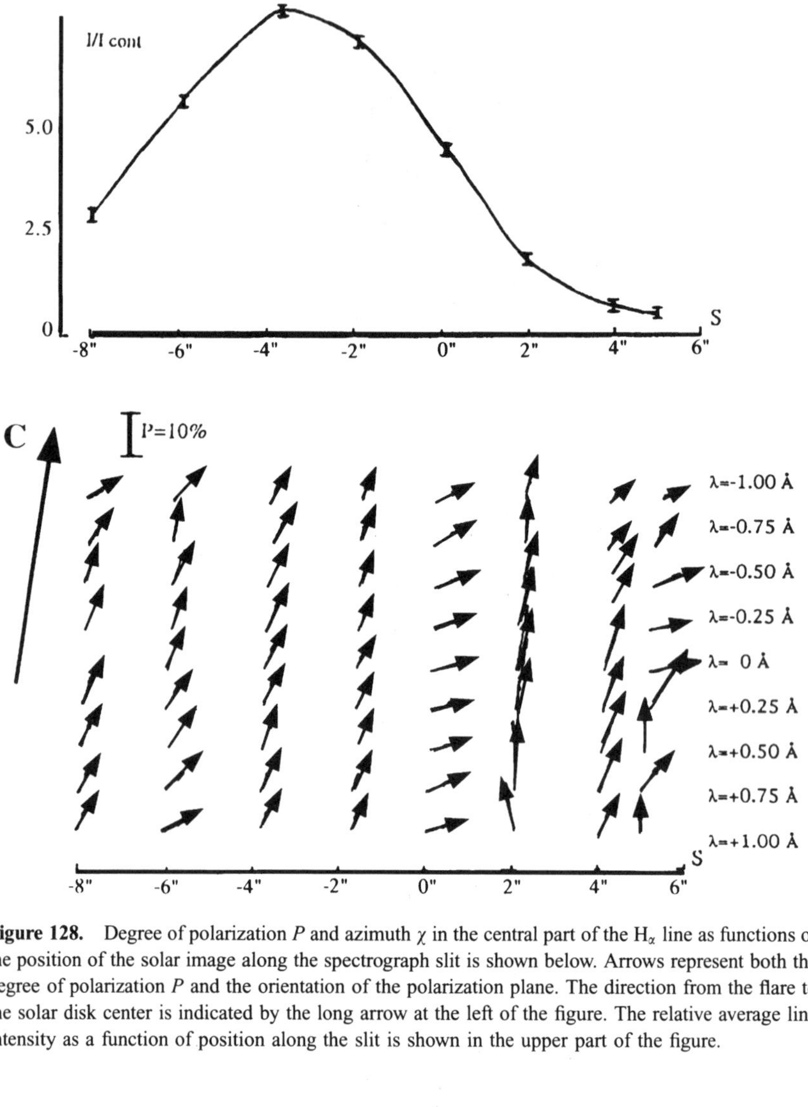

Figure 128. Degree of polarization P and azimuth χ in the central part of the H_α line as functions of the position of the solar image along the spectrograph slit is shown below. Arrows represent both the degree of polarization P and the orientation of the polarization plane. The direction from the flare to the solar disk center is indicated by the long arrow at the left of the figure. The relative average line intensity as a function of position along the slit is shown in the upper part of the figure.

account the method used to compute the values of I_c, P_c, and χ_c, it is possible to regard $\Delta I/I$, $\Delta P/P$, and $\Delta\chi/\chi$ as the maximum errors in the determination of the intensity and polarization. The two last columns give the angle $\chi_\gamma = \gamma - \chi$ between the direction to the disk center and the direction of maximum polarization.

Thus, in order to study linear polarization in seven areas of a third class flare on 16.05.91, 51 H_α line profiles were analyzed of which 37 with the highest

TABLE 22
Observational Data

Cutting (″)	U (22.5°)	$-Q$ (45°)	Q (0°)	U (22.5°)	$-Q$ (45°)
5	00303	02202	00403	02503	00603
4	03203	03303	03403		
			04404		
2	00302	00202, 03302	00402, 03402	00502	00602
	03202	02201	04402	02502	
0	00301	00201	00401, 03401	00501	00601
	03201	03301	04401	0250	
−1.75	00304	02203	00404	02504	00604
	03204	03304	03404		00605
			04405		
−3.75	00305	02204	00405	02505	00606
	03205	03305	04406		
6	03206	03306	04407		
8	03207	03307	04408		

quality were used to obtain the results. Linear polarization of the central part of the H_α line within the range ± 1 Å was evidently detected. Stokes parameters Q/I and U/I within a range of ± 1 Å from the H_α line center in the emitting flare region was 5–6% on average. In a region of mustaches these Stokes parameters increase, changing their signs. The degree of polarization for a flare was 7–10% on average, being significantly higher (up to 20%) in regions with lower emission and in an adjacent region of the nonperturbed chromosphere. The least-squares error for the latter regions is higher than for regions with increased emission. Orientation of the polarization plane was the same as the solar center disk direction, except that part where the H_α line profile resembles mustaches.

Numerical estimates of energy transport to the chromospheric region of a solar flare are based on the typical concept of impact polarization by fast

TABLE 23
Data on Parameters I_c, P_c, and χ_c

Cutting,	I	I_c	$\Delta I/I$	P	P_c	$\Delta P/P$	χ^0	χ_c^0	$\Delta\chi/\chi$	χ_j^0	χ_{jc}^0
4	0.5	0.5	0	0.126	0.07	0.57	40			11	
2	1.5	1.5	0	0.13	0.13	0	18	30	0.5	33	20
0	4.45	4.5	0.011	0.08	0.07	0.13	−18	−20	0.1	69	70
−1.75	7.0	7.3	0.042	0.04	0.05	0.22	50	40	0.22	1	10
−3.75	8.15	8.0	0.018	0.067	0.075	0.11	50	50	0	1	1
−6	5.7	5.7	0	0.077	0.075	0.03	45	40	0.12	6	11
−8	3.2	2.85	0.115	0.086	0.09	0.05	36	31	0.15	15	20

particles. A high-velocity proton beam, propagating from the corona, causes hydrogen atoms to collide with chaotically distributed velocity vectors, which form an optically active region of the solar chromosphere. Due to the axial symmetry of the collision process, ordering of velocity vectors in the beam is transmitted into ordering of angular momenta of the ensemble of hydrogen atoms, which in turn results in linear polarization of the H_α hydrogen line.

Suppose a frame of reference is introduced as Section 13.2. The determination of the polarization characteristics of H_α radiation is reduced to computing the cross section of angular momenta ordering of an excited hydrogen ensemble, $\tilde{\tilde{\sigma}}_q^k(ll'|v)$.

To solve this problem within the Glauber approximation, it is possible to use formula (12.5.4) for the scattering amplitudes and then formula (12.5.6) for the cross sections. Stokes parameter $\tilde{Q}$ and finally the polarization characteristics of H_α radiation within the reference frame (X, Y, Z) can be calculated, using formulas (12.4.10).

Results of these calculations with the velocity distribution function of protons within the magnetic loop of a solar flare, (5.4.12) and (13.2.9), are plotted in Figure 104. Comparison of these results with the data of Figure 103, presenting the computation in the Born approximation, shows that at small velocities the Glauber approximation gives a much smaller degree of polarization than the Born approximation, and this difference decreases with rising velocity. Theoretical calculations of the degree of polarization for the Glauber approximation, calculated in the (X, Y, Z) coordinate system for inclination angles of $\vartheta = 60^\circ$, 63°, 68.55°, 49.20°, are presented in Table 24. Observed values of the degree of polarization for some of these angles are given in Table 25.

The seventh column of Table 24 contains theoretical degrees of polarization for an inclination angle of 21.55°. The corresponding observational data in Table

TABLE 24

Theoretical Values of the Degree of Polarization of H_α Radiation for the Excitation of an Ensemble of Hydrogen Atoms in the Chromosphere by a Proton Beam (Glauber Approximation)

v (10^8 cm/sec)	E (keV)	60.0°	63.0°	68.5°	49.2°	21.55°
1.0	5.2	5.9	4.8	3.1	10.5	23.9
2.0	20.8	4.7	3.8	2.5	8.3	18.3
3.0	46.7	3.2	2.6	1.7	5.5	11.8
4.0	83.0	2.7	2.2	1.5	4.7	10.1
5.0	129	2.3	1.9	1.2	4.0	8.4
6.0	187	2.0	1.5	1.0	3.2	6.8
7.0	254	1.7	1.4	0.9	2.8	5.9
8.0	332	1.2	1.0	0.6	2.1	4.2

TABLE 25
Estimate of the Proton Energy Range by Comparing Observations of the Degree of Polarization of the Hydrogen H_α Line for Optimal Angular Parameters and Computations with the Glauber Approximation

	49.2°	21.55°
Polarization degree (%)	2.2 ± 0.9	6.5 ± 2.1
Proton beam energy (keV)	360 ± 160	210 ± 90

23, averaged over the sections and wavelengths, are given in Table 25. The bottom row of Table 25 presents the kinetic energy for the relative motion of colliding particles for angles 49.2° and 21.55° at which the theoretical and experimental degrees of polarization agree within the limits of experimental error. The kinetic energy of the relative motion was estimated to be much lower for the first three angles of inclination than for the rest. This may be explained by the complicated structural features of the object under study. The influence of these structural features can be illustrated by the scatter of data from the fifth column of Table 23. Indeed, the wavelength-averaged values of the degree of polarization taken along the cutting show that the degree of polarization is changing along the cutting by more than a factor of 3 (from 4 to 13%). Apart from the structure of the studied entity, an explanation of this data scatter may also be related to imperfection of the theoretical model.

These computations, and the possibility of using the Glauber approximation for impact spectral sensing, were examined. We note first that, following the general conclusions of Section 12.8, a decrease in the degree of polarization with increasing energy of the relative motion takes place mainly owing to the behavior of the multipole moments of velocity distribution function (5.4.12). Second, there exists a special restriction of the Glauber approximation that must be taken into consideration when describing spectropolarimetric phenomena. Impact-induced ordering of angular momenta is a finer effect than the ordinary impact population of atomic states. A major role in impact polarization is played by impact excitation of the magnetic sublevels of the state for some orbital moment. There exists a specific selection rule of the Glauber method that has no physical meaning, but forbids transitions when the absolute value of the magnetic quantum number changes from 2. This special selection rule may have a stronger influence on the impact alignment cross section $\sigma_0^2(ll')$ than on $\sigma_0^0(ll')$ and thereby affects computations of spectropolarimetric parameters.

These results show that, if we take into consideration the multichannel problem of the theory of scattering as well as the more adequate description of the structural and spatial properties of the object under study, then a more accurate description of the collision process seems to be desirable for improved reliability of impact spectropolarimetric sensing.

13.4. Impact Spectropolarimetric Sensing for the Flare on 15.09.1981

The reliability of spectropolarimetric sensing as applied to solar physics problems was improved by simultaneously observing the polarization, with corresponding computation of the spectropolarimetric parameters, of the two hydrogen lines H_α and H_β. A flare was observed on September 15, 1981 in the Halle 17842 group (No. 18E22) with an automated solar telescope of Sayany Observatory, affiliated to the Institute of Solar and Terrestrial Physics at Irkutsk, Russia.[(194)]

A Wollaston prism mounted after the slit of the spectrograph was used for spectropolarimetric observations. The deflection angle of the Wollaston prism and the distance from the slit were chosen to obtain equal filling of the collimator by the ordinary and extraodinary beams. An chromatic half-wave plate was fixed in front of the slit of the spectrograph. For a position of the $\lambda/2$ plate corresponding to 0°, the oscillations of the electric field vector for the first spectral band were parallel to the slit and the grooves of the grating. Oscillations of the electric vector for the second spectral band were parallel to the dispersion. For the first position parameter Q was recorded. Rotation of the $\lambda/2$ plate through 22.5° gave parameter U.

In order to record the H_α and H_β lines simultaneously, the grating was inclined such that, at the cassette plane, the H_α line was in the third order of the grating while the H_β line was in the fourth order. The mutual influence of light from both orders was excluded by mounting the filters in the cassette part. Therefore every spectrogram taken at a particular moment consisted of four spectral bands: the spectra of the H_α and H_β lines in two orthogonal linear polarizations.

The flare started at 0009 UT and continued until 0130. Due to weather conditions, spectropolarimetric observations were performed during the period 0044 until 0102 UT, which corresponded to the decaying stage of the flare. Fourteen spectrograms were obtained, seven with 0° orientation of the $\lambda/2$ plate and the other seven with 22.5° orientation. The profiles of two spectrograms recorded every 30 s for the position of the $\lambda/2$ plate at 0° as well as the profiles of two spectrograms for the 22.5° position were averaged. Finally, 12 spectrograms recorded in the interval 0044 to 0102 UT were used.

The spectrograms were processed by a JFO 451 microphotometer with slit height chosen equal to 1/3 of the emission structure in the Balmer line (1 s approximately). The intensity of the spectral lines J_λ was reduced to that of the continuum. Figure 129 gives typical contours of the H_α and H_β lines for this package of data, the intensity of which was averaged over two spectral bands. The shape of these contours is not very typical for the flares: they have narrow spectral emission with self-reversal in the center. Such contours are more typical of flares situated above spots.

Spectropolarimetric parameters were measured in different parts of the H_α and H_β contours, marked in Figure 129 by 1–5:

(1) the total half-width for H_α, $\Delta\lambda = 1.576$ Å, and for H_β, $\Delta\lambda = 1.103$ Å;

(2) the narrower central part of the profile, the boundary of which is located at the red and blue peaks approximately, for H_α, $\Delta\lambda = 1.103$ Å and for H_β, $\Delta\lambda = 0.610$ Å;

(3) the narrow central hole with half-width for H_α, $\Delta\lambda = 0.20$–0.25 Å, and for H_β, $\Delta\lambda = 0.12$–0.18 Å;

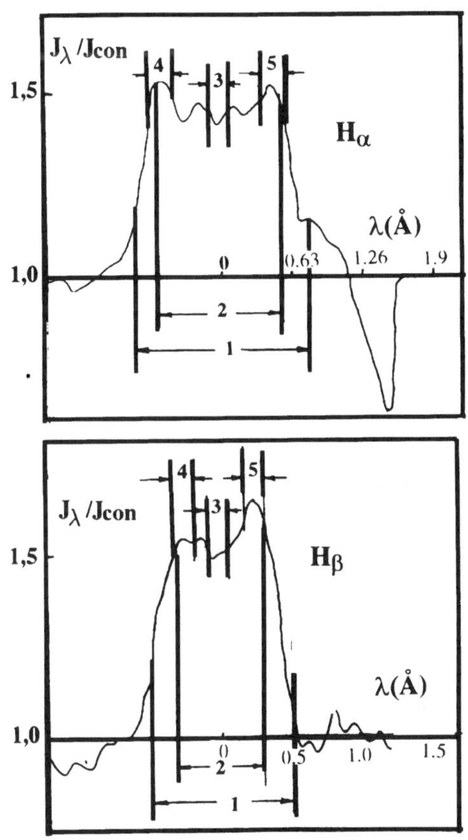

Figure 129. Typical H_α and H_β profiles for the flare on 15.09.1981.

(4) the blue peak;
(5) the red peak.

The half-width of these peaks for H_α was $\Delta\lambda = 0.20–0.25$ Å, and for H_β, $\Delta\lambda = 0.12–0.18$ Å.

Figures 130a and 131a represent the temporal behavior of H_α and H_β, intensity-averaged over every chosen spectral band. The behavior of both lines is mostly similar: a growth in intensity at the end of the observation period and small intensity oscillations in the middle part. At the initial moment the intensity in H_α grows sharply, while the increase in H_β intensity is only at the beginning.

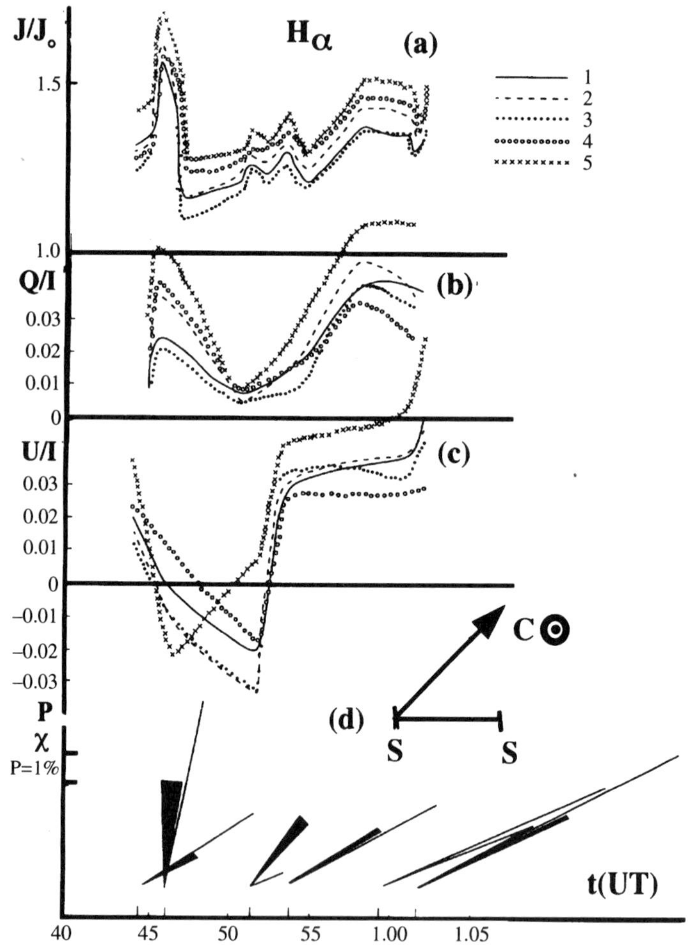

Figure 130. Temporal variation in J/J_c, Q/I, U/I, P, and χ for different parts of the H_α profile in Figure 129 and the geometry of observation (d).

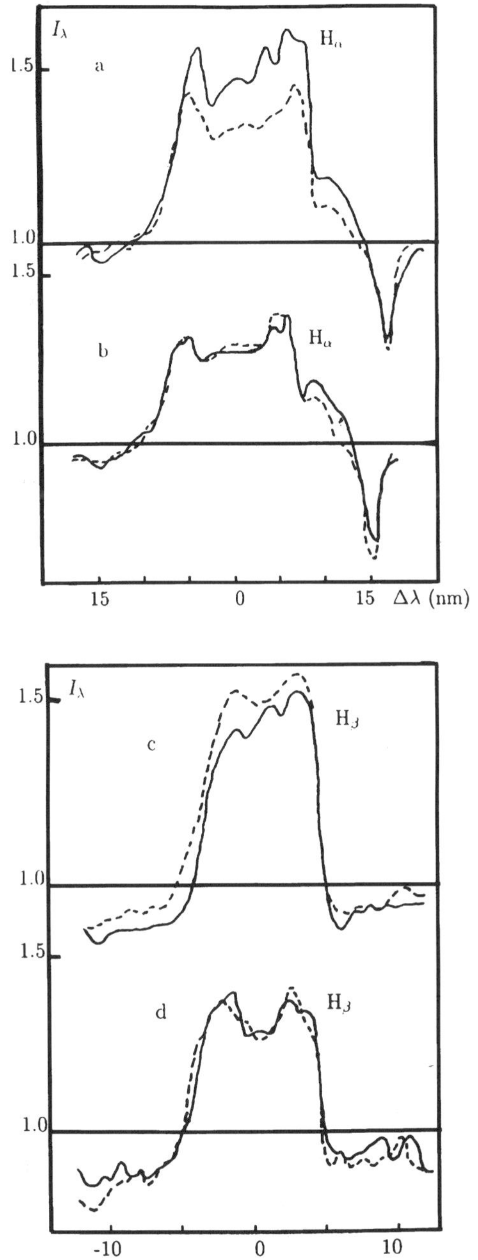

Figure 132. Illustration of the different behavior of $I_{\lambda 1,2} = J_{\lambda 1,2}/J_{con1,2}$ and $J_{\lambda 2}/J_{con2}$ profiles for the H_α and H_β lines within two orthogonally polarized spectral bands $(1, 2)$ at different moments of observation (a, b).

TABLE 26
Stokes Parameters, Degree of Polarization P, Azimuth of the Polarization Plane χ for the H_α Hydrogen Line Averaged over Four Parts of the Profile

UT	$\tilde{Q}/I$	$\Delta Q/I$	UT	$\tilde{U}/I$	$\Delta U/I$	UT	P	ΔP	χ^0	$\Delta\chi^0$
0045.00	0.010	0.003	0044.00	0.017	0.003	0044.30	0.020	0.004	29.8	3.1
0045.30	0.031	0.011	0046.45	−0.004	0.006	0046.00	0.032	0.010	86.3	13.5
0050.50	0.006	0.001	0052.00	−0.025	0.006	0051.30	0.026	0.006	51.9	5.4
0054.15	0.015	0.003	0053.15	0.028	0.003	0053.45	0.033	0.004	30.9	2.2
0058.00	0.041	0.004	0101.00	0.034	0.004	0059.30	0.053	0.006	20.0	0.6
0101.20	0.034	0.004	0102.00	0.041	0.006	0101.40	0.053	0.007	25.2	0.8

TABLE 27
Same as Table 26, for the H_β Line Averaged over Five Parts of the Contour

UT	$\tilde{Q}/I$	$\Delta Q/I$	UT	$\tilde{U}/I$	$\Delta U/I$	UT	P	ΔP	χ^0	$\Delta\chi^0$
0045.00	0.017	0.003	0044.00	0.001	0.003	0044.30	0.017	0.004	1.7	9
0045.30	0.021	0.003	0046.45	0	0.004	0046.00	0.021	0.003	0	11
0050.50	−0.013	0.006	0052.00	−0.035	0.010	0051.30	0.037	0.012	34.7	1
0054.15	−0.020	0.004	0053.15	0.001	0.004	0053.45	0.020	0.004	89.0	12
0058.00	−0.008	0.007	0101.00	0.014	0.005	0059.30	0.016	0.001	59.8	30
0101.20	−0.005	0.002	0102.00	−0.025	0.004	0101.40	0.026	0.004	39.3	3

TABLE 28
Same as Table 26, for the Red Peak of the H_α Line

UT	Q/I	UT	U/I	UT	P	χ^0
0045.00	0.017	0044.00	0.037	0044.30	0.041	32.6
0045.30	0.052	0046.45	−0.021	0046.00	0.057	78.2
0050.50	0.007	0052.00	0.008	0051.30	0.011	24.4
0054.15	0.029	0053.15	0.043	0053.45	0.052	28.0
0058.00	0.058	0101.00	0.052	0059.30	0.078	21.0
0101.20	0.059	0102.00	0.075	0101.40	0.095	26.0

contours of H_α and H_β and did not depend on intensity. An exception was the "red peak," especially for the H_α line.

Data on parameters Q/I and U/I for near times were used to compute the degree of polarization P and the azimuth of polarization χ with respect to the entrance slit of the spectrograph S. These data are presented in the last block of columns of Tables 26–28 and in Figures 130d and 131d. The error bars ΔP and $\Delta\chi$ were estimated from errors in the observational data on ΔQ and ΔU. The degree of polarization is represented by the length of the respective lines, and the inclination of these lines gives the azimuth χ of the polarization plane with error

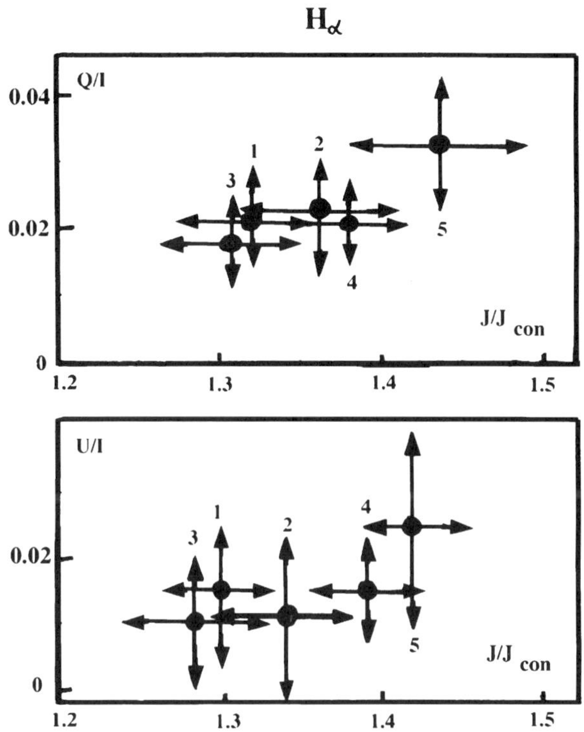

Figure 133. Time-averaged values of Q/I and U/I over the whole observation period for different parts (1–5) of the H_α profile (Figure 129).

fan $\Delta\chi$; the angle between the direction to the solar disk center and the entrance slit of the spectrograph is also given in both figures. Values of P and χ for the "red peak" are given in Figure 130d by separate solid lines.

The result of such a substantial analysis of spectropolarimetric observables for solar flare emission lines allowed us to infer that linear polarization during the simultaneous observation of spectropolarimetric effects in hydrogen lines is more clearly detectable for the H_α line than for the H_β line. The degree of polarization for the H_α line was 3–5%, and for the H_β line it was a bit less around 3.7%. The degree of polarization in the "red peak" attained 9.5%. For the H_α line the orientation of the polarization plane was more often close to the direction to the solar disk center; for the H_β line the temporal variation in the azimuth χ was more chaotic. It can be proposed that impact polarization was more clearly represented in the core of the H_α line, which is formed in higher layers of the chromosphere than the core of the H_β line.

The considerable scatter of data obtained with the approximate Born and Glauber methods of atomic collisional calculation suggest that a more accurate

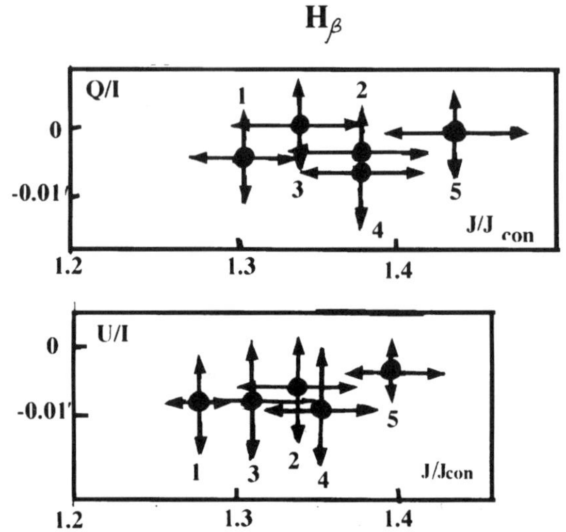

Figure 134. Time-averaged values of Q/I and U/I over the whole observation period for different parts (1–5) of the H_β profile (Figure 129).

description of the collisional process be sought. So for these series of observational data the collisional variation in density matrix elements $\rho^{ll'}_{mm'}$ or polarization moments of the atomic density matrix $\rho^{(\kappa)}_q(ll')$ were calculated by means of numerical integration of the system of equations governing the impact parameter method within the approximation of straight line trajectories.

A general description of this technique was presented in more detail in Section 12.6. This approach was utilized to examine spectropolarimetric effects on the H_α and H_β lines, corresponding to transitions from excited states of hydrogen atom with $n = 3, 4$, respectively, with account for the ground state $1s$. Within the subsequent frame of reference $(\tilde{\tilde{X}}, \tilde{\tilde{Y}}, \tilde{\tilde{Z}})$, for which the $O\tilde{\tilde{Z}}$ axis is directed at the projectile and $(\tilde{\tilde{Z}}, \tilde{\tilde{X}})$ is the hydrogen atom–precipitating proton scattering plane, the wave function of the hydrogen atom–proton system is as follows:

$$\tilde{\tilde{\Psi}}(\mathbf{r}, t) = \tilde{\tilde{c}}_{100}(t)\tilde{\tilde{\psi}}_{100}(\mathbf{r})e^{-i\frac{E_1 t}{\hbar}} + \sum_{l,m} \tilde{\tilde{c}}_{nm}(t)\tilde{\tilde{\psi}}_{3lm}(\mathbf{r})e^{-i\frac{E_n t}{\hbar}}, \qquad (13.4.2)$$

where $n = 3, 4$, and

$$\tilde{\tilde{\psi}}_{nlm}(\mathbf{r}) = \tilde{\tilde{R}}_{nl}(r)\tilde{\tilde{Y}}_{lm}(\theta, \varphi) \qquad (13.4.3)$$

is the wave function of the hydrogen atom in energy state E_n.

The radial part of the wave function is represented by the function

$$\tilde{\tilde{R}}_{nl}(r) = \sqrt{\frac{(2\varepsilon_n)^{2l+3}(n-l-1)!}{2n(n+l)!}} r^l e^{-\varepsilon_n r} L_{n-l-1}^{2l+1}(2\varepsilon_n r) \qquad (13.4.4)$$

where $\varepsilon_n = 1/na_0$, $L_n^\alpha(x)$ is the Lager polynomial, and $\tilde{\tilde{Y}}_{lm}(\theta\varphi)$ is a spherical function.

The time dependence of function $\tilde{\tilde{\psi}}_{nlm}$ comes only from the rotation of the subsequent system of reference.

For the particular case $n = 3, 4$, the radial part of the wave function takes the form

$$\tilde{\tilde{R}}_{10} = \sqrt{4\varepsilon_1} e^{-x_1}, \qquad \tilde{\tilde{R}}_{30} = \sqrt{\frac{4}{9}\varepsilon_3^3}(3 - 6x_3 + 2x_3^2)e^{-x_3}$$

$$\tilde{\tilde{R}}_{31} = \sqrt{\frac{8}{9}\varepsilon_3^3} x_3(2 - x_3)e^{-x_3}, \qquad \tilde{\tilde{R}}_{32} = \sqrt{\frac{8}{45}\varepsilon_3^3} x_3^2 e^{-x_3}$$

$$\tilde{\tilde{R}}_{40} = \sqrt{\frac{2}{3}\varepsilon_4^3}(3 - 9x_4 + 6x_4^2 - x_4^3)e^{-x_4}, \qquad \tilde{\tilde{R}}_{41} = \sqrt{\frac{4}{15}\varepsilon_4^3} x_4(5 - 5x_4 + x_4^2)e^{-x_4}$$

$$\tilde{\tilde{R}}_{42} = \sqrt{\frac{4}{15}\varepsilon_4^3} x_4^2(3 - x_4)e^{-x_4}, \qquad \tilde{\tilde{R}}_{43} = \frac{1}{6}\sqrt{\frac{16}{35}\varepsilon_4^3} x_4^3 e^{-x_4}$$

where $x_n = \varepsilon_n R$.

By substituting Eq. (13.4.3) into the time-dependent Schrödinger equation and introducing a new variable angle ϑ, related to the impact parameter b by

$$\vartheta = \arctan(b/vt)$$

a dimensionless Massey parameter

$$\eta = e^2/v\varepsilon_n b\hbar = v_0/v\varepsilon_n b$$

and a dimensionless frequency

$$\omega_{mm'} = \frac{(E_{n'} - E_n)b}{\hbar v} \qquad (n, n' = 1, 3)$$

We obtain a system of differential equations of tenth order for $n = 3$ and seventeenth order for $n = 4$:

$$\frac{d}{d\tau}\tilde{\tilde{c}}_{nlm} = -i\eta \sum_{n'l'} \langle nlm|V(r,R)|n'l'm\rangle \tilde{\tilde{c}}_{n'l'm} e^{-i\omega_{nn'}\cot\vartheta} + \frac{\delta_{n,n'}\delta_{l,l'}}{2}[\sqrt{(l-m)(l+m+1)}\tilde{\tilde{c}}_{nlm+1} - \sqrt{(l+m)(l-m+1)}\tilde{\tilde{c}}_{nlm-1}] \quad (13.4.5)$$

Here, function $V(r, R)$ characterizes interaction of an atomic electron with the incident proton:

$$V(r,R) = \frac{\varepsilon_3 R^2}{|\mathbf{r}-\mathbf{R}|} = \frac{\varepsilon_3 R^2}{r_>}\sum_{\chi=0}^{\infty}\left(\frac{r_<}{r_>}\right)^k P_k(\cos\vartheta) \quad (13.4.6)$$

where $r_{>,<}$ are values bigger or smaller than r and R.

The last expression for $V(r, R)$ enables the matrix element on the right hand side of Eq. (13.4.5) to be written as

$$V_{nln'l'}^{(m)} = \langle nlm|V(r,R)|n'l'm'\rangle = \sum_k f_{nl;n'l'}^{(k)}(\varepsilon_n R)F_{l,l'}^{(k)}(m), \quad (13.4.7)$$

where

$$F_{l,l'}^{(k)}(m) = \sqrt{\frac{(2l'+1)}{(2l+1)}}\begin{bmatrix} l' & k & l \\ 0 & 0 & 0 \end{bmatrix}\begin{bmatrix} l' & k & l \\ m & 0 & m \end{bmatrix}$$

and

$$f_{nl;n'l'}^{(k)}(\varepsilon R) = \frac{1}{\varepsilon_n^2}\left[\frac{1}{(\varepsilon_n R)^{k+1}}\int_0^{\varepsilon_n R} dx\, x^{k+2}\tilde{\tilde{R}}_{nl}\left(\frac{x}{\varepsilon_n}\right)\tilde{\tilde{R}}_{n'l'}\left(\frac{x}{\varepsilon_n}\right) + (\varepsilon_n R)^k \int_{\varepsilon_n R}^{\infty} dx x^{k+2}\tilde{\tilde{R}}_{nl}\left(\frac{x}{\varepsilon_n}\right)\tilde{\tilde{R}}_{n'l'}\left(\frac{x}{\varepsilon_n}\right)\right] \quad (13.4.8)$$

Matrix elements of interaction (13.4.7) can be represented in terms of linear combinations of quantities $f_{nln'l'}^{(k)}$ given in Eq. (13.4.8). The coefficients of these linear combinations can be determined only with the help of the momenta coupling rules. These linear combinations for the computation of matrix elements of interaction between the atomic states of hydrogen with quantum numbers not exceeding 4 are given below:

$$V^{(0)}_{n0n'l}(R) = \frac{1}{\sqrt{2l+1}} f_{n0n'l}(R), \qquad l = 0, 1, 2, 3$$

$$V^{(0)}_{n1n'1}(R) = f^{(0)}_{n1n'1}(R) + \frac{2}{5} f^{(2)}_{n1n'1}(R)$$

$$V^{(1)}_{n1n'1}(R) = f^{(0)}_{n1n'1}(R) - \frac{1}{5} f^{(2)}_{n1n'1}(R)$$

$$V^{(0)}_{n1n'2} = \sqrt{\frac{4}{15}} f_{n1n'2}(R) + \frac{3\sqrt{3}}{7\sqrt{5}} f_{n1n'2}(R)$$

$$V^{(1)}_{n1n'2}(R) = \frac{1}{\sqrt{5}} f^{(1)}_{n1n'2}(R) - \frac{3}{7\sqrt{5}} f^{(2)}_{n1n'2}(R)$$

$$V^{(0)}_{n1n'3}(R) = \frac{3\sqrt{7}}{5\sqrt{7}} f^{(2)}_{n1n'3}(R) - \frac{4}{3\sqrt{21}} f^{(4)}_{n1n'3}(R)$$

$$V^{(0)}_{n2n'2}(R) = f^{(0)}_{n2n'2}(R) + \frac{2}{7} f^{(2)}_{n2n'2}(R) + \frac{2}{7} f^{(4)}_{n2n'2}(R)$$

$$V^{(1)}_{n2n'2}(R) = f^{(0)}_{n2n'2}(R) + \frac{1}{7} f^{(2)}_{n2n'2}(R) - \frac{4}{21} f^{(4)}_{n2n'2}(R)$$

$$V^{(2)}_{n2n'2} = f^{(0)}_{n2n'2}(R) - \frac{2}{7} f^{(2)}_{n2n'2}(R) + \frac{1}{21} f^{(4)}_{n2n'2}(R)$$

$$V^{(0)}_{n2n'3}(R) = \frac{3}{\sqrt{35}} f^{(1)}_{n2n'3}(R) + \frac{4}{3\sqrt{35}} f^{(2)}_{n2n'3}(R) + \frac{1}{33}\sqrt{\frac{5}{7}} f^{(4)}_{n2n'3}(R)$$

$$V^{(1)}_{n2n'3}(R) = \frac{2\sqrt{2}}{\sqrt{35}} f^{(1)}_{n2n'3}(R) + \frac{\sqrt{2}}{3\sqrt{35}} f^{(2)}_{n2n'3}(R) - \frac{5\sqrt{10}}{33\sqrt{7}} f^{(4)}_{n2n'3}(R)$$

$$V^{(2)}_{n2n'3}(R) = \frac{1}{\sqrt{7}} f^{(1)}_{n2n'3}(R) - \frac{2}{3\sqrt{7}} f^{(2)}_{n2n'3}(R) + \frac{5}{33\sqrt{7}} f^{(4)}_{n2n'3}(R)$$

$$V^{(0)}_{n3n'3}(R) = f^{(0)}_{n3n'3}(R) + \frac{4}{15} f^{(2)}_{n3n'3}(R) + \frac{2}{11} f^{(4)}_{n3n'3}(R) + \frac{100}{429} f^{(6)}_{n3n'3}(R)$$

$$V^{(1)}_{n3n'3}(R) = f^{(0)}_{n3n'3}(R) + \frac{1}{5} f^{(2)}_{n3n'3}(R) + \frac{1}{33} f^{(4)}_{n3n'3}(R) - \frac{25}{143} f^{(6)}_{n3n'3}(R)$$

$$V^{(2)}_{n3n'3}(R) = f^{(0)}_{n3n'3}(R) - \frac{7}{33} f^{(4)}_{n3n'3}(R) + \frac{10}{143} f^{(6)}_{n3n'3}(R)$$

$$V^{(3)}_{n3n'3}(R) = f^{(0)}_{n3n'3}(R) - \frac{1}{3} f^{(2)}_{n3n'3}(R) + \frac{1}{11} f^{(4)}_{n3n'3}(R) - \frac{5}{429} f^{(6)}_{n3n'3}(R)$$

The system, consisting of matrix elements between an odd combination of functions, has only a trivial solution. Therefore, in order to determine the cross sections of impact excitation of hydrogen, the system corresponding to an even combination of functions should be solved for the initial condition of an atomic hydrogen at the ground state before collision.

As the hydrogen atom before collision was in the ground state, the initial condition for system (13.4.5) is

$$\tilde{\tilde{c}}_{nlm} = \delta_{n1}\delta_{l0}\delta_{m0}$$

Symmetry of collisions with respect to reflection in the collision plane, namely

$$\tilde{\tilde{c}}_{nlm} = \tilde{\tilde{c}}_{nl-m} \tag{13.4.9}$$

allows one to reduce the order of system (13.4.5), which should be resolved in order to obtain the excitation matrix.

If, instead of coefficients $\tilde{\tilde{c}}_{nlm}$, one uses linear combinations of these coefficients having special parity with respect to reflection in the collision plane, in case $n = 3$ it is possible to construct 7 even and 3 odd combinations of this type, and the right hand side of Eq. (13.4.5) will be split into blocks of orders 7 and 3, respectively. The third-order system, determining the temporal development of the odd functions, has only a trivial solution, because this system should be solved for zero initial conditions.

In case $n = 4$, a similar situation arises: one can construct 11 even combinations of functions $\tilde{\tilde{\psi}}_{nlm}$ and 6 odd combinations. The system of differential equations will be split into two independent systems of orders 11 and 6. The system of sixth order has only a trivial solution. The system of even combinations should be solved for initial conditions similar to the case $n = 3$.

The system for even linear combinations was solved numerically 400 times for impact parameters lying within the interval $1 \le \varepsilon_3 b \le 40$ and 10 times for the interval $0.1 \le \varepsilon_3 b \le 1$. As a result, the products $\tilde{\tilde{c}}_{n,l,m}(\pi)\tilde{\tilde{c}}^*_{n,l',m'}(\pi)$ were calculated. Then, the cross sections of impact ordering of momenta $\tilde{\sigma}_0^k(l, l'|v)$, introduced by formula (12.6.10), were obtained by computing the integral

$$\tilde{\sigma}^{ll'}_{mm'}(v) = \frac{2\pi}{\varepsilon_3^2}\int_0^\infty d\eta\, \eta^{-3}\tilde{c}_{3lm}\tilde{c}^*_{3l'm'} \tag{13.4.10}$$

On averaging with account for axial symmetry of the collisional process, the elements of the matrix of cross sections of the collisional creation of atomic coherence within an ensemble of hydrogen atoms in the state with $n = 3, 4$ will take the form

$$\tilde{\sigma}^k_q(ll') = \delta_{q0}\sum_{m,m'}(-1)^{l'-m'}\begin{bmatrix} l & l' & k \\ m & -m' & q \end{bmatrix}\int d\mathbf{v}|\mathbf{v}|\,\mathbf{f}(\mathbf{v})\mathbf{P}_\mathbf{k}(\cos\beta)\sigma^{\mathbf{ll'}}_{\mathbf{mm'}}(v) \tag{13.4.11}$$

Collisional variation in the polarization moment $\tilde{\rho}_q^k(l, l')$ can be expressed by an equation in terms of the cross section of collisional polarization of the excited state:

$$\frac{d}{dt}\tilde{\rho}_q^k(l, l') = -\frac{\gamma_l + \gamma_{l'}}{2}\tilde{\rho}_q^k(l, l') + n_0 v \tilde{\sigma}_q^k(l, l'|v) \tag{13.4.12}$$

where γ_l is the radiation damping constant for the $n = 3$ state and n_0 is the density of the precipitating protons.

A stationary solution of this system has the form

$$\tilde{\rho}_q^k(l, l') = -\frac{2n_0}{\gamma_l + \gamma_{l'}}\tilde{\sigma}_q^k(l, l'; v) \tag{13.4.13}$$

This expression enables one to calculate the degree of polarization for the transition $nl \rightarrow n'l'$ within a system of reference with the OZ axis directed along the proton beam:

$$\tilde{P} = \frac{I_z - I_y}{I_z + I_y} = -\frac{3\sum_{ll'} W^{(2)}(l, l')\tilde{\rho}_q^2(l, l')}{\sum_{ll'}[2\sqrt{2}W^{(0)}(l, l')\tilde{\rho}_0^0(l, l') - W^{(2)}(l, l')\tilde{\rho}_q^2(l, l')]} \tag{13.4.14}$$

Here, coefficients $W_{ll'}^{(k)}$ are represented by expressions (3.6.17) and (3.6.18), while matrix elements $\Gamma_{nln_0l_0}$ from Eq. (3.6.18) were taken as the square roots of the experimental data of oscillator strengths for transitions $n = 3 \rightarrow n' = 2$ (for the H_α line) and $n = 4 \rightarrow n' = 2$ (for the H_β line).[164]

Results of calculations for the degree of polarization by the impact parameter technique, using velocity distribution function $f(\mathbf{v})$, proposed elsewhere,[51] are given in Table 29 for the particular conditions of observation

TABLE 29

Theoretical Degrees of Polarization of the H_α Hydrogen Line for Different Energies of a Proton Beam and Geometrical Conditions of Observation for the Impact Parameter Method and the Velocity Distribution of Protons Presented Elsewhere[51]

v (10^8 cm/sec)	E (keV)	$\tilde{P}$	60.0°	63.0°	68.5°	49.2°	21.55°	61.87°
2.0	20.8	5.5	1.43	1.194	0.78	2.42	4.79	1.28
3.0	46.7	8.3	2.21	1.83	1.2	3.72	7.26	1.09
4.0	83.0	10.3	2.79	3.31	1.52	4.67	9.04	2.5
5.0	129	11.3	3.09	2.56	1.68	5.16	9.93	2.77
6.0	187	13.6	3.79	3.14	2.07	6.3	12	3.4
7.0	254	15.6	4.42	3.67	2.42	7.31	13.7	3.96
8.0	332	19.1	5.57	4.64	3.07	9.16	16.9	5.01

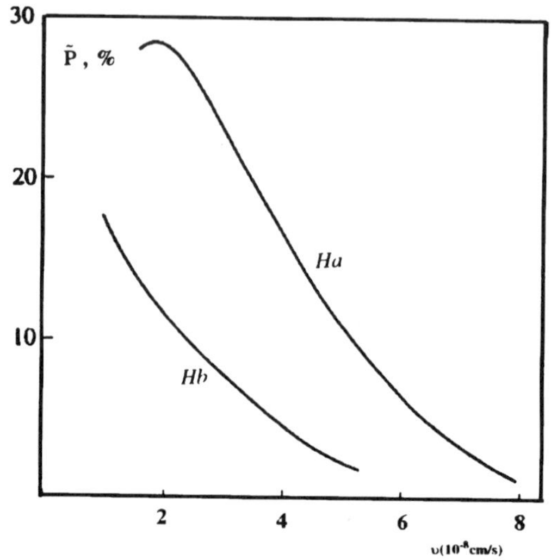

Figure 135. Theoretical dependencies of the degree of polarization $\tilde{P}$ of the H_α and H_β hydrogen lines in the collision reference frame for the impact parameter approximation and a proton velocity distribution function.[(51)]

corresponding to Section 13.3 and illustrated by Figure 135, where the theoretical dependencies of the degree of polarization of hydrogen lines are plotted.

The second row of Table 30 represents the resultant energies of the proton beam, which provide agreement with observational data presented in Section 13.3 (first row of Table 30) and calculations for the impact parameter technique.

The structure of Table 31 is similar to that discussed earlier. Theoretical data on the degree of polarization, corresponding to collisional and observation reference frames, are given. Comparison of theoretical and observational data for the flare on 15.09.1981 shows that the energy intervals of protons, providing agreement between computations and observation and given in the first row of Table 32 for the flare on 15.09.1981, is lower than that for the flare described in Section 13.3.

TABLE 30
Observational Data on the Degree of Polarization of the H_α Hydrogen Line (Section 13.3) and Determination of the Energy of a Proton Beam Bombarding the Upper Atmosphere with Comparison to Theoretical Values Obtained by the Impact Parameter Technique

	60.0°	63.0°	68.5°	49.2°	21.55°	61.87°
Polarization (%)	4.5 ± 1.6	5.5 ± 1.7	3.5 ± 0.9	2.3 ± 0.9	6.5 ± 2.0	3.6 ± 0.6
Beam energy (keV)	280 ± 70	> 280	> 300	50 ± 30	50 ± 30	220 ± 80

TABLE 31
Theoretical Values of the Degree of Polarization of Hydrogen Lines in a Collision Frame of Reference, $\tilde{P}$, and in the Frame of Reference Connected to a Ground-Based Observer, P, for the Angle 61.87° Between the OZ and $O\tilde{Z}$ Axes for the Impact Parameter Approximation Technique with a Velocity Distribution of Protons Proposed Elsewhere(51)

v (10^8 cm/sec)	E (keV)	$\tilde{P}H_\beta$ (%)	P_{H_β} (%)	$\tilde{P}H_\alpha$ (%)	P_{H_α} (%)
2.0	28.5	12	1.6	5.5	1.28
3.0	46.7	7.5	0.9	8.3	1.09
4.0	83.0	4.2	0.54	10.3	2.5
5.0	129	2.5	0.22	11.3	2.72
6.0	187	1	0.1	13.6	3.4
7.0	254			15.6	3.96
8.0	332			19.1	5.01

In conclusion, Figure 136 shows the velocity dependence of Stokes parameter $\tilde{Q}$ of hydrogen lines calculated in the collisional reference frame; it could be used for other problems of spectropolarimetric sensing. Any improvement in the impact spectropolarimetric sensing of solar flares based on hydrogen lines should include a more adequate solution of the collision problem. This implies a separate solution for different values of the principal quantum number n, an account of the cascade transitions from higher-lying levels in Eq. (13.4.5), a more detailed concept of charged particle kinetics in a solar plasma, and finally, improvement of the basic solar flare model itself.

The impact excitation of hydrogen atoms in the chromosphere by beams of charged fast particles propagating preferentially along the radial direction was the principal supposition for the creation of the observed polarization. The fact that nonzero Q values were observed when the angle between the spectrograph slit and the direction to the solar disk center was 45° could be explained by assuming

TABLE 32
Estimate of the Proton Energy Range by Comparing of Observations of the Degree of Polarization of Hydrogen Lines for the Optimal Angular Parameter of Observation of the Flare on 15.09.1981 (61.87°) and Computations Under the Impact Parameter Approximation Technique

	P_{H_β} (%)	P_{H_α} (%)
Observation	2.3 ± 0.5	3.6 ± 0.6
Proton energy (keV)	30 ± 10	200 ± 80

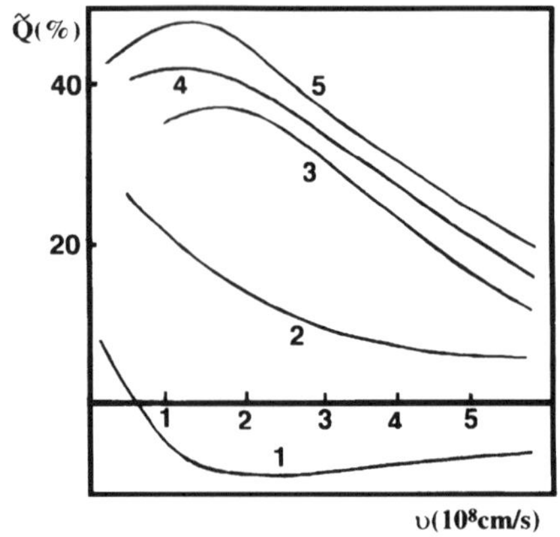

Figure 136. Theoretical data on Stokes parameter $\tilde{Q}$ within the collisional frame of reference as a function of proton velocity: (1) H_α, by the impact parameter method, (2) H_α, by the Glauber approximation, (3) H_β, by the impact parameter method, (4) H_α, by the Born approximation, (5) H_β, by the Born approximation. 1, 3—numerical model of proton kinetics within the magnetic loop;(198) 2, 4, 5—velocity distribution function of protons.(51)

that impact polarization of hydrogen is created by a beam with a nonradial preferential direction of propagation. The local magnetic field may be responsible for beam deflection effects; on the other hand, the magnetic field could directly affect the polarization of atomic ensembles and the polarization characteristics of observed radiation, according to the general equations and computed functions presented in Section 12.3.

13.5. Impact Spectropolarimetric Sensing of Mustaches

Ellerman bombs or mustaches represent in themselves an interesting phenomenon in the solar atmosphere, but they are not yet understood. This phenomenon is recognized as a fine-structure short-living formation with a lifetime of around 10 min.(208,209) The principal feature of mustaches is the specific profile of the Balmer lines: bright, long, emission wings and a self-reversed profile within the line core. The emission contours of more that 100 mustaches have been measured(210) and showed that the wings of contours could reach up to 3–10 Å from the center of the H_α line. The characteristic profile of Balmer lines in mustaches led to the conclusion(211) that heating of mustaches takes place in the lower chromosphere, while upper layers are mostly heated in the course of flares. One work(200) represents an extensive review of the characteristic features of Ellerman bombs. In connection with this work it is worth emphasizing that, according to many studies, Ellerman bombs are directly linked to the extraordinary behavior of the magnetic field(211–215) and the development of regions of solar activity.(200,208,213)

Special studies(216) showed that the emission line profiles of mustaches cannot be attributed to Stark broadening. Doppler broadening of hydrogen lines in mustaches, caused by collision processes of the hydrogen atoms with plasmoids from the lower chromosphere,(217) is quite dubious owing to the lack of a clear understanding of the acceleration mechanism of plasmoids.(200) Some published evidence(200) allows us to assert that linear polarization of the line center created by impacts with accelerated protons could not be applied to explain spectropolarimetric effects of mustache emission.

Detailed spectropolarimetric measurements of mustaches, based on the idea of suprathermal excitation of Balmer lines by beams of energetic particles, were undertaken for two main reasons. The first was unquestionable observational evidence of the presence of linear polarization in mustaches. Weak linear polarization was detected in mustaches in early work,(218) already. In extensive observations presented elsewhere(189–191) a very high degree of polarization (exceeding 20%) was reported in a number of cases. It was of special interest and significance that the preferential direction of the polarization plane, estimated by processing 40 spectrograms of mustaches, was tangential.(190) The degree of polarization, measured(200) from observations of 32 Ellerman bombs, was much lower than other work,(189–191) being at a level of 3–4%. Most important is that the linear polarization of hydrogen line emission was beyond doubt.

The second reason for substantial spectropolarimetric analysis of mustaches was linked to work(219) in which H_α contours were calculated on the basis of semiempirical models $F1$ and $F2$ of a solar flare(205) with account for nonthermal excitation and ionization of hydrogen atoms by electron impact. These calculated profiles could be very effectively used for the diagnostics of a flare plasma, especially at the initial phase. There usually exists rather good agreement between calculated profiles using data from Fang *et al.*(219) and observations of contours of mustaches. The wings of these profiles are well expressed and rather broadened, and the center of the line is self-reversed.

Polarization data on mustaches used for spectropolarimetric sensing were obtained with the help of the solar telescope of Sayany Observatory of the Institute of Solar and Terrestrial Physics at Irkutsk (Russia)(192) using the simultaneous observations of two H_α and H_β Balmer lines of atomic hydrogen. In order to insure reliability of the spectropolarimetric observations, a Wollaston prism was placed just behind the spectrograph slit. Oscillations of the electric vector in the ordinary beam were parallel to the vertical axis, identifying the direction of both the slit and the grating ruling, and orthogonal to the vertical axis for the extraordinary beam. An achromatic $\lambda/2$ plate was mounted in front of the spectrograph entrance slit with one axis either coinciding with the doubling axis or making an angle of 22.5° with it in the course of observations.

In order to obtain the picture at a given time in both H_α and H_β hydrogen lines, the grating was oriented such that, when it was possible to observe the H_α line in the plane of the photoplate in the third order, the H_β line was seen simultaneously in the fourth order. Special light filters to eliminate light originating from the orders of the grating were fixed in front of the photoplate holder. Four spectra were recorded within a single exposure: the H_α and H_β lines in two orthogonal polarizations. A special photoelectric guide was used to provide exact adjustment of the studied location at the entrance slit plane of the spectrograph.

Mustache spectrograms pertaining to the large composite sunspot group located at 40° W, 9° N on September 13, 1981 were very rich and included strips of the continuum emission, mustaches, ejections, and pores. The mustaches were picked out after subsequent treatment and were observed against the background of the penumbra edge.

The spectrograms were processed by means of a self-recording microphotometer and a computer. The microphotometer slit length was taken equal to 1/3 of the emitting object size. Records of the quiet area of the neighboring chromosphere were obtained with the same slit. The wavelength range for the H_α line region was ± 3 Å from the line center, and ± 2.5 Å for the H_β line. Special processing of spectrograms revealed that emission was no longer observed at these distances from the line centers.

In order to assess instrumental errors, H_α and H_β line profiles were recorded by this technique in a quiet region. The profiles corresponding to the unperturbed photosphere were recorded with the same width of the entrance slit as for mustaches. In addition, using a very long microphotometer slit, the region of the spectrum with no mustaches was also recorded. The H_α line profiles agreed reasonably within two strips; the H_β line showed a difference in the line center, attaining 5%. This difference resulted from the slight defocusing of two spectral strips in this spectral region.

If polarized light passes through the optical reception channel when the phase plate axes coincide with the axes (both vertical and horizontal) of the Wollaston prism, then the following emission parameters are recorded within the first and second strips: $\frac{1}{2}(J+Q)$ and $\frac{1}{2}(J-Q)$, respectively.

Addition and subtraction of data within the first and second strips would yield $(J_1+J_2)=J$ and $(J_1-J_2)=Q$.

By rotating the phase plate through 22.5°, the emission intensities within the first and second strips are, respectively, $\frac{1}{2}(J+U)$ and $\frac{1}{2}(J-U)$.

Addition and subtraction would give parameters J and U, respectively.

Seven spectrograms of the same emission in mustaches have been obtained with the aid of a photoguide. Table 33 presents the numbers of spectrograms and the time when these pictures were taken, the positions of the phase plate, and the Stokes parameters determined from a given spectrogram. A disadvantage of this

TABLE 33
Observational Data for Mustaches

Spectrum number	UT	$\lambda/2$ (°)	Stokes parameter
100 (I)	0125.00	22.5	U, J
101 (II)	0125.30	0	Q, J
102 (III)	0126.30	22.5	U, J
104 (IV)	0128.00	0	Q, J
105 (V)	0142.00	0	Q, J
106 (VI)	0142.40	22.5	U, J
107 (VII)	0144.00	22.5	U, J

technique is that some errors could be introduced because the pictures, taken at different positions of the phase plate, are not obtained simultaneously.

The profiles of the H_α and H_β hydrogen lines observed for mustaches are shown in Figure 137 for two spectral bands, corresponding to orthogonally linear polarizations. The wings of the profiles in Figure 137 corresponding to the quiet region were used to compare two profiles in order to determine the line core polarization. The slight difference in the two profiles corresponding to the two spectral bands is natural, because these profiles of the H_α and H_β lines were obtained at the earliest times of observation, when the mustaches had maximal intensity. For the H_β line, this difference is opposite for the core of the line and for the wing region, corresponding to the quiet location. Such behavior of the line profiles could not be explained by instrumental polarization errors, but only if linear polarization was present in the light emitted by the object under investigation. Of all the spectrograms taken at 0142 and 0144 UT, only the profiles of spectrogram No. 106, shown in Figure 137, indicated the existence of linear polarization. Therefore, it was concluded that at times of maximal intensity (initial phase of mustache brightening or the first 2–3 minutes of their lifetime), linear polarization was detected at the line core. No linear polarization was observed in the decaying phase, close to the disappearance of mustaches.

The difference in intensities within two spectral bands divided by the sum of the same intensities, or the quantity $(J_1 - J_2)/(J_1 + J_2)$, gives the values of Stokes parameters Q/J or U/J.

Let us assume that the time intervals between spectrograms 100 and 101 (30 s) and between 102 and 104 (1 min 30 s) is not sufficient for substantial variations in the studied object. Then, it is possible to determine the degree of polarization and the azimuth of the linear polarization from the first and second pairs of spectrograms as follows:

$$P = \sqrt{\frac{Q^2 + U^2}{J^2}}, \qquad \text{cotan } 2\chi = \frac{\mathrm{Q}}{\mathrm{U}}$$

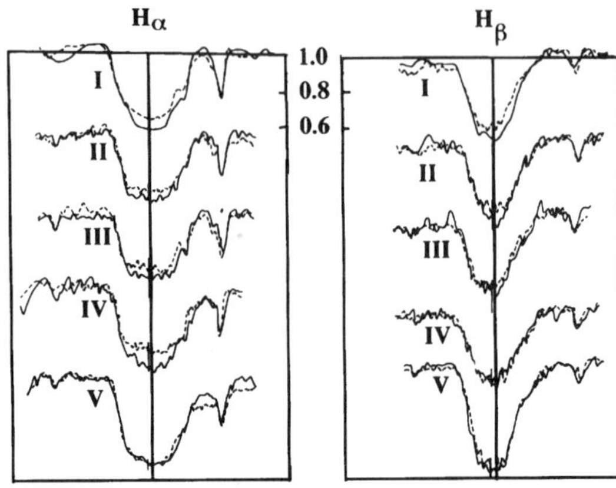

Figure 137. H_α and H_β hydrogen line profiles for two orthogonal linear polarizations in mustaches.

Figure 138 shows profiles $(J_1 - J_2)/(J_1 + J_2)$ for the first four spectrograms. The specific behavior at the center of lines is well pronounced for some H_α and H_β profiles.

The final observational results for the degree of polarization and the azimuth of polarization taken from these spectrograms are presented in Table 34. Since a single spectrogram provides only Q/J or U/J, the first column indicates two moments, corresponding to two successive spectrograms.

The difference between the data in Table 34 and other work[(192)] is connected to the fact that, in the latter work,[(192)] the data on parameters Q and U were averaged over the spectral interval ± 1 Å with respect to the line center while the data on χ^* and χ^c in Table 34 were obtained exactly for the line center. Concerning the orientation of the azimuth of polarization, it should be noted that

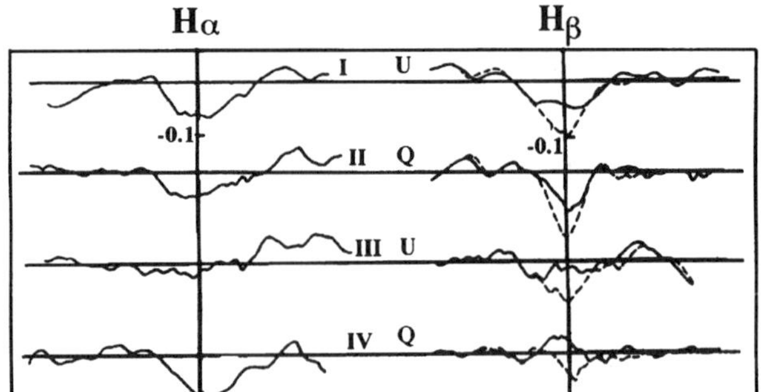

Figure 138. $(J_1 - J_2)/(J_1 + J_2)$ profiles for the first four spectrograms (I–IV). For the H_β line, dashed curves represent the difference between the studied profile and the profile for the quiet region.

TABLE 34
Time Variation of the Degree Polarization P, Azimuth of Polarization χ with Respect to the Slit Direction for an Interval of ± 1 Å of the Central Region of the H_α and H_β Hydrogen Emission Lines, and the Azimuth Angles χ^* and χ^c for the Center of H_α and H_β Hydrogen Emission Lines, with Angle χ^c Measured with Respect to the Direction to the Solar Disk Center

UT	P_{H_α} (%)	P_{H_α} (°)	$\chi^*_{H_\alpha}$ (°)	$\chi^c_{H_\alpha}$ (°)	P_{H_β} (%)	χ_{H_β} (°)	$\chi^*_{H_\beta}$ (°)	$\chi^c_{H_\beta}$ (°)
0125.00–25.30	7.8	20	27	75	9	28	15	87
0125.30–26.30	5.2	37	13	85	7.6	39	7	95
0126.30–28.00	8.1	40	8	91	2.4	72	72	30

azimuth χ in Table 34 was measured from the direction of the entrance slit, which was oriented at 102° to the solar disk center direction. In calculating the azimuth of polarization with respect to the solar center disk direction (χ_c), it is easy to obtain for these observational data that the polarization of mustaches is preferentially tangential with respect to the solar center disk direction.

A comparative analysis of the line profiles presented for these series of observations in mustaches, together with calculations performed elsewhere,[219] allows one to conclude that linear polarization of mustaches is caused mainly by electron impact or by electron beams precipitating into the chromospheric location.

The polarization characteristics of an excited ensemble of hydrogen atoms are determined by polarization moments $\rho^k_q(ll')$. These describe the ordering of angular momenta of an excited atomic ensemble and are related to the density matrix of this ensemble in the representation of magnetic quantum numbers given by relation (12.2.5). Under stationary conditions the variation in these polarization moments is described by the matrix equation

$$\Gamma\rho = N \tag{13.5.1}$$

where N is the matrix of rate constants of the process of collisional ordering of angular momenta, and Γ is the matrix describing the variation in population of the ensemble of excited atoms of hydrogen due to radiative transitions between its different states.

The determination of the polarization moments of an ensemble of excited hydrogen atoms is connected to the solution of the collisional problem of the computation of the matrix of rate constants of collisional ordering of an ensemble and the solution of Eq. (13.5.1) with account for excited state population changes due to radiative transitions from the upper states. A limited number of states is usually accounted for in particular computations. In an analysis of the polarization characteristics of the H_α hydrogen line,[220] ten basic functions were used: one $1s$ ground state function and nine functions of the excited state with principal quantum number $n = 3$. Subject to these restrictions, Eq. (13.5.1) could be written in the form

$$\frac{\gamma_l + \gamma_{l'}}{2} \rho_q^k(ll') = n_0 \langle v\sigma_q^k(ll') \rangle \tag{13.5.2}$$

where $\gamma_{l,l'}$ are the radiative damping constants of state $3l$, n_0 is the density of the exciting particles, and $\langle v\sigma_q^k(ll') \rangle$ are rate constants for collisional ordering of states with angular momenta l and l' of the hydrogen state with $n = 3$.

As Eq. (13.5.2) does not account for population changes due to radiative transitions from states with $n > 3$ populated also by electron impact, its solution only approximately describes a real polarization process even if its right hand side is computed by a rather precise technique. Taking into consideration that, under these restrictions, the solution of Eq. (13.5.2) will have an approximate nature, we shall not resolve the collisional problem accurately because, for a solution of the multichannel problem with account for an excitation channel and channels describing transitions between magnetic sublevels of different momenta of state $n = 3$ with allowance for s, p, and d partial waves, a basis of more than twenty functions is needed.

A criterion for the validity of any approximation could be the ratio[44]

$$s = \frac{\bar{U}ka}{mv^2} \approx \frac{v_0}{v}$$

where $\bar{U}$ is the mean interaction energy, r is the effective radius of interaction, and $v_0 = e^2/\hbar$ is the atomic velocity.

If $s \gg 1$, which means that the uncertainty of the scattering angle is much less than the value of this angle, then the concept of the trajectory of motion can be exploited. For the opposite case, $s \ll 1$, the interaction can be treated as a perturbation and collisions described under the Born approximation. For the intermediary case $s \approx 1$, which is realized for an electron energy typical for the excitation of hydrogen lines in mustaches, the eikonal approximation can be used. Therefore, to solve the collisional problem we used the Glauber approximation, which results from the eikonal approximation in second order by resolving the exponent in terms of powers of U/k^2, following Grien.[39,44]

The rate constants are expressed in terms of amplitudes of the inelastic scattering of an electron on the hydrogen atom with excitation of the latter from the ground to the excited state with quantum numbers $n = 3$, l, and m and distribution function of the relative velocities $\Phi(\mathbf{v})$:

$$\langle v\sigma_q^k(ll') \rangle = \frac{128\pi^2 v_0^4}{9(2k+1)} \delta_{q0} \sum_m (-1)^{l'+m'} \begin{bmatrix} l & l & k \\ m & m' & q \end{bmatrix} \cdot \int_0^\infty dx\, x\Phi_k(x) \int_0^{3x} \frac{dz}{z^3} f_{lm}(x, z) f_{l'm'}(x, z) \tag{13.5.3}$$

Here $x = v/v_0$, $v_0 = e^2/\hbar$ is the atomic velocity, z is a dimensionless parameter related to the transmitted impulse q by $z = 4/3a_0q$, and $\Phi_k(v)$ is the multipole moment of the relative velocity distribution function:

$$\Phi_k(v) = \frac{2k+1}{4\pi}\int d\Omega\Phi(\mathbf{v})\mathbf{P_n}(\cos\theta) \tag{13.5.4}$$

The scattering amplitude was calculated within the Glauber approximation. For scattering resulting in transition $1s$–$3l$ and characterized by wave number k and transmitted impulse q, the scattering amplitude assumes the form

$$f_{lm} = \frac{ik}{2\pi}\int \langle 1s0|[1 - \exp(2i\chi(\rho', r))]|3lm\rangle e^{i(\mathbf{q},\rho')}\, d^2\rho' \tag{13.5.5}$$

where quantity $\chi(\rho', r)$ is determined by the interaction potential of the incoming and valent electrons, $V(\mathbf{r}, \mathbf{r})$:

$$\chi(\rho', r) = \frac{1}{\hbar v}\int_{-\infty}^{\infty} V(\mathbf{r}, \mathbf{r})\, dz' \tag{13.5.6}$$

Integration in expressions (13.5.5) and (13.5.6) is carried out in a cylindrical reference frame over the coordinates of the incoming electron. The scattering amplitude (13.5.5) can be expressed in terms of the hypergeometric function $F(a, b; c; -z^2)$, described elsewhere.[76,136] The following scattering amplitudes will be nonzero:

$$\begin{aligned}
f_{00}(x,z) &= -\frac{9\sqrt{3}}{1024}a_0\Gamma(1 - i/x)\Gamma(1 + i/x)z^s \\
&\quad \times \left[s(s^2 - 9s + 14)\Psi_1 + 6(s^2 + 8s + 12)z^2\Psi_1' \right. \\
&\quad \left. + 12(s - 5)z^4\Psi_1'' + 8z^6\Psi_1'''\right] \\
f_{1,\pm1} &= \mp i\frac{27}{256}a_0 e^{\pm\varphi}\Gamma(1 + i/x)\Gamma(2 - i/x)z^{1+s} \\
&\quad \times \left[\frac{s(s-7)}{4}\Psi_2 + \frac{2s-9}{2}z^2\Psi_2' + z^4\Psi_2' + z^4\Psi_2''\right] \\
f_{2,0} &= -\frac{27\sqrt{6}}{256}a_0\Gamma(i/x)z^s[(s+2)\Psi_3 + 2z^2\Psi_3'] \\
f_{2,\pm2} &= -\frac{27}{2048}e^{\pm 2i\varphi}\Gamma(1 + i/x)\Gamma(3 - i/x)\Gamma(1 - i/x)z^{2+s}[3\Psi_4 + 2z^2\Psi_4']
\end{aligned} \tag{13.5.7}$$

where $s = 2 - 2i/x$; Ψ_k $(k = 1, 2, 3, 4)$ are the following combinations of hypergeometric functions:

$$\Psi_1 = F(1 - i/x, 1 + i/x; 1; -z^2)$$

$$\Psi_2 = F(2 - i/x, 1 - i/x; 2; -z^2) - (2 - i/x)F(3 - i/x, 1 - i/x; 2; -z^2)$$

$$\Psi_3 = \sum_{j=0}^{3} B_j \frac{\Gamma(1 + j - i/x)}{j!} F(j + 1 - i/x, -i/x; 1; -z^2) \qquad (13.5.8)$$

$$B_0 = 1, \qquad B_1 = -6, \qquad B_2 = 4.5, \qquad B_3 = -2/3$$

$$\Psi_4 = 3F(3 - i/x, 2 - i/x; 3; -z^2) - (3 - i/x)F(4 - i/x, 2 - i/x; 3; -z^2)$$

and Ψ' denotes the derivative with respect to the argument.

For characteristic parameter $s \approx 1$, apart from the Glauber approximation, the impact parameter method with curving of the trajectory can also be used to solve the collisional problem. In using this method, we introduce two systems of reference: $(\tilde{X}, \tilde{Y}, \tilde{Z})$ with $O\tilde{Z}$ axis antiparallel to the relative velocity vector of colliding particles and subsequent system $(\tilde{\tilde{X}}, \tilde{\tilde{Y}}, \tilde{\tilde{Z}})$ with $O\tilde{\tilde{Z}}$ axis directed towards the incident electron, which moves along a linear trajectory characterized by impact parameter b and relative velocity v (Figure 83). At time $t_0 = -b/v \operatorname{cotan} \vartheta_0$, which is determined by a known classical relation,[36,40,175] a break in the trajectory takes place at an angle given by

$$\vartheta_0 = \arcsin \frac{1}{\sqrt{1 + \left(\frac{e^2}{v^2 mb}\right)^2}} \qquad (13.5.9)$$

The wave function of an atomic electron in the field of two centers can be sought as an expansion in terms of a system of ten functions, one of which is the eignfunction of the $1s$ hydrogen state while the other nine describe the excited state with principal quantum number $n = 3$:

$$\tilde{\tilde{\Psi}}(\mathbf{r}, \mathbf{R}) = \tilde{\tilde{c}}_{1,0,0}\tilde{\tilde{\psi}}_{1,0,0}(\mathbf{r}, t) \exp \frac{iE_1 t}{h} + \sum_m \tilde{\tilde{c}}_{3,l,m}\tilde{\tilde{\psi}}_{3,l,m}(\mathbf{r}, t) \exp \frac{iE_3 t}{h} \qquad (13.5.10)$$

If the last expression is substituted into the Schrödinger equation, integration over time is transformed to integration over the angle $\vartheta_0 = -\arctan(b/vt)$, and dimensionless parameter $\eta = (v_0/v)/b\varepsilon_3$ is introduced where $\varepsilon_3 = 1/3a_0$, then we obtain a system for the expansion coefficients in expression (13.5.10):

$$\frac{d}{d\vartheta}\tilde{\tilde{c}}_{n,l,m} = -i\eta \sum_{n',l'} \langle n,l,m|W(\mathbf{r},R)|n',l',m\rangle \tilde{\tilde{c}}_{n,l,m} \exp(-\omega_{nn'} \cot\vartheta) + \frac{\delta_{n,n'}\delta_{ll'}}{2}[\sqrt{(l-m)(l+m+1)}\tilde{\tilde{c}}_{n,l,m+1} - \sqrt{(l+m)(l-m+1)}\tilde{\tilde{c}}_{n,l,m-1}] \tag{13.5.11}$$

where $W(\mathbf{r}, R)$ is related to the interaction energy of atomic and incident electrons:

$$W(\mathbf{r},R) = \frac{\varepsilon_3 R^2}{|\mathbf{r}-\mathbf{R}|} \quad \text{and} \quad \omega_{n,n'} = \frac{n'^2 - n^2}{2nn'^2}\frac{v}{v_0}\varepsilon_n b$$

Matrix elements from expression (13.5.11) can be computed by representing function $W(\mathbf{r}, R)$ in terms of Legendre polynomials:

$$\langle n,l,m|\frac{1}{|\mathbf{r}-\mathbf{R}|}|n',l',m\rangle = \varepsilon_3\sqrt{\frac{2l'+1}{2l+1}}\sum_k f^{(k)}_{n,ln',l'}(R)\begin{bmatrix} l' & k & l \\ 0 & 0 & 0 \end{bmatrix}\begin{bmatrix} l' & k & l \\ m & 0 & m \end{bmatrix} \tag{13.5.12}$$

where

$$f^{(k)}_{n,ln',l'}(R) = \frac{1}{\varepsilon_3}\left[\frac{1}{R^{k+1}}\int_0^R dr\, r^{k+2}\tilde{\tilde{R}}_{n,l}(r)\tilde{\tilde{R}}_{n',l'}(r) + R^k\int_R^\infty dr\, r^{1-k}\tilde{\tilde{R}}_{n,l}(r)\tilde{\tilde{R}}_{n',l'}(r)\right] \tag{13.5.13}$$

and $\tilde{\tilde{R}}_{n,l}(r)$ is the radial part of the atomic electron wave function.

Before collision, the hydrogen atom was in the ground state, so the initial condition for system (13.5.11) takes the form

$$\tilde{\tilde{c}}_{n,l,m} = \delta_{n,1}\delta_{l,0}\delta_{m,0}$$

Due to symmetry of the collisional process with respect to reflection in the collision plane, it is possible to employ ten functions $\tilde{\psi}_{n,l,m}$ in order to construct seven even and three odd combinations, with respect to reflection in this plane. As the odd combinations have zero initial conditions, the system describing its time variation will have only a zero solution. Therefore, the symmetry condition for coefficients $\tilde{\tilde{c}}_{n,l,m}$ is

$$\tilde{\tilde{c}}_{n,l,m} = (-1)^m \tilde{\tilde{c}}_{n,l,-m} \tag{13.5.14}$$

The values of coefficients $\tilde{\tilde{c}}_{3,l,m}(\pi)$ for a particular impact parameter b are determined by solving system (13.5.11).

If we transform to the collision reference frame with the help of rotation matrix $D^{(2)}(0, \vartheta_0, 0)$, it is possible to determine the cross section for transition between the levels of the state with $n = 3$:

$$\tilde{\sigma}^{l,l'}_{m,m'} = \frac{2\pi v_0^2}{\epsilon_3 v^2} \int_0^\infty \frac{d\eta}{\eta^3} \tilde{c}_{3,l,m}(\pi) \tilde{c}_{3,l',m'}(\pi) \tag{13.5.15}$$

Finally, we can compute the rate constant for the creation of coherence between states with different angular moments of this state:

$$\langle v\tilde{\sigma}^{(k)}_q(l, l')\rangle = \delta_{q,0} \sum_{m,m'} (-1)^{l'+m'} \begin{bmatrix} l & l' & k \\ m & -m' & q \end{bmatrix} \frac{4\pi}{2k+1} \int_0^\infty v\, v^3 \Phi_k(v) \tilde{\sigma}^{l,l'}_{m,m'} \tag{13.5.16}$$

where the multipole moment of the velocity distribution function is determined by Eq. (12.4.5).

This multipole moment can be obtained using an approach[51] in which a distribution function of protons was proposed for an ionized medium that accounted for scattering of protons on hydrogen atoms. An expression of the energy loss of a probe particle while scattering on an ensemble of field particles[183] was used. In the present case, assuming that the probe and field particles are electrons, the distribution function can be found[51] in the form

$$\Phi(v') = \left(v'^4 + \frac{16\pi e^4 N\Lambda}{v_0 m^2 \cos\vartheta'} \right)^{-2.5} \delta(v - v') \tag{13.5.17}$$

where N is the linear density of the chromosphere which, according to the $F2$ model,[205] is equal to 3.8×10^{13} cm; Λ is the Coulomb logarithm, which is close to unity, and ϑ is the pinch angle.

The density in the optically active chromospheric region is quite high, $n \sim 10^{16}$ cm^{-3}. Hence it is possible to conclude that the scattering of electrons on field particles is mostly of a diffuse nature and the distribution of relative velocities is represented by the drift velocity distribution function, or the Maxwell distribution function with drift. The multipole moment of this velocity distribution function takes the form

$$\Phi_k(x) = (2k+1)Ge^{-\delta x^2}\frac{I_{k+1/2}(2\delta\xi x)}{\sqrt{x}}$$

where $\delta = v_0^2/kT_{eff}$, $\xi = u/v_0$, u is the drift velocity, v_0 the atomic velocity, k the Boltzmann constant, T_{eff} the effective temperature of the chromospheric plasma, $I_{k+1/2}(z)$ is the Bessel function of pure imaginary argument, and G is a constant.

Spectropolarimetric data given in Table 34 are connected to the observer system of reference (X, Y, Z); see Figure 83. Stokes parameter Q in the observer system of reference is determined by the polarization moment $\rho_q^{(k)}(l, l')$, referred to the same reference system:

$$\begin{aligned} Q &= \frac{I_z - I_y}{I} = -\frac{\sum_{l,l'} w_{l,l'}^{(2)}[3\rho_0^{(2)}(l,l') + \mathrm{Re}(\rho_2^{(2)})]}{I\sqrt{6}} \\ U &= \frac{I_{zy^+} - I_{zy^-}}{I} = -2\frac{\sum_{l,l'} w_{l,l'}(2)\mathrm{Im}[\rho_1^{(2)}(l,l')]}{I} \\ I &= I_z + I_y = \frac{1}{\sqrt{6}}\sum_{l,l'}\{2\sqrt{2}w_{l,l'}^{(0)}\rho_0^{(0)}(l,l') + w_{l,l'}^{(2)}[\rho_0^{(2)}(l,l') + \sqrt{6}\,\mathrm{Re}(\rho_2^{(2)})]\} \end{aligned} \quad (13.5.18)$$

where I_z and I_y are intensities of radiation with linear polarization along axes OY and OZ, respectively, while I_{zy^+} and I_{zy^-} are the same for polarization along the diagonals in the YZ plane.

Designating $r_{l,l'}$ the radial part of the matrix element for the dipole transition $1s - 3l$, it is possible to obtain the following expression for the coefficients $w_{l,l'}^{(k)}$:

$$w_{l,l'}^{(k)} = \sum_{l_0}\sqrt{(2l+1)(2l'+1)}r_{l,l_0}r_{l',l_0}\begin{bmatrix} l & 1 & l_0 \\ 0 & 0 & 0\end{bmatrix}\cdot\begin{bmatrix} l' & 1 & l_0 \\ 0 & 0 & 0\end{bmatrix}\begin{Bmatrix} 1 & 1 & k \\ l & l' & l_0\end{Bmatrix} \quad (13.5.19)$$

If the projection of the polarization vector in the ZY plane with respect to the entrance slit of the spectrometer is characterized by angle χ and the direction of polarization vector corresponds to the symmetry axis of the collisional process, it is easy to compute Euler angles (α, β, γ), specifying the position of the (X, Y, Z) system with respect to the $(\tilde{X}, \tilde{Y}, \tilde{Z})$ reference system. Polarization moments in these two reference systems are related via the Wigner rotation matrix:

$$\rho_q^{(k)}(l,l') = \sum_{q'} D_{q',q}^{(k)}(\alpha,\beta,\gamma)\tilde{\rho}_q^{(k)}(l,l') = D_{0,q}^{(k)}(\alpha,\beta,\gamma)\tilde{\rho}_0^{(k)}(l,l') \quad (13.5.20)$$

because, due to axial symmetry of the collisional process, only polarization moments with $q = 0$ are not zero in the system $(\tilde{X}, \tilde{Y}, \tilde{Z})$. Euler angles (α, β, γ) characterize the relative position of these two reference systems.

Rate constants of the angular momenta ordering of an ensemble of hydrogen atoms in a state with $n = 3$ have been calculated using the technique described above. The hypergeometric functions and Γ functions were computed using algorithms published elsewhere.[207,136] Results of computations are presented in Figure 139 and Table 35.

The third column of Table 35 contains calculated values of the degree of polarization for the monokinetic beam approximation (distribution function δ) in the collision reference system $(\tilde{X}, \tilde{Y}, \tilde{Z})$, calculated for a particular value of

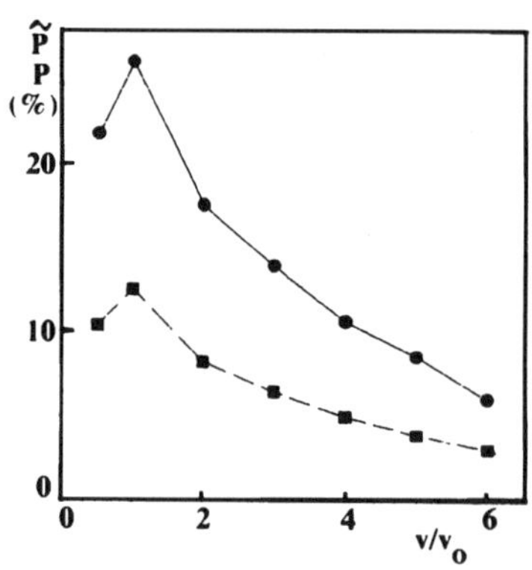

Figure 139. Degree of polarization of the H_α hydrogen line as a function of parameter $x = v/v_0$ ($v_0 = e^2/\hbar$ is the atomic velocity); (●) and (■) refer to the collision system of reference and the system of reference of a ground-based observer, respectively.

TABLE 35
Calculated Degree of Polarizations P and $\tilde{P}$ as a Function of Parameter $x = v/v_0$, Corresponding to Some Energy of Relative Motion

x	E (eV)	$\tilde{P}$ (%)	P (%)
0.5	3.4	21.9	10.4
1.0	13.6	26.4	12.4
2.0	54.5	17.7	8.26
3.0	122	14.0	6.49
4.0	217	10.7	5.02
5.0	340	8.58	3.92
6.0	490	6.08	3.09

parameter $x = v/v_0$ indicated in the first column. The degree of polarization, computed within the observer system of reference (X, Y, Z) for velocity distribution function (13.5.17), is given in the last column of Table 35. Calculations with the drift velocity distribution function for reasonable temperatures of the upper chromospheric layer ($T \sim 10^4$ K) gave very low values of the degree of polarization, confirming that the drift velocity distribution function is not suitable for a description of the kinetics of momenta ordering of an excited atomic hydrogen by electron impact.

Comparison of observational data in Table 34 with computations for different energies of electrons showed that theoretical evaluation is in rather good agreement with observation for electron energies of about 100 eV. Electrons with such energies were proposed to be responsible for the linear polarization of mustache emission in the framework of known astrophysical models. The error in observational data on the degree of polarization was estimated at about 30%, and the error in theoretical estimates of the exciting electron energy was evaluated at 50%.

The implementation of spectropolarimetric sensing of mustaches could be significantly improved by using more precise methods of collisional theory than the Glauber approximation and employing a more realistic velocity distribution function.

13.6. Concluding Remarks

This chapter summarizes results of quantitative spectropolarimetric sensing of energy input in the chromospheric region of the solar atmosphere for particular nonthermal events. The final data were obtained as a result of sophisticated observations of the linear polarization of hydrogen emission lines and computations with the help of the developed methodology. The approaches and results presented in this chapter show that the impact spectropolarimetric technique can be used to estimate the energy transported by precipitating charged particles for many natural or artificial ionized media.

14
Conclusions

Optical spectroscopy is a continuously developing field which provides new practical methods for the remote sensing of different natural and artificially ionized entities. This book presents a detailed description of a new spectro-polarimetric method of sensing based on the phenomenon of the polarization of ensembles of atomic particles resulting from collisional processes in a diluted ionized media.

The quantum mechanical apparatus governing the irreducible representation of a density operator was shown to be the most effective means for describing the collisional interaction of atomic particles inside a real ionized medium, resulting in the polarization of atomic ensembles and the emission of polarized light. This optimal theoretical approach provides a comprehensive account of the macroscopic symmetry and allowed us to show that polarization reflects the structural features of an ionized medium. Quadrupole ordering of momenta, or self-alignment of atoms and ions, created by collisions with fast, light, precipitating particles is analyzed in this book in detail with regard to many real diagnostic applications. Special attention was paid for the first time to the polarization of ensembles under collisions of slow, heavy, atomic particles when collisional relaxation in a system of closely spaced levels and charge exchange play a principal role.

Methodologically, the main new diagnostic opportunities are connected to the fact that the macroscopic quadrupole orientation of momenta of atomic particles and the emission of linearly polarized light reflect the structural features of the ionized entity under study and the environmental effects responsible for sustaining an energy balance. Spectropolarimetric effects resulting from the intrusion of external fast beams of charged particles, precipitating to a known diluted medium and forming an ionized entity in itself, have been studied in more detail because such physical situations arise in many real cases.

The spectropolarimetric sensing technique is illustrated in this book by ground-based observations and quantitative diagnostics of nonthermal entities in the solar atmosphere and, more specifically, by studies of solar flares. A solar flare as one of the most physically rich phenomena was taken as an example of the implementation of impact spectropolarimetric sensing, because optically active chromospheric location is formed by the bombardment of fast charged particles precipitating from the higher shells of the solar atmosphere. Due to the complicated nature of the ionized medium in the optically active upper chromosphere, many collisional spectropolarimetric effects occur.

Different observational techniques from eminent scientific groups in this field have been described together with results of spectropolarimetric observations for a number of solar flare events. This material was presented in order to illustrate to the reader the practical approaches to spectropolarimetric sensing and the type of instrumental equipment. All results for the most practically significant hydrogen emission lines are examined and used to implement impact spectropolarimetric sensing by comparing observations with theoretical computations of the polarization of hydrogen lines employing appropriate astrophysical models.

Impact spectropolarimetric methodology, straightforward theoretical and practical illustrations, and estimates of the energy transported by particles to the chromospheric region of a solar flare enable methods of impact spectropolarimetric sensing to be applied also to many other ionized gas objects under natural and artificial conditions. These developments and new fundamental results can be expected in the near future.

References

1. S. A. Kazantsev and J.-C. Hénoux, *Polarization Spectroscopy of Ionized Gases* Kluwer Academic Publishers, Dordrecht, London, Boston (1995).
2. S. A. Kazantsev, *Sov. Phys. Usp. (USA)* **26**, 328 (1983).
3. J.-C. Hénoux and M. Semel, Proceedings of Solar Maximum Year Workshop, *Izmiran* **1**, 207 (1981).
4. J.-C. Hénoux, D. Herischi, G. Chambe, M. Machado, B. Woodgate, R. Shine, and J. Bakers, *Astron. Astrophys.* **119**, 233 (1983).
5. J.-C. Hénoux and G. Chambe, *JQSRT* **44**, 193 (1990).
6. E. Vogt and J.-C. Hénoux, *Solar Phys.* **164**, 345 (1996).
7. S. A. Kazantsev, N. Feautrier, J.-C. Hénoux, V. V. Luchinkina, and A. V. Liaptsev, *Astron. Astrophys. Rev.* **6**, 1 (1994).
8. S. A. Kazantsev, *Opt. Spectrosc. (USA)* **52**, 509 (1982).
9. S. A. Kazantsev, N. Ya. Polynovskaya, L. N. Pyatnitsky, and S. A. Edelman, *Sov. Phys. Usp. (USA)* **31**, 785 (1988).
10. S. A. Kazantsev, *JETP Lett. (USA)* **37**, 159 (1983).
11. S. A. Kazantsev and A. V. Subbotenko, *Sov. J. Plasma Phys. (USA)* **10**, 78 (1984).
12. S. A. Kazantsev and A. V. Subbotenko, *J. Phys. D* **20**, 741 (1987).
13. A. I. Drachev, S. A. Kazantsev, A. G. Rys, and A. V. Subbotenko, *Opt. Spectrosc. (USA)* **69**, 585 (1990).
14. A. G. Petrashen, V. N. Rebane, and T. K. Rebane, *Opt. Spectrosc. (USA)* **58**, 785 (1985).
15. S. A. Kazantsev, N. T. Polezhaeva, and V. N. Rebane, *Opt. Spectrosc. (USA)* **63**, 15 (1987).
16. S. A. Kazantsev, A. G. Petrashen, V. N. Rebane, and T. K. Rebane, *J. Exp. Theor. Phys. JETP (USA)* **79**, 384 (1994).
17. S. A. Kazantsev, A. G. Petrashen, V. N. Rebane, T. K. Rebane, V. N. Funtov, L. Windholz, and Ch. Neureiter, *Phys. Scr.* **52**, 572 (1995).
18. S. A. Kazantsev, A. V. Mikhalev, and A. G. Petrashen, *Opt. Spectrosc. (USA)* **85**, N6 (1998).
19. J. W. Chamberlain, *Planet. Space Sci.* **2**, 73 (1959).
20. R. A. Duncan, *Planet Space Sci.* **1**, 112 (1959).
21. J. W. Chamberlain, *J. Geophys. Res.* **9**, 1239 (1974).
22. E. G. Berezhko, *Geomagn. Aeron. (USSR)* **21**, 1131 (1981).
23. Q. Harang, *Inst. Theor. Astrophys. Blinder* **No. 11**, 15 (1960).
24. L. M. Fishkova, *Night Emission of the Middle Latitude High Atmosphere of the Earth*, Mitsniereba Publ., Tbilisi, 271 pp. (1983).
25. Yu. A. Nadubovich, "Polarization effects during optical flares, aurorae and twilights," in *Physical Phenomena in the Atmosphere of High Latitudes*, p. 40, Yakutsk Division of the Siberian Department of the Academy of Science of the USSR (1977).

26. I. A. Khvostikov, *Dokl. Acad. Sci. USSR* **27**, 219 (1940) (in Russian).
27. A. Yutsys and A. Banzaitis, *Theory of Momentum in Quantum Mechanics*, Nauka, Vilnius, 622 pp. (1965) (in Russian).
28. U. Fano, *Phys. Rev.* **90**, 577 (1953).
29. M. I. Diakonov and V. I. Perel, *JETP (USA)* **47**, 1483 (1964).
30. M. I. Diakonov and V. I. Perel, *JETP (USA)* **48**, 345 (1965).
31. A. Omont, *J. Phys.* **26**, 26 (1965).
32. E. El Baas and B. Castel, *Graphic Methods in Spin Algebra*, Nauka, Moscow 1980 (in Russian).
33. V. E. Berestetsky, E. M. Lifshits, and L. P. Pitaevsty, *Relativistic Quantum Theory*, Nauka, Moscow, 929 pp. (1963) (in Russian).
34. A. G. Petrashen, V. N. Rebane, and T. K. Rebane, *JETP (USA)* **77**, 187 (1993).
35. S. A. Kazantsev, N. Ya. Polynovskaya, L. N. Piatnitsky, and S. A. Edelman, *Opt. Spectrosc. (USA)* **58**, 28 (1985).
36. L. D. Landau and E. M. Lifshits, *Course of Theoretical Physics*, Nauka, Moscow (1963) (in Russian).
37. Y. Itikawa, *Phys. Rep.* **143**, 70 (1986).
38. L. A. Burkova and V. I. Ochkur, *JETP (USA)* **76**, 76 (1979).
39. J. T. Gien, *Phys. Rep.* **160**, 123 (1988).
40. G. F. Drukarev, *Collisions of Electrons with Atoms and Molecules*, Nauka, Moscow, 253 pp. (1978) (in Russian).
41. M. Ya. Amusia and I. A. Chernyshov, "Many electron correlations and scattering processes," *Case Stud. At. Phys.* **5**, 50 (1975).
42. E. J. Tweed and J. Larglois, *J. Phys. B* **24**, 1779 (1991).
43. D. H. Jakubassa-Amundsen, *J. Phys. B* **23**, 3335 (1992).
44. J. T. Gien, *J. Phys. B* **9**, 3203 (1976).
45. E. Gerjoy and B. H. Tomas, *Rep. Prog. Phys.* **37**, 1325 (1974).
46. B. H. Tomas and E. Gerjoy, *J. Math. Phys.* **12**, 1567 (1971).
47. E. I. Dashevskaya and E. E. Nikitin, *Opt. Spectrosc. (USA)* **63**, 472 (1987).
48. E. I. Dashevskaya and E. E. Nikitin, *J. Chem. Phys. (USA)* **4**, 1171 (1985).
49. V. L. Zagrebin and N. A. Pavlovskaya, *Opt. Spectrosc. (USSR)* **62**, 27 (1988).
50. S. A. Kazantsev, A. G. Petrashen, N. M. Firstova, and J.-C. Hénoux, *Opt. Spectrosc. (USA)* **75**, 382 (1993).
51. J.-C. Hénoux, G. Chambe, D. Smith, B. Tamres, N. Feautrier, and M. Rivira, *Astrophys. J., Suppl.* **73**, 303 (1990).
52. M. Lombardi, *C.R. Acad. Sci. Paris* **265**, 191 (1967).
53. V. N. Rebane, *Opt. Spectrosc. (USA)* **24**, 309 (1968).
54. E. Chamon, M. Lombardi, M. Carre, and M. Gaillard, *J. Phys.* **38**, 581 (1973).
55. A. G. Petrashen, V. N. Rebane, and T. K. Rebane, *Opt. Spectrosc. (USA)* **58**, 785 (1985).
56. A. G. Petrashen, V. N. Rebane, and T. K. Rebane, *Opt. Spectrosc. (USA)* **55**, 819 (1984).
57. A. G. Petrashen, V. N. Rebane, and T. K. Rebane, *JETP (USA)* **94**, 46 (1988).
58. S. A. Kazantsev, A. G. Petrashen, N. T. Polezhaeva, V. N. Rebane, and T. K. Rebane, *JETP Lett. (USA)* **45**, 17 (1987).
59. S. A. Kazantsev, A. G. Petrashen, N. T. Polezhaeva, and V. N. Rebane, *Opt. Spectrosc. (USA)* **68**, 740 (1990).
60. A. G. Petrashen, V. N. Rebane, and T. K. Rebane, *Opt. Spectrosc. (USA)* **67**, 6 (1989).
61. A. G. Petrashen, V. N. Rebane, and T. K. Rebane, *Opt. Spectrosc. (USA)* **55**, 819 (1983).
62. A. G. Petrashen, V. N. Rebane, and T. K. Rebane, *Opt. Spectrosc. (USA)* **57**, 200 (1984).
63. S. A. Kazantsev, A. G. Petrashen, N. T. Polezhaeva, and V. N. Rebane, *Opt. Spectrosc. (USA)* **68**, 544 (1990).
64. C. Cohen-Tannoudji, *Commun. Atom and Mol. Phys.* **2**, 24 (1970).

65. L. A. Vainstein, I. I. Sobelman, and E. A. Yukov, *Excitation of Atoms and Spectral Line Broadening*, Nauka, Moscow 319 pp. (1978) (in Russian).
66. V. N. Rebane and T. K. Rebane, *Opt. Spectrosc. (USA)* **54**, 701 (1983).
67. K. Ida, S. Hidekuma, Y. Miura, T. Fujia, M. Mozi, K. Hoshizo, N. Suzuki, T. Yamauchi, and JFT-2M Group, *Phys. Rev. Lett.* **65**, 1364 (1990).
68. K. Ida and S. Hidekuma, *Rev. Sci. Instrum.* **60**, 867 (1989).
69. B. H. Bransden, *Adv. At. Mol. Phys.* **1**, 85 (1965).
70. J. Jackson and H. Schiff, *Phys. Rev.* **89**, 359 (1953).
71. L. M. Brodsky, V. S. Potapov, and P. V. Tolmachev, *JETP (USA)* **58**, 264 (1970).
72. A. V. Vinogradov and V. P. Shevelko, *JETP (USA)* **59**, 193 (1970).
73. V. P. Shevelko, *Z. Phys.* **287 A**, 19 (1978).
74. Dž. Belkić, R. Gayet, J. Hanssen, and A. Salin, *J. Phys. B* **19**, 2945 (1986).
75. R. Hippler, in: *Electronic and Atomic Collisions*, pp. 241–253, North-Holland Publishing Company, Amsterdam, New York, Oxford (1984).
76. E. Gerjoy, B. H. Tomas, and V. S. Sheorey, *J. Phys. B*, **5**, 321 (1972).
77. L. J. Lembo, D. Danzman, Ch. Stoller, W. E. Meyerhof, and T. Häusch, *Phys. Rev. Lett.* **55**, 1874 (1985).
78. L. Lembo, K. Danzmann, Ch. Stoller, W. E. Meyerhof and T. Häusch, *Phys. Rev. A* **37**, 1141 (1988).
79. A. Salin, *J. Phys.* **45**, 671 (1984).
80. C. Geussorgues, C. Plaentine, and A. Sain, *J. Comput. Phys.* **10**, 223 (1975).
81. A. Salin, *Comput. Phys. Commun.* **14**, 121 (1978).
82. I. V. Komarov, L. I. Ponomarev, and S. Yu. Slavianov, *Spherical and Coulomb Spherical Functions*, Nauka, Moscow, 319 pp. (1976) (in Russian).
83. I. V. Komarov, *Problems of the Theory of Atomic Collisions*, Issue 1, St. Petersburg Univ. Publ., 319 pp. (1976) (in Russian).
84. N. Ya. Truskova, "Determination with given accuracy of the eigenfunctions and eigenvalues of the two-center problem in quantum mechanics," Preprint PN-10207, Institute of Nuclear Physics, Dubna (1976) (in Russian).
85. A. I. Baz, Ya. B. Zeldovich, and L. N. Perelomov, *Scatterings, Reactions and Decays in Quantum Mechanics*, Nauka, Moscow, 467 pp. (1966) (in Russian).
86. N. Andersen, J. W. Gallagher, and I. V. Hertel, in: *Electron and Atom Collisions* (D. C. Lorents, W. E. Meyerhoff, and J. R. Peterson, eds.), pp. 57–126, Elsevier, Amsterdam, (1986).
87. E. Landi Del Innocenti, V. Bommier, and S. Sahal-Brechot, *Astron. Astrophys.* **235**, 469 (1990).
88. V. Bommier, E. Landi Del Innocenti, and S. Sahal-Brechot, *Astron. Astrophys.* **244**, 383 (1991).
89. V. Bommier, E. Landi Del Innocenti, and S. Sahal-Brechot, *Astron. Astrophys.* **244**, 391 (1991).
90. E. Landi Del Innocenti, V. Bommier, and S. Sahal-Brechot, *Astron. Astrophys.* **244**, 401 (1991).
91. P. Hammond, W. Kanas, A. G. MacConkey, and J. W. MacConkey, *Phys. Rev. A* **40**, 1804 (1990).
92. R. Hippler, in: *Electronic and Atomic Collisions* (H. B. Gibody and W. R. Newell, F. H. Read and A. C. H. Smith, eds.), North-Holland Publishing Company, Amsterdam, Oxford, New York, Tokyo (1988).
93. T. Fujimoto and S. A. Kazantsev, *Plasma Phys. Contr. Fusion.* **39**, 1 (1997).
94. A. G. Petrashen, V. N. Rebane, and T. K. Rebane, *Opt. Spectrosc. (USA)* **78**, 579 (1995).
95. A. G. Petrashen, V. N. Rebane, and T. K. Rebane, *Opt. Spectrosc. (USA)* **80**, 713 (1996).
96. S. A. Kazantsev, A. G. Petrashen, N. M. Firstova, and J.-C. Hénoux, *Opt. Spectrosc. (USA)* **78**, 729 (1995).
97. T. G. Winter, *Phys. Rev. A* **25**, 679 (1982).
98. T. G. Winter and C. D. Lin, *Phys. Rev. A* **29**, 3071 (1984).
99. J. P. Faroux, Thèse, Paris (1969).
100. N. M. Kabachnik, *J. Phys. B* **26**, 3803 (1993).

101. W. Fritsch and C. D. Lin, *Phys. Rep.* **202**, Nos. 1–2 (1991).
102. K. Blum, *Density Matrix Theory and Applications*, Second Edition, Plenum Press, New York, London (1998).
103. K. Sandhya Devi and J. D. Garcia, *J. Phys. B* **16**, 2837 (1983).
104. J. P. Beijers, S. J. Doornenbal, J. van Eck, and H. G. Heideman, *J. Phys. B* **20**, 6617 (1987).
105. A. Crowe, J. C. Nogueira, and Y. C. Liew, *J. Phys. B* **16** 481 (1983).
106. G. G. King, S. Trajmar, and J. W. MacConkey, *Commun. At. Mol. Phys.* **23**, 229 (1989).
107. G. S. Panev, N. Andersen, T. Andersen, and P. Dalby, *Z. Phys.* **5**, 331 (1987).
108. S. Wang, D. J. van Der Burgt, J. J. Corr, J. M. MacConkey, and B. H. Madison, *J. Phys. B* **27**, 329 (1994).
109. N. Andersen and S. E. Nilsen, *Adv. At. Mol. Phys.* **18**, 265 (1982).
110. R. E. Clark, J. Abdallah, G. Csanak, and S. P. Kramer, *Phys. Rev. A* **40**, 2835 (1989).
111. A. K. Katiyar and R. Srivastava, *Phys. Rev. A* **40**, 1289 (1989).
112. K. Bartschat and D. Madison, *J. Phys. B* **20**, 5839 (1987).
113. K. Bartschat and D. Madison, *J. Phys. B* **21**, 153 (1988).
114. D. Detleffsen, M. Anton, A. Werner, and K.-H. Schartner, *J. Phys. B* **27**, 4195 (1994).
115. O. Scholler, J. S. Briggs, and R. M. Dreizler, *J. Phys. B* **19**, 2505 (1986).
116. S. Kumar and M. Srivastava, *J. Phys. B* **9**, 1911 (1976).
117. T. H. Chang and R. Roe, *Phys. Rev. Lett.* **31**, 1097 (1978).
118. Y. Alhassid and B. Shao, *Phys. Rev.* **46**, 3978 (1992).
119. J. M. Laming, *Astrophys. J.* **357**, 275 (1990).
120. J. M. Laming, *Astrophys. J.* **362**, 219 (1990).
121. M. Craeter, *Solar Phys.* **130**, 337 (1990).
122. A. Hirabayashy, Y. Nambu, M. Hasuo, and T. Fujimoto, *Phys. Rev. A* **37**, 83 (1988).
123. A. G. Petrashen, V. N. Rebane, and T. K. Rebane, *Opt. Spectrosc. (USA)* **85**, 333 (1998).
124. J. Bogdorfer, *Z. Phys. A* **309**, 285 (1983).
125. J. Bogdorfer, *Phys. Rev. Lett.* **52**, 2225 (1984).
126. S. A. Kazantsev and A. G. Petrashen, *Opt. Spectrosc.* **77**, 898 (1994).
127. J. Laming, *Nucl. Instrum. Methods Phys. Res.* **43**, 359 (1989).
128. S. A. Kazantsev, V. V. Luchinkina, A. D. Mezentsev, A. P. Mustafaev, V. N. Rebane, A. G. Rys, and Yu. A. Stepanov, *Opt. Spectrosc.* **76**, 909 (1994).
129. N. Shimakura and M. Kimura, *Phys. Rev. A* **44**, 1659 (1991).
130. C. Forster, P. L. Cooper, A. Prichinson, D. Tlower, and L. Mendez, *J. Phys. B* **24**, 3933 (1991).
131. B. Zuyelman, P. Cooper, M. Ford, A. Palgarni, J. Gerra, and M. Raimondi, *Phys. Rev. A* **46**, 3846 (1992).
132. T. H. McLaughlin, H. Tanuma, J. Hodgkimson, R. McCullough, and H. B. Gilbody, *J. Phys. B* **26**, 3871 (1993).
133. P. W. Heddle, *J. Phys. B* **12**, 489 (1979).
134. S. A. Kazantsev and A. V. Subbotenko, *Spectropolarimetric Diagnostics of Gas Discharges*, St. Petersburg State University Publ. (1983) (in Russian).
135. S. A. Kazantsev, A. G. Rys, and M. P. Chaika, *Opt. Spectrosc. (USA)* **54**, 421 (1983).
136. M. A. Abramovitz and I. Stegun (eds.), *Handbook of Mathematical Functions*, NBS, p. 829 (1964).
137. V. Bommier and S. Sahal-Brechot, *Ann. Phys.* **16**, 555 (1991).
138. V. Bommier, *Ann. Phys.* **16**, 598 (1991).
139. L. D. Landau and E. M. Lifshits, *Quantum Mechanics*, Nauka, Moscow (1963) (in Russian).
140. J. C. Percival and M. J. Seaton, *Philos. Trans. R. Soc. London, Ser. A* **251**, 113 (1958).
141. U. Fano and Ch. Macek, *Rev. Mod. Phys.* **45**, 553 (1973).
142. A. I. Drachev, S. A. Kazantsev, and A. V. Subbotenko, *Opt. Spectrosc. (USA)* **69**, 585 (1990).

143. A. I. Drachev, S. A. Kazantsev, A. G. Rys, and A. V. Subbotenko, *Opt. Spectrosc. (USA)* **70**, 277 (1991).
144. A. Werner and K.-H.Schartner, *J. Phys. B* **29**, 125 (1996).
145. H. Kleinpoppen and E. Kraiss, *Phys. Rev. Lett.* **20**, 361 (1968).
146. A. Mahan, A. Gallagher, and S. Smith, *Phys. Rev. A* **13**, 156 (1976).
147. R. F. Syms, M. R. McDowell, and P. W. Edmunds, *J. Phys. B*, **20**, 1587 (1975).
148. F. T. Chang and C. H. Chang, *Phys. Rev. A* **15**, 118 (1977).
149. H. Kleinpoppen and A. Scharmann, in: *Progress in Atomic Spectroscopy*, Part A, p. 239 (1978).
150. J. Aboudarham, K. Berrington, J. Callway, N. Feautrier, J.-C. Hénoux, G. Peach, and H. Saraph, *Astron. Astrophys.* **262**, 302 (1992).
151. R. Hippler, H. Madeheim, W. Harpich, H .Kleinpoppen, and H. Lutz, *Phys. Rev. A* **38**, 1662 (1988).
152. W. Fritsch and C. B. Lin, *Phys. Rev. A* **27**, 3361 (1983).
153. J. T. Park, J. E. Alday, J. M. George, and J. L. Peacher, *Phys. Rev. A* **14**, 608 (1976).
154. J. T. Park, *Physics of Electronic and Atomic Collisions*, pp. 109–121 North Holland Publishing Comapny, Amsterdam, New York, London, Oxford (1982).
155. S. Chwirot and J. Slevin, *J. Phys. B* **20**, 6139 (1987).
156. A. K. Katryar and R. Srivastava, *Phys. Rev. A* **40**, 1289 (1989).
157. S. Chwirot and J. Slevin, *J. Phys. B* **18**, L881 (1985).
158. J. McConkey, *J. Phys. B*, **21**, 317 (1988).
159. N. Andersen, I. V. Hertel, and H. Kleinpoppen, *J. Phys. B* **17**, L901 (1984).
160. N. Andersen, T. Andersen, J. S. Dahler, S. E. Nielsen, and G. Nienhius, *J. Phys. B*, **16**, 817 (1983).
161. N. Andersen, T. Andersen, H.-P. Neitzke, and E. H. Pedersen, *J. Phys. B* **18**, 2243 (1985).
162. V. A. Fock, "Multielectron problem in quantum mechanics and atomic structure," Jubilee USSR Academy of Sciences Publication, Part 1, pp. 255–287 (1947).
163. M. I. Petrashen and E. D. Trifonov, *Application of Group Theory in Quantum Mechanics*, Nauka, Moscow (1967) (in Russian).
164. A. G. Radtsig and B. M. Smirnov, *Parameters of Atoms and Atomic Ions*, Nauka, Moscow, pp. 243 (1986) (in Russian).
165. W. Hanle, *Z. Phys.* **30**, 93 (1924).
166. E. Jacquet, P. Boduch, and M. Chantepie, *Phys. Scr.* **49**, 154 (1994).
167. P. Boduch, M. Chantepie, D. Hennecart, X. Husson, and D. Lecler, *Phys. Scr.* **46**, 337 (1992).
168. M. Chantepie, E. Jacquet, S. A. Kazantsev, S. Laulhier, and A. G. Petrashen, *Opt. Spectrosc. (USA)* **81**, 38 (1995).
169. L. D. Landau and E. M. Lifshits, *Theoretical Mechanics*, Nauka, Moscow (1963) (in Russian).
170. U. Fano, *Phys. Rev.* **133**, B828 (1964).
171. H. A. Bethe and E. E. Solpiter, *Quantum Mechanics of One- and Two-Electron Atoms*, Academic Press, New York (1957).
172. J. D. Jackson, *Classical Electrodynamics*, John Wiley & Sons Inc. New York, London, Sydney (1962).
173. E. A. Duman, B. M. Smirnov, and M. I. Chibisov, *JETP (USA)* **53**, 311 (1967).
174. C. Harel and A. Salin, *J. Phys. B* **10**, 3511 (1977).
175. N. F. Mott and H. S. W. Massey, *The Theory of Atomic Collisions*, Third Edition, Oxford University Press, London (1965).
176. J.-C. Hénoux, G. Chambe, M. Semel, S. Sahal, B. Woodgate, D. Shine, J. Beckers, and M. Mochado, *Astron. Astrophys.* **265**, 1066 (1983).
177. W. Jr. Henze, *The Ultraviolet Spectrometer and Polarimeter Experiment on the Solar Maximum Mission*, Marshal Space Flight Center Report MSFC-2323, p. 179.
178. G. Chambe and J.-C. Hénoux, *Astron. Astrophys.* **80**, 123 (1979).
179. B. Woodgate, *Sol. Phys.* **65**, 73 (1980).
180. J. Calvert, D. Gringer, J. Montenegro, R. Nola, R. Rutledge, E. Tandberg-Hanssen, C. L. Wyman, and J. M. Beckers, *Opt. Eng.* **18**, 287 (1979).

181. M. Chantepie, E. Jacquet, S. A. Kazantsev, S. Laulhier, and A. G. Petrashen, *Opt. Spectrosc. (USA)* **81**, 186 (1995).
182. J.-C. Hénoux, G. Chambe, D. Heristchi, M. Semel, B. Woodgate, D. Shin, and J. Beckers, *Sol. Phys.* **86**, 115 (1983).
183. V. A. Trubnikov, *Sov. J. Plasma Phys. (USA)* **1**, 105 (1965).
184. J.-C. Hénoux, in: *Solar Polarimetry*, Proc. 11th National Sacramento Peak Observatory Workshop (L. H. November, ed.), pp. 285–295 (1990).
185. A. N. Babin and A. N. Koval, *Izv. Crim. Observ.* **66**, 89 (1983) (in Russian).
186. A. N. Babin and A. N. Koval, *Izv. Crim. Observ.* **70**, 3 (1985) (in Russian).
187. A. N. Babin and A. N. Koval, *Izv. Crim. Observ.* **72**, 142 (1985) (in Russian).
188. A. N. Babin and A. N. Koval, *Solar. Phys.* **98**, 159 (1985).
189. A. N. Babin and A. N. Koval, *Izv. Crim. Observ.* **73**, 3 (1985) (in Russian).
190. A. N. Babin and A. N. Koval, *Izv. Crim. Observ.* **77**, 9 (1987) (in Russian).
191. A. N. Babin and A. N. Koval, *Izv. Crim. Observ.* **75**, 52 (1986) (in Russian).
192. N. M. Firstova, *Sol. Phys.* **103**, 11 (1986).
193. A. V. Bulatov, J.-C. Hénoux, S. A. Kazantsev, N. M. Firstova, and A. G. Petrashen, *Opt. Spectrosc. (USA)* **78**, 729 (1995).
194. N. M. Firstova and P. Kotrc, *Bull. Astron. Inst. Czech.* **38**, 257 (1987).
195. J.-C. Hénoux, S. A. Kazantsev, N. M. Firstova, and A. G. Petrashen, *Opt. Spectrosc. (USA)* **80**, 532 (1996).
196. J.-C. Hénoux, S. A. Kazantsev, N. M. Firstova, and A. G. Petrashen, *Opt. Spectrosc. (USA)* **80**, 635 (1996).
197. S. A. Kazantsev, N. M. Firstova, N. A. Lankevich, and A. B. Gubin, *Opt. Spectrosc. (USA)* **70**, 580 (1991).
198. M. Kralicky, J.-C. Hénoux, and D. Smith, *Astron. Astrophys.* **310**, 629 (1996).
199. A. Werner, D. Hasselkampf, O. Wilhelmi, and K.-H. Schartner, Proc. ICPEAC, Vienna (1997).
200. D. M. Rust and S. L. Keil, *Sol. Phys.* **140**, 55 (1992).
201. J.-C. Hénoux, G. Chambe, M. Semel, S. Sahal, B. Woodgate, D. Shine, J. Beckers, and M. Machado, *Astrophys. J.* **265**, 1066 (1983).
202. V. I. Skomorovsky, *Geomagn. Aeron. Sol. Phys.* **91**, 97 (1991) (in Russian).
203. N. M. Firstova, N. A. Lankevich, and A. V. Gubin, *Geomagn. Aeron. Sol. Phys.* **91**, 164 (1991) (in Russian).
204. V. N. Obridko, *Solar Spots and Centers of Activity*, Nauka, Moscow (1985) (in Russian).
205. M. E. Machado, E. H. Avrett, J. E. Vernazza, and R. W. Noyes, *Astrophys. J.* **242**, 336 (1980).
206. B. Thomas and J. Gerjuoy, *J. Math. Phys.* **19**, 1561 (1971).
207. Y. Luke, *Algorithms for Computation of Mathematical Functions*, Physics-Mathematics Publisher, Moscow (1977) (in Russian).
208. J. Vorpahl and T. Pope, *Sol. Phys.* **25**, 347 (1972).
209. H. Kurokawa, J. Kawaguchi, Y. Funakoshi, and Y. Nakai, *Sol. Phys.* **79**, 77 (1982).
210. O. Engvold and P. Maltby, in: *Mass Motions in Solar Flares* (H. Ohman, ed.), p. 109, Wiley, London (1968).
211. R. Kitai, *Sol. Phys.* **87**, 135 (1983).
212. A. N. Koval, *Izv. Crim. Astrophys. Observ.* **43**, 278 (1965) (in Russian).
213. A. Bruzek, *Sol. Phys.* **2**, 451 (1967).
214. D. M. Rust, in: Structure and Development of Solar Active Regions (K. O. Kiepenheuer, ed.), JAU Symp. **35**, 77 (1968).
215. J.-R. Roy and H. Leparskas, *Sol. Phys.* **30**, 449 (1973).
216. A. B. Severny and A. N. Koval, *Izv. Crim. Astrophys. Observ.* **26**, 3 (1961) (in Russian).
217. J. S. Shklovsky, *Ann. Rev. Astron. Astrophys.* **2**, 363 (1958).
218. A. B. Severny and V. L. Khokhlova, *Izv. Crim. Astrophys. Observ.* **20**, 68 (1958) (in Russian).

219. C. Fang, J.-C. Hénoux, and W. Q. Gan, *Astron. Astrophys.* **274**, 917 (1993).
220. S. A. Kazantsev, A. G. Petrashen, N. M. Firstova, and J.-C. Hénoux, *Opt. Spectrosc. (USA)* **82**, 659 (1997).
221. J.-C. Hénoux, G. Chambe, M. Semel, S. Sahal-Bzechot, B. Woodgate, D. Shine, J. Beckers and M. Macado, *Astrophys. J.* **265**, 1066 (1983); *Astron. Astrophys.* **119**, 133 (1983).
222. G. M. Simnett, *Sol. Phys.* **106**, 165 (1986).
223. D. H. Tamres, D. B. Melrose, and R. C. Canfield, *Astrophys. J.* **342**, 576 (1989).
224. V. E. Stepanov, V. G. Banin, and V. J. Kruglov, *Novaya Tekhnika v Astronomii* **6**, 42 (1979) (in Russian).
225. V. I. Skomorovsky and N. M. Firstova, *Sol. Phys.* **163**, 209 (1996).
226. A. N. Koval, *Izv. Crim. Astrophys. Observ.* **57**, 133 (1977) (in Russian).
227. P. Bakshi and G. Kalman, *Sol. Phys.* **47**, 307 (1977).
228. G. V. Sholin, *Doklady Akademii Nauk USSR* **195**, 589 (1970) (in Russian).
229. N. M. Firstova, *Sol. Phys.* 90, 269 (1984).
230. N. M. Firstova and A. V. Bulatov, *Sol. Phys.* **164**, 361 (1996).
231. F. Ellerman, *Astrophys. J.* **45**, 298 (1917).
232. A. B. Severny, *Astron. Zh.* **33**, 74 (1956) (in Russian).
233. A. B. Severny, 22nd IAU Congress, Rottoch-Eggern., pp. 1–10 (1963).
234. A. N. Koval, *Izv. Crim. Astrophys. Observ.* **34**, 278 (1985) (in Russian).

Index

COLLISIONS OF ELECTRONS WITH ATOMS AND MOLECULES
G. F. Drukarev

THEORY OF MULTIPHOTON PROCESSES
Farhad H. M. Faisal

PROGRESS IN ATOMIC SPECTROSCOPY, Parts A, B, C, and D
Edited by W. Hanle, H. Kleinpoppen, and H. J. Beyer

RECENT STUDIES IN ATOMIC AND MOLECULAR PROCESSES
Edited by Arthur E. Kingston

QUANTUM MECHANICS VERSUS LOCAL REALISM: The Einstein-Podolsky-Rosen Paradox
Edited by Franco Selleri

ZERO-RANGE POTENTIALS AND THEIR APPLICATIONS IN ATOMIC PHYSICS
Yu. N. Demkov and V. N. Ostrovskii

COHERENCE IN ATOMIC COLLISION PHYSICS
Edited by H. J. Beyer, K. Blum, and R. Hippler

ELECTRON–MOLECULE SCATTERING AND PHOTOIONIZATION
Edited by P. G. Burke and J. B. West

ATOMIC SPECTRA AND COLLISIONS IN EXTERNAL FIELDS
Edited by K. T. Taylor, M. H. Nayfeh, and C. W. Clark

ATOMIC PHOTOEFFECT
M. Ya. Amusia

MOLECULAR PROCESSES IN SPACE
Edited by Tsutomu Watanabe, Isao Shimamura, Mikio Shimizu, and Yukikazu Itikawa

THE HANLE EFFECT AND LEVEL CROSSING SPECTROSCOPY
Edited by Giovanni Moruzzi and Franco Strumia

ATOMS AND LIGHT: INTERACTIONS
John N. Dodd

POLARIZATION BREMSSTRAHLUNG
Edited by V. N. Tsytovich and I. M. Ojringel

INTRODUCTION TO THE THEORY OF LASER–ATOM INTERACTIONS (Second Edition)
Marvin H. Mittleman

ELECTRON COLLISIONS WITH MOLECULES, CLUSTERS, AND SURFACES
Edited by H. Ehrhardt and L. A. Morgan

THEORY OF ELECTRON–ATOM COLLISIONS, Part 1: Potential Scattering
Philip G. Burke and Charles J. Joachain

POLARIZED ELECTRON/POLARIZED PHOTON PHYSICS
Edited by H. Kleinpoppen and W. R. Newell

INTRODUCTION TO THE THEORY OF X-RAY AND ELECTRONIC SPECTRA OF FREE ATOMS
Romas Karazija

VUV AND SOFT X-RAY PHOTOIONIZATION
Edited by Uwe Becker and David A. Shirley

DENSITY MATRIX THEORY AND APPLICATIONS (Second Edition)
Karl Blum

SELECTED TOPICS ON ELECTRON PHYSICS
Edited by D. Murray Campbell and Hans Kleinpoppen

PHOTON AND ELECTRON COLLISIONS WITH ATOMS AND MOLECULES
Edited by Philip G. Burke and Charles J. Joachain

COINCIDENCE STUDIES OF ELECTRON AND PHOTON IMPACT IONIZATION
Edited by Colm T. Whelan and H. R. J. Walters

PRACTICAL SPECTROSCOPY OF HIGH-FREQUENCY DISCHARGES
Sergei A. Kazantsev, Vyacheslav I. Khutorshchikov, Günter H. Guthöhrlein, and Laurentius Windholz

IMPACT SPECTROPOLARIMETRIC SENSING
S. A. Kazantsev, A. G. Petrashen, and N. M. Firstova